Introduction to Mobile Communications Engineering

For a complete listing of the *Artech House Mobile Communications Library*, turn to the back of this book.

Introduction to Mobile Communications Engineering

José M. Hernando
F. Pérez-Fontán

Artech House
Boston • London

Library of Congress Cataloging-in-Publication Data
Hernando, José M.
  Introduction to mobile communications engineering / José M.
Hernando, F. Pérez-Fontán.
    p.   cm.— (Artech House mobile communications library)
  Includes bibliographical references and index.
  ISBN 0-89006-391-5 (alk. paper)
  1. Mobile communication systems.   I. Pérez-Fontán, F.   II. Title.
III. Series
  TK6570.M6H42   1999
  621.3845—dc21   99-23714
                                                                      CIP

British Library Cataloguing in Publication Data
Hernando, Jose M.
  Introduction to mobile communciations engineering. -
  (Artech House mobile communications library)
  1. Mobile communications sytesms
  I. Title II. Perez-Fontan, F.
  621.3'845

  ISBN 0-89006-391-5

Cover by Lynda Fishbourne

© 1999 Artech House
685 Canton Street
Norwood, MA 02062

All rights reserved. Printed and bound in the United States of America. No part of this book may be reproduced or utilized in any form or by any means, electronic or mechanical, including photocopying, recording, or by any information storage and retrieval system, without permission in writing from the author
  All terms mentioned in this book that are known to be trademarks or service marks have been appropriately capitalized. Artech House cannot attest to the accuracy of this information. Use of a term in this book should not be regarded as affecting the validity of any trademark or service mark.

International Standard Book Number: 0-89006-391-5
Library of Congress Catalog Card Number: 99-23714

10 9 8 7 6 5 4 3 2 1

Contents

	Preface	xv
I	**Basic Concepts of Mobile Communications**	**1**
1	**Introduction: The Propagation Channel in Mobile Communications**	**3**
1.1	Introduction	3
1.2	Received Field Strength Variations	6
1.3	Characterization of Signal Variations	8
1.4	Path Loss and Signal Variations: Characterization of the Slow Variations	10
1.5	Fast Variations	14
	References	24
2	**Introduction to Mobile and Private Mobile Radio (PMR) Systems**	**25**
2.1	Introduction to Private Mobile Radio (PMR) Systems	25
2.2	Channel Types in Mobile Communications	27
2.3	Types of Radio Station Control	39
2.4	Signaling in PMR Systems	44

2.5	Area Coverage Techniques	52
2.6	Typical Example of a Dispatch System	58
2.7	Data Transmission in PMR	59
	References	62

II Propagation Modeling in Mobile Communications 63

3 Multipath Propagation 65

3.1	Introduction	65
3.2	Angle of Incidence and Spectrum of the Received Signal	65
3.3	Received Signal Envelope Statistics	69
3.4	Received Signal Phase	75
3.5	Second-Order Statistics	76
3.6	Random Frequency Modulation	81
3.7	Fading in the Ricean Case	83
3.8	Spatial Correlation of the Field Components	85
3.9	Frequency Selective Fading	87
3.10	Wideband Channel Characterization	95
3.11	Global Wideband Characterization of the Land Mobile Channel	106
3.12	Land Mobile Channel Measurements	111
3.13	Cost 207 Channel Model	115
	References	119

4 Propagation Path Loss 121

4.1	Introduction	121
4.2	The Okumura Model	124
4.2.1	Terrain Features	124

4.2.2	Environment Types	128
4.2.3	Calculating the Received Median Field Strength Value	130
4.2.4	Correction Factors	132
4.2.5	Quantification of Built-Up Area Losses	136
4.2.6	Locations Variability	141
4.2.7	Calculations Using the Okumura Model	146
	Appendix 4A: Review of the Plane Earth Model	**150**
4.3	Hata Model	157
4.4	Modified JRC Model	159
4.4.1	Additional Losses in Built-Up Areas	165
4.5	Physical-Geometrical Models	169
4.5.1	Ikegami Model	170
4.5.2	COST 231 Model Walfish-Ikegami	176
4.5.3	Ray-Tracing Based Models	179
	References	189
5	**Mobile Network System Engineering**	**191**
5.1	Introduction	191
5.2	Coverage Computations	192
5.3	Mobile Station Coverage Evaluation	207
5.4	Interference-Limited Systems	212
5.5	Intermodulation	213
5.6	Frequency Planning	225
5.7	Cochannel Interference Compatibility	229
	References	233
	Selected Bibliography	233
6	**Propagation in New Scenarios**	**235**
6.1	Introduction	235
6.2	Microcell Scenarios	235
6.2.1	Microcell Propagation Modeling	239

6.2.2	Empirical Propagation Models for Urban Microcells	244
6.3	Indoor Propagation	248
6.3.1	Indoor Picocell Propagation Models	248
6.4	Indoor Coverage Using Outdoor Base Stations—Penetration Losses	252
	References	255
7	**Base Station Engineering**	**257**
7.1	Introduction: The Antenna System	257
7.2	Transmitter Combiners	257
7.3	Antenna Multicouplers	267
7.4	Duplexers	269
7.5	Feeders	269
7.6	Complete Antenna Connection System	272
	References	274
	Selected Bibliography	274
III	**Description of Different Mobile Standards**	**275**
8	**Trunked Systems**	**277**
8.1	Introduction	277
8.2	Elements in a Trunked Radio Network	280
8.2.1	Network Control Center (NCC)	282
8.2.2	Base Stations	282
8.2.3	Antenna System	288
8.2.4	Support Systems	288
8.2.5	Interconnection Network	290
8.2.6	Frequency Plan	292
8.3	MPT13xx Standards	292
8.3.1	Introduction	292
8.3.2	Types of Calls and Facilities	295

8.3.3	Characteristics of MPT-13xx Systems	296
8.3.4	Call Processing	297
8.3.5	Multisite Systems	297
8.3.6	Access Protocol	298
8.3.7	Types of Signaling Messages	298
8.3.8	Principles of the Random Access Protocol	301
8.3.9	Terminal Addressing	303
8.3.10	Examples of Signaling Sequences	304
8.4	Trunked Network Dimensioning	317
	References	321
9	**The Cellular Concept**	**323**
9.1	Introduction	323
9.2	Keys to the Cellular Concept	324
9.3	Cellular Geometry	327
9.4	Analog Cellular Systems	332
9.4.1	AMPS System Parameter Selection Criteria	334
9.4.2	Operation of an Analog Cellular System: AMPS/TACS	336
9.4.3	Control System	338
9.4.4	Signaling Message Formats	346
9.4.5	Specific Features of AMPS and TACS	347
9.5	Rolling Out an AMPS Network	351
9.5.1	Adjacent Channel Interference Reduction	351
9.5.2	Cell Splitting	354
9.5.3	Overlaid Cell Concept	355
9.5.4	Example of an AMPS Network Rollout Plan	358
9.6	Structure and Basic Functions of a Cellular Network	360
9.7	Dimensioning of Cellular Networks	362

9.8	Cochannel Interference Issues	367
	References	371

Appendix 9A: Summary of Parameters Used — **371**

Appendix 9B: Erlang-B Table — **373**

10	**The GSM System**	**375**
10.1	Introduction	375
10.1.1	Origin and Evolution of the GSM System	376
10.2	Basic Specifications of GSM	378
10.3	Elements and Architecture of GSM	381
10.3.1	Identifiers and Contents of the Different Network Databases	385
10.3.2	Registration and Location Update	387
10.3.3	Authentication	388
10.3.4	Equipment Identification	391
10.3.5	Roaming	391
10.3.6	Call Hand-Over	391
10.3.7	Stages in a Call	393
10.3.8	Call Setup and Routing Protocol	393
10.4	The Radio Interface	397
10.4.1	TDMA Structures	398
10.4.2	Channel Types	401
10.4.3	Setting Up Calls	403
10.4.4	Voice Link	408
10.4.5	Coding the Signaling Bits	412
10.5	Other Features of the GSM Radio Interface	413
10.5.1	Adaptive Time Alignment	413
10.5.2	Power Control	415
10.5.3	Call Hand-Over	416

10.5.4	Discontinuous Transmission and Reception	417
10.5.5	Slow Frequency Hopping	418
10.6	Link Budget Data	419
10.7	GSM Services	422
10.7.1	Teleservices	423
10.7.2	Carrier/Bearer Services	424
10.7.3	Supplementary Services	424
	References	425
	Selected Bibliography	425
11	**Other Mobile Radio Systems**	**427**
11.1	Introduction	427
11.2	The North American Cellular Digital System (NADC)	427
11.2.1	Overview	427
11.2.2	Description of the NADC System	430
11.2.3	Logical Channels	431
11.2.4	NADC Radio Interface	432
11.2.5	Modulation	436
11.2.6	Speech Processing (Full-Rate Channel)	438
11.3	The Mobile Telephone CDMA System	439
11.3.1	Introduction to CDMA	439
11.3.2	Diversity Mechanisms in CDMA	444
11.3.3	Power Control	447
11.4	The CDMA IS-95 Standard	448
11.4.1	Introduction	448
11.4.2	Speech Coding	449
11.4.3	The Walsh Matrix	450
11.4.4	Coding on the Base-to-Mobile Link	450
11.4.5	Coding on the Mobile-to-Base Link	453
11.4.6	CDMA System Operation	456
11.4.7	Comparison of Capacities in Different FDMA, TDMA, and CDMA Systems	458

11.5	Cordless Systems	459
11.5.1	Introduction	459
11.5.2	Advanced Cordless Telecommunications Systems	460
11.5.3	Cordless Telecommunications and Cellular Radio	463
11.5.4	The Digital Enhanced Cordless Telecommunication (DECT) System	465
11.5.5	Technical Principles of the DECT System	472
11.5.6	Basic Operation of the DECT System	477
11.5.7	Other Information of Interest	478
11.6	The Digital Trunking System TETRA	479
11.6.1	Introduction	479
11.6.2	TETRA Telecommunications Services	481
11.6.3	Harmonized TETRA Frequencies	481
11.6.4	TETRA Functional Configuration	483
11.6.5	TETRA Signal Processing	484
	References	488
	Selected Bibliography	489
12	**Future Mobile Communications Systems**	**491**
12.1	Introduction	491
12.2	IMT-2000	494
12.2.1	Introduction	494
12.2.2	Frequency Bands	495
12.2.3	Basic Objectives of IMT-2000	495
12.2.4	Radio Interfaces	497
12.2.5	Traffic Densities and Spectrum Requirements	500
12.2.6	Radio Transmission Technology Evaluation	501
12.3	The UMTS Concept	503
12.3.1	Introduction	503

12.3.2	UMTS Objectives	506
12.3.3	Spectrum Requirements	507
12.3.4	Radio Interface	508
	Selected Bibliography	514
	About the Authors	**515**
	Index	**517**

Preface

The last 20 years have been referred to as the "information explosion" era. The huge development of information technologies has driven a parallel expansion of telecommunications. More recently, in the past five years, a similar process has been registered in mobile communications. The mobile communications area, whose market objective is to supply a mobile counterpart to all fixed telecommunications applications, is currently the most dynamic sector of the telecommunications business, with growth rates that would have been unthinkable some years ago.

An increasing number of people work in mobile radio, and others intend to join this activity. Some of these individuals need to receive appropriate training to access this booming labor market. This book's primary aim is to meet this training need. In the literature, there are many specialized books on mobile communications for readers with some previous experience in the field. We believe, however, that there is also a need for general textbooks that provide an overview of mobile communications engineering for undergraduate students and engineers who want to receive training in a new field. With this objective in mind, the book provides a firsthand treatment of all the topics related to mobile communications, from the basic dispatching radio networks to the most advanced cellular mobile telephone systems. In addition, the book describes currently existing systems and gives insight into the future of mobile technology.

The book covers a wide range of topics to provide a complete introductory view of mobile communications issues. This will allow readers to acquire a general understanding of these systems so that they can delve more deeply, as needed, into specific matters covered in more specialized texts. Furthermore, the book presents a great deal of practical data necessary for the practicing engineer. In fact, most engineering aspects needed in planning tasks can be addressed with the basic and simple formulations outlined in the book. Accordingly, this book would be useful to undergraduate students and engineers wanting to undertake studies in mobile communications by self-study, in seminar and masters courses, in factory training, or in consultancy activities. The only prerequisites for readers are a general background in electromagnetic theory and telecommunication principles.

The text is divided into three main parts: Part One deals with the basic concepts of mobile communications. It describes the fundamental characteristics of mobile radio channels and offers a general view of private mobile radio networks. Part Two is devoted to propagation modeling and engineering issues. It is probably the core body of the book and provides information relevant to mobile systems projects and engineering. Part Three presents different mobile standards.

We hope that this book will be useful to interested readers. We would kindly appreciate any reader suggestions and feedback.

José M. Hernando and Fernando Pérez-Fontán,
June 1999

Basic Concepts of Mobile Communications

1

Introduction: The Propagation Channel in Mobile Communications

1.1 Introduction

This chapter stresses the peculiarities of propagation in mobile communications in a simple manner [1] and presents the differences between it and other radio communication systems using the same frequency bands, such as fixed radio links and sound and television broadcasting. The chapter focuses mainly on conventional land mobile communications, i.e., those with coverages ranging over a few kilometers and transmission antennas located on elevated sites. Other channels such as land mobile satellite (LMS) and indoor and urban microcell communications are also widely used in mobile communication applications. Similar methods to those described in this chapter for conventional mobile communications are applied to study and characterize these channels.

The engineering involved in land mobile systems also has many similarities with that used in other area coverage systems (point-to-multipoint), as is the case of sound and television broadcasting. These similarities are primarily found in the computation of path losses and in the way in which coverage calculations are carried out.

Man-made structures, such as buildings or small houses in suburban areas with sizes ranging from a few meters to tens of meters, exert a decisive influence on mobile communications. In urban environments, the size of structures is even greater. In rural environments, features such as isolated trees or groups of trees may have similar dimensions. These environmental features are similar or greater in size than the transmitted wavelength (metric waves (VHF) and decimetric waves (UHF)) and may both block and scatter radio signals, causing specular and/or diffuse reflections. These contributions may reach the mobile receiver via multiple paths in addition to that of the direct signal. In many cases, these echoes are responsible for a certain amount of energy reaching the receiver, thus making communications possible, especially when the direct signal is blocked by environmental elements found in the transmitted signal path. Thus, two main effects are then characteristic in mobile propagation, the *shadowing* and *multipath effects*.

While base station heights are on the order of 30m [1] and are normally set on elevated sites with no blocking/scattering elements in their vicinity, mobile antenna heights will usually be smaller than those of the natural and man-made features found in the surrounding area. Typical values range from 1.5 to 3m, normal heights for several types of vehicles (cars, vans, trucks, etc.). For other radio communication services such as television broadcasting operating in the same frequency bands, VHF and UHF, the propagation channel will present different characteristics, since the receiving antennas are directive and are located well above the ground. In this case, both the shadowing effect on the direct signal and multipath are considerably reduced. In some instances, however, they may be present and thus cause, for example, double imaging on conventional analog television receivers.

The frequencies used in mobile communications are above 30 MHz, and the maximum length of the links does not exceed 25 km. It must be taken into account that mobile communications are bidirectional and that the uplink (mobile-to-base) is power-limited. This is especially so in the case of portable, hand-held terminals. Furthermore, mobile radio coverage ranges are short due to the screening effects of the terrain and buildings in urban areas. This makes frequency reuse possible at relatively short distances. This is also an important feature in mobile networks that require a great deal of frequency efficiency to accommodate a large number of users (cellular telephony networks).

Two extreme scenarios may be considered: 1) that in which a strong direct signal, referred usually as a line-of-sight (LOS) signal, is available together with a number of weaker multipath echoes and 2) that in which a number of weak multipath echoes is received and no direct signal is available.

The first case occurs on open land or in very specific spots in city centers, in places such as crossroads or squares with a good visibility of the transmitter. An example of this situation would be found in densely built-up urban areas where there is a large building that is clearly visible both from the transmitter and the receiver. This situation may be modeled by a Rice distribution for variations in the received RF signal envelope (Rice case). In this case, the received signal will be strong and quite steady, with small slow and fast variations due to shadowing and multipath effects. See Figure 1.1(a).

Introduction: The Propagation Channel in Mobile Communications 5

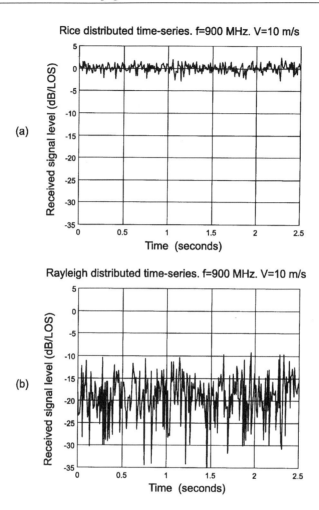

Figure 1.1 (a) Rice distributed time-series, (b) Rayleigh distributed time-series relative to LOS, and (c) Rayleigh environment in a rural, tree-shadowed area.

Case 2 is typically found in highly built-up urban environments. This is the worst of all situations, since the received signal will be weak and subject to marked variations due to shadowing and multipath effects—see Figure 1.1(b). This kind of situation may also occur in rural environments where the signal is obstructed by trees (woods or tree alleys). The received signal amplitude variations in this situation can normally be modeled with a Rayleigh distribution (Rayleigh case). Figure 1.1(c) illustrates a Rayleigh scenario in a rural environment. For stretches of the road sketched in Figure 1.1(c) where the transmitter is visible, the Rice case would be applicable.

(c)

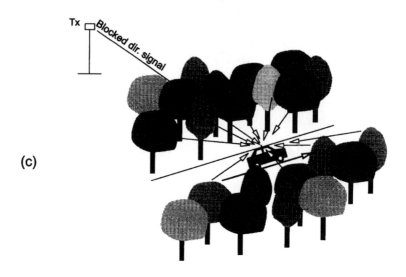

Figure 1.1 (continued).

1.2 Received Field Strength Variations

The received field strength may be graphically represented in terms of time, $e(t)$, or as a function of the traveled distance, $e(x)$ [1]. Figure 1.2 shows a typical mobile communication scenario with a base station and a mobile station running along a radial route relative to the base station so that the terrain profile fits that of the route profile. Figure 1.2 also shows a diagram of the received signal envelope as a function of the distance from the transmitter. The first thing to note is the fact that the RF carrier envelope is subject to strong oscillations as the mobile travels away from the transmitter.

If ideal conditions are assumed (i.e., no changes in propagation conditions occur), when a mobile receiver traverses a given environment at speed V—Figure 1.3(a)—and later traverses the same environment at half the speed $V/2$—Figure 1.3(b)—the very same series of received signal values would be observed except for the fact that the time axes would be scaled. See Figure 1.3(a, b). This would be so if propagation conditions did not change in time, and, thus, the same echoes from the same scatterers were received in the very same places with the same magnitudes and phases. This brings up the standing wave concept, which is produced by the combination/interference of the different rays present at every route position. As it travels along its route, the mobile just "sees" the same standing wave, i.e., it receives the signal time series with the same shape but with a different scaling on the time axis.

The ideal conditions mentioned above will rarely occur in reality, and usually, time variations will be present. However, the use of the standing wave

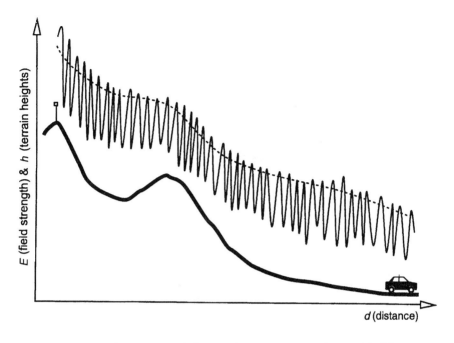

Figure 1.2 Variations in the received signal with the movement of the mobile [1].

notion to visualize the mobile propagation channel is still valid, because the multipath ray combination effects will still exist and a standing wave pattern somehow moving in time will be present.

During measurements, the mobile speed $V(t)$ should remain constant. If this were so, it would be possible to make direct conversions between the representations in the time $e(t)$ and traveled distance domains $e(x)$ ($t = x/V$). In practice, however, it is not possible to keep the same speed throughout the measurements due to such factors as traffic lights and traffic density conditions. For this reason, it is necessary not only to record the field intensity but also the instantaneous speed or the traveled distance by means of a *fifth wheel* that sends reference impulses with each revolution. The received time series $e(t)$ must be processed together with the instantaneous velocity record to produce a new time-series $e'(t)$, referred to a constant mobile speed or, alternatively, a series in terms of the traveled distance $e(x)$. The variable x may either be expressed in meters or in wave lengths. In order to make these conversions, a *velocity-weighted* algorithm must be used. Based on these preprocessed signal records in the distance domain, it is possible to separate and study individually both the fast and slow variations due to multipath and shadowing phenomena, respectively. Figure 1.4(a) shows an example of a received signal and an instantaneous mobile speed [2].

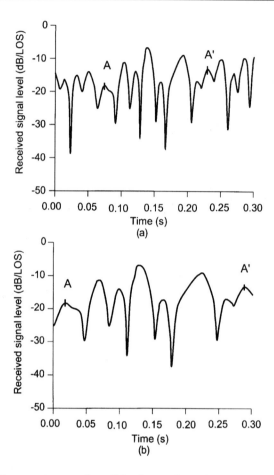

Figure 1.3 Standing wave pattern "seen" by the receiver when traveling at speeds (a) V and (b) $V/2$ [1].

1.3 Characterization of Signal Variations

Generally, the received signal variations $e(t)$ or $e(x)$ may be broken down, in a more or less artificial way, into two terms [1]:

- The slow or long-term variations: $m(t)$ or $m(x)$;
- The fast or short-term variations: $r_o(t)$ or $r_o(x)$.

The received signal may, therefore, be described as the product of these two terms

$$e(t) = m(t) \cdot r_o(t) \text{ or, alternatively, } e(x) = m(x) \cdot r_o(x) \quad (1.1)$$

Introduction: The Propagation Channel in Mobile Communications 9

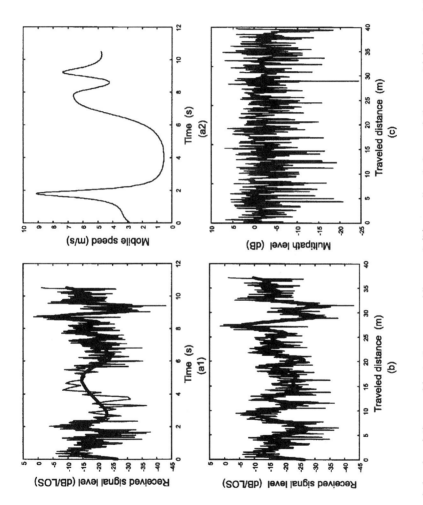

Figure 1.4 (a) Received signal variations and instantaneous mobile speed, (b) overall and slow variations, and (c) fast variations after separating the slow variations [2].

The fast variations will be superimposed on the slow signal variations. Figure 1.4(b) illustrates a signal variation time series (once the abscissa axis with converted to traveled distance, as described in Section 1.2); the figure also shows the slow variations. Figure 1.4(c) shows the fast variations after removing (filtering out) the slow variations in the time series in Figure 1.4(b). The slow signal variations are extracted from the overall variations by low-pass filtering. This is equivalent to compute the signal average over a few samples within a stretch of route with a length $2L$ of some tens of wavelengths

$$m(x_i) = \frac{\sum_{k=-N}^{N} e_{i+k}}{2N+1} \text{ for } e_{i-N} \ldots e_i \ldots e_{i+N} \in x_i - L < x < x_i + L$$

(1.2)

Typically, values of 20–40 λ are used. For example, for the 900-MHz (λ = 0.33m) band used in cellular mobile communications in Europe, the averaging length will be $2L \approx$ 7–14m. The average value $m(x_i)$ computed for a given route position x_i is usually called the *local mean* at x_i.

It has been observed experimentally that the slow variations of the received signal, that is, the variations of the local mean $m(x)$, follow a log-normal distribution when expressed in linear units or, alternatively, a normal distribution when expressed in logarithmic units $M(x_i) = 20 \log m(x_i)$ (see Chapter 3).

The length ($2L$) of route considered for the computation of the local mean—i.e., that used to separate the fast from the slow variations—is usually called the *small area*. It is within a small area where the fast variations of the received signal are studied, since they can be described there with well-known distributions (Rayleigh). Over larger route stretches ranging from 100m to 1 or 2 km, the variations of the local means are studied. These zones are usually called *larger areas*. Typically, standard propagation models do not attempt to predict the fast received signal variations. Instead they predict the median $\tilde{M}(x) = \tilde{E}$ and the standard deviation (or location variability) σ_L of the local mean variations (normal distribution in logarithmic units) within the larger area. Figure 1.5 illustrates these concepts.

1.4 Path Loss and Signal Variations: Characterization of the Slow Variations

As shown in Figure 1.2, as the mobile travels along its route, strong received signal variations are observed. These variations may be classified into *fast* or

Introduction: The Propagation Channel in Mobile Communications 11

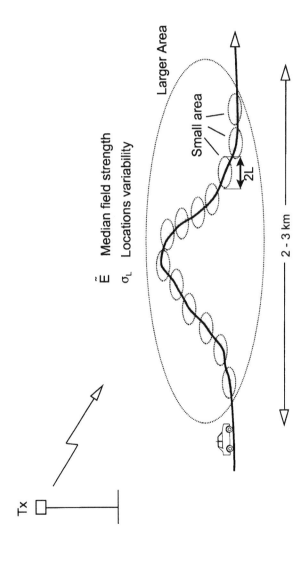

Figure 1.5 Illustration of the different areas used when studying propagation in mobile communications.

short-term variations and *slow* or *long-term* variations. The fast variations are superimposed on the slow variations. The slow variations are due to shadowing/blockage by natural or man-made structures along the propagation path and in the vicinity of the mobile. To study the slow variations, it is first necessary to separate them from the fast variations by low-pass filtering of the overall signal.

The first slow variations component is due to the free-space losses, L_{fs} (dB)

$$l_{fs} = \frac{p_t}{p_r} = \left(\frac{4\pi d}{\lambda}\right)^2 \text{ or } L_{fs} = 32.4 + 20 \log f(\text{MHz}) + 20 \log d (\text{km})$$

(1.3)

where p_t and p_r are the transmitted and received powers, respectively, and L_{fs} and l_{fs} are the free-space losses in decibels and in linear units, respectively. Throughout this book, variables in capital letters will denote magnitudes expressed in logarithmic units (dB) and in lowercase variables magnitudes expressed in linear units.

Free-space losses give rise to a drop in the received signal power with distance following a power law with exponent $n = 2$ (Figure 1.6)

$$p_1 \alpha \frac{1}{d_1^2} \text{ and } p_2 \alpha \frac{1}{d_2^2} \text{ then } \Delta_p = 10 \log \frac{p_2}{p_1} = 20 \log \frac{d_1}{d_2}$$

(1.4)

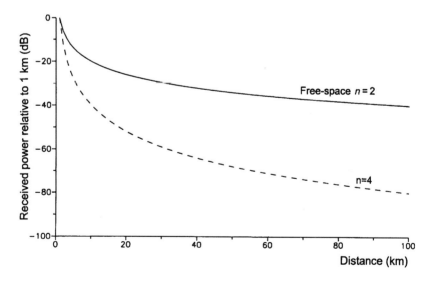

Figure 1.6 Received signal decay with distance: $n = 2$ and $n = 4$ laws.

where α means "proportional to" and Δ_p is the received power/propagation loss increment in decibels.

The above expressions show a 20 dB/decade decay rate of the received signal power with distance. These variations are steeper in the first few kilometers of the radio path and become more gentle over greater distances. For example, using (1.4), from km 1 to km 2 of the radio path, a 6-dB loss increment due to free-space propagation is observed. However, the same 6-dB loss increment is observed from km 10 to km 20.

The second component is mainly observed some distance away from the transmitter. The most important variations are those due to the irregularity of the terrain (obstruction by hillsides) and those due to natural and/or man-made features (houses, buildings, trees, etc.) in the vicinity of the mobile.

However, it has been experimentally observed that, in typical mobile propagation paths, the signal loss variations as a function of the distance from the transmitter do not decrease with the square of the distance (as in free-space conditions), but, rather, they present higher exponent values. The signal decay with distance is usually modeled with a $l \propto d^n$ law, i.e., a law of variation of losses l with distance to the power of n. The values of n will typically be somewhere near 4, i.e., 40 dB/decade (Figure 1.6). For example, in the case of plane earth propagation, $n = 4$ due to the interaction of a direct and a reflected ray as explained in detail in Chapter 4.

The propagation loss expressions normally provided by mobile propagation models (see Chapter 4) have forms similar to the one in (1.5).

$$L(\text{dB}) = A + B \log d \text{ (km)} = A + 10 \, n \log d \text{ (km)} \qquad (1.5)$$

Several other factors greatly influence path losses and are taken into consideration by the different propagation models. These factors are listed as follows:

- Frequency;
- Distance;
- Height of the base station antenna;
- Height of the mobile station antenna;
- Height of the base station relative to the surrounding terrain (effective height);
- Terrain irregularity (undulation: Δh, roughness: σ_t) (see Chapter 4);
- Land-usage (morphography) in the vicinity of the mobile terminal.

When calculating the total path losses, all the different factors must be taken into account

$$L_{\text{Total}} = L_{\text{Basic}} = L_{\text{Reference}} + L_{\text{Terrain Irregularity}} + L_{\text{Environment}} \quad (1.6)$$

The *basic loss,* defined as the loss between isotropic antennas, is made up of three main components. The first component represents the *reference* losses, typically the free-space loss, although some models like Hata's [3] take urban area loss as a reference. Another element is the loss due to *terrain* blockage, and the final element comprises the loss due to the *environment* in the vicinity of the mobile (urban, suburban, rural, open, woodland, etc.).

1.5 Fast Variations

The fast signal variations [1], superimposed on the slow variations, are due to a number of multipath echoes reaching and combining at the receiver antenna. These echoes are generated by specular and/or diffuse reflections from environmental features, generally located near the mobile, such as houses, buildings, lamp posts, or any other artificial structures or natural elements, such as trees, bushes, or the ground itself.

The study of the multipath phenomenon is performed for small sections ($2L$) of the mobile route (small area) 20λ to 40λ long. For these short stretches of route, the powers of the different multipath contributions reaching the receiver do not change from a practical point of view. However, as the mobile runs, the relative phases of the multipath components change, giving rise to variations in the total received signal power. The following simple examples illustrate how the multipath phenomenon is generated [1].

Case 1

Figure 1.7 illustrates a situation in which a stationary mobile receiver is in the direct line-of-sight (LOS) of a base station and other mobiles are passing by. Also found in Figure 1.7 is a possible record of the received signal variations. In this case, the received signal will basically consist of the direct path and will be practically constant except for some small variations due to occasional echoes from the vehicles passing close by.

Case 2

It is assumed here that a mobile is traveling at a constant speed V in an open area. In this case, it receives a strong direct signal, whereas the powers of

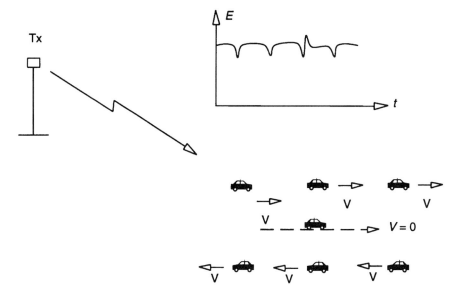

Figure 1.7 Case 1: Stationary mobile terminal [1].

multipath echoes are assumed to be negligible. This situation is illustrated in Figure 1.8(a). The received signal may be expressed as

$$e_r = e_0 e^{j2\pi f_t t} e^{-j\frac{2\pi}{\lambda} x \cos\theta} \quad (1.7)$$

where x is the transmitter-to-mobile distance and f_t is the transmitted frequency.

Replacing the distance x with $x = tV$ in (1.7) yields

$$e_r = e_0 e^{j2\pi\left(f_t - \frac{V}{\lambda}\cos\theta\right)t} \quad (1.8)$$

In (1.8), it can be observed how the received carrier frequency f_r is offset from the original transmitted frequency f_t due to the Doppler shift effect

$$f_r = f_t - \frac{V}{\lambda}\cos\theta \quad (1.9)$$

The magnitude of Doppler shift f_D caused by the relative mobile-base movement is

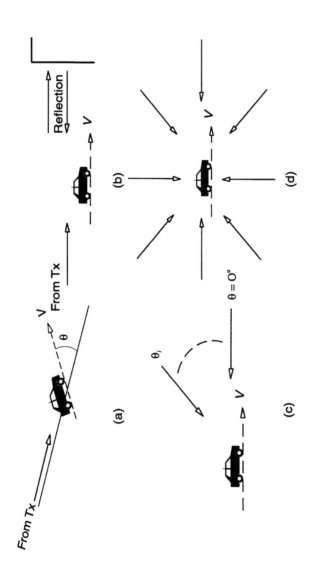

Figure 1.8 (a) Case 2: Mobile moving at speed V with direct visibility and negligible multipath. (b) Case 3: Direct plus reflected signal. (c) Case 4: Multipath with only two echoes. (d) Case 5: Multiple echoes with no direct signal [1].

$$f_D = \frac{V}{\lambda}\cos\theta \qquad (1.10)$$

It may also be noted that the magnitude of the received signal $|e_r| = e_o$ does not change. In addition, it can be observed how, if the mobile moves away from the transmitter ($\theta = 0$ degrees), the received frequency decreases to a minimum value

$$f_{r\min} = f_t - \frac{V}{\lambda} \qquad (1.11)$$

Conversely, when it travels in the direction of the transmitter ($\theta = 180$ degrees), the received frequency increases to a maximum value

$$f_{r\max} = f_t + \frac{V}{\lambda} \qquad (1.12)$$

A simulation [4] has been performed with a 1,500-MHz carrier ($\lambda = 0.2$m) and a mobile speed of 80 km/h. Values were obtained for the magnitude and phase of the received signal at equally spaced points along the mobile route. It can be observed that the signal amplitude does not vary. However, the phase does vary—Figure 1.9(a)—causing a shift in the frequency initially transmitted (i.e., it causes a Doppler shift of the transmitted carrier frequency—Figure 1.9(b)—of $-V/\lambda = -(80 \times 1000)/(3600 \times 0.2) = -110$ Hz). When the mobile travels perpendicular to the mobile-to-base direction ($\theta = 90$ degrees), the received frequency will be the same as the transmitted frequency.

Case 3

It will be assumed here that the mobile travels at a constant speed V along a road running between the transmitter and a reflecting surface, e.g., a large building—Figure 1.8(b). The received signal will follow the expression

$$e_r = e_o\exp\left[j2\pi\left(f_t - \frac{V}{\lambda}\cos\theta\right)t\right] - e_o\exp\left[j2\pi\left(f_t + \frac{V}{\lambda}\cos\theta\right)t\right] \qquad (1.13)$$

where it was assumed that the reflection coefficient of the building face is $R = -1$. The resulting received signal amplitude $|e_r|$ will oscillate between 0 and $2e_o$ (twice the free space amplitude).

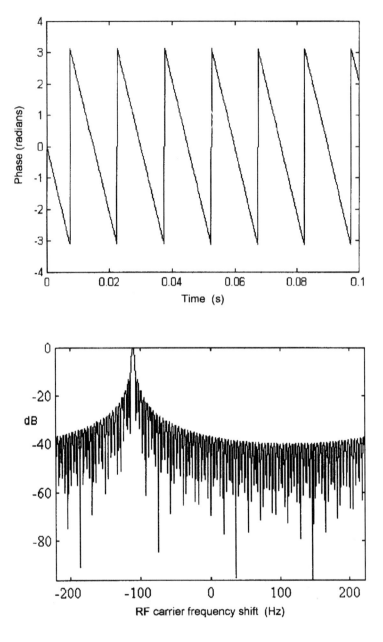

Figure 1.9 (a) Case 2: Phase variations. (b) Case 2: Doppler shift.

Introduction: The Propagation Channel in Mobile Communications

$$|e_r| = 2e_0 \sin\left(2\pi\frac{V}{\lambda}t\right) \qquad (1.14)$$

This phenomenon is the same as that observed on a short-circuited transmission line where a standing wave is generated by the interference effect of a direct and a reflected wave. Figure 1.10 shows the standing wave observed in this mobile communication scenario in linear units. Alternatively, in Figure 1.11(a) the received signal is shown in logarithmic units (decibels) with values between +6 dB and very marked negative ones corresponding to signal magnitudes of $2e_0$ (twice the transmitted amplitude) and 0, respectively.

The phenomenon of fast fading may be explained using the concept of a standing wave that the mobile receiver "sees" as it travels. Obviously, in more realistic situations, the standing wave pattern produced by N echoes from N scatterers will give rise to much more complex signal variations, as shown in Fig. 1.13, Case 5, that can only be described statistically and not in a deterministic way as in this simple case.

Case 4

Assuming now [1] that two echoes of equal magnitudes reach the mobile receiver, and further assuming that these echoes do not come exactly from opposite directions as in Case 3, e.g. ($\theta = 0$ degrees) and ($\theta = \theta_i$)—Figure 1.8(c)—the received signal would be

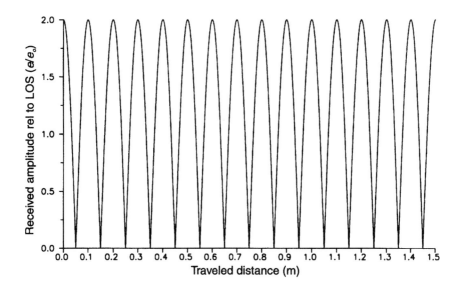

Figure 1.10 Case 3: Standing wave traversed by the mobile.

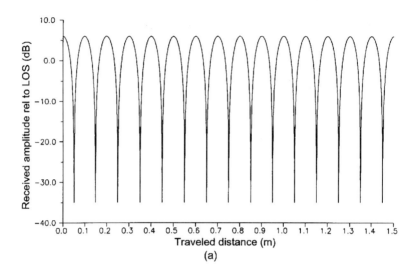

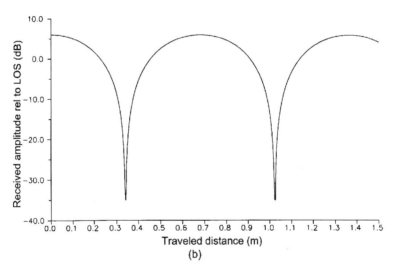

Figure 1.11 (a) Case 4: θ_i = 180 degrees. (b) Case 4: θ_i = 45 degrees.

$$e_r = e_o e^{j2\pi f_t t}\left(e^{+j\frac{2\pi}{\lambda}x} + e^{+j\frac{2\pi}{\lambda}x\cos\theta_i}\right) \quad (1.15)$$

$$= 2e_o e^{j2\pi f_t t} e^{j\frac{2\pi x}{\lambda}\frac{1}{2}(1+\cos\theta_i)} \cos\left(\frac{2\pi}{\lambda}\frac{x}{2} - \frac{2\pi}{\lambda}\frac{x}{2}\cos\theta_i\right)$$

Replacing x with $x = tV$ and rearranging the expression gives

$$e_r = 2e_o\cos\left(\frac{2\pi}{\lambda}\frac{Vt}{2}(1-\cos\theta_i)\right)e^{j2\pi f_t t}e^{j\frac{2\pi}{\lambda}\frac{Vt}{2}(1+\cos\theta_i)} \quad (1.16)$$

In (1.16), both the transmitted frequency shift and the amplitude fluctuations can be clearly seen. The fading frequency f_{fading} [1] (i.e., the frequency of variation of the received signal amplitude) is

$$f_{\text{fading}} = \frac{V}{2\lambda}(1-\cos\theta_i) \quad (1.17)$$

When $\theta = 0$ degrees, as in Case 2, the fading frequency f_{fading} is zero (i.e., no amplitude fading occurs if all echoes come from the same direction) [1]. When the echoes come from opposite directions, i.e., when $\theta = 180$ degrees, the maximum fading frequency case occurs, with a value

$$f_{\text{fading max}} = \frac{V}{\lambda} \quad (1.18)$$

To illustrate the variation in the fading frequency with the angle of arrival (AoA) of the echoes, two examples are now presented. Assume two scatterers and no direct ray. In all the cases, one of the scatterers is fixed facing the mobile, and the other scatterer varies its angle of arrival θ_i between 180 degrees and 45 degrees (Figure 1.11). Additionally, Doppler shift spectra are shown for $\theta_i = 180$ degrees and $\theta_i = 45$ degrees (Figure 1.12). In Figure 1.12, the two spectral lines corresponding to the two echoes are clearly seen.

Case 5

Finally, the general case where multipath echoes arrive from multiple directions is shown (Figure 1.8(d))

$$e_r = \sum_{i=1}^{N} e_i e^{j2\pi f_t t} e^{j\frac{2\pi}{\lambda}Vt\cos\theta_i} \quad (1.19)$$

Figure 1.13(a) illustrates a simulated sequence [4] of received signal level values in decibels relative to the direct LOS signal obtained for a scenario such as that shown in Figure 1.1 (a mobile traveling through a tree alley where the direct signal is obstructed). Figure 1.13(b) illustrates the variations observed on the phase.

It is noted that this case resembles more closely the conditions observed in real environments where the number of echoes is very large. These situations

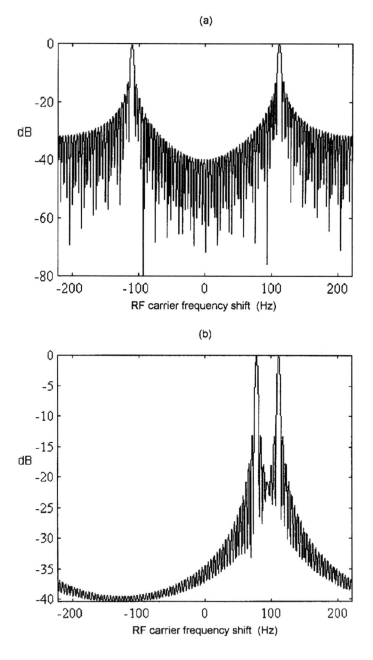

Figure 1.12 (a) Case 4: Doppler spectrum for θ_i = 180 degrees. (b) Case 4: Doppler spectrum for θ_i = 45 degrees.

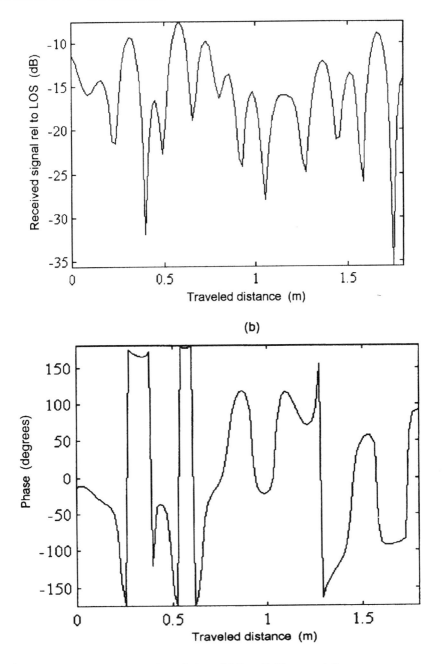

Figure 1.13 (a) Case 5: Amplitude variations. (b) Case 5: Phase variations.

are extremely complex. To study these, statistical distributions, such as the Rice and Rayleigh distributions, must be used. These statistical distributions are examined in detail in Chapter 3.

References

[1] Lee, W. C. Y., *Mobile Communications Design Fundamentals,* New York, NY: John Wiley & Sons, Inc., 1993.

[2] Lecours, M., et al., "Statistical Modeling of a Land Mobile Radio Channel," *IEEE Vehicular Technology Conference,* 1986, pp. 232–238.

[3] Hata, M., "Empirical Formula for Propagation Loss in Land Mobile Radio Services," *IEEE Transactions on Vehicular Technology,* VT-29, August 1980, pp. 317–325.

[4] Fontan, F. P., A. Seoane, and M. A. V. Castro, "Matlab for Windows Software Aid in a Mobile Communications Course," *Int. Journal of Electrical Engineering Education,* Vol. 32, No. 4, October 1995, pp. 341–349.

2

Introduction to Mobile and Private Mobile Radio (PMR) Systems

2.1 Introduction to Private Mobile Radio (PMR) Systems

Mobile radio plays an extremely important role in commercial, business, and security sectors in the more developed countries. Of course, public telephone applications (cellular systems) are even more important. The growth in demand has been very fast since the beginnings of mobile radio and calls for adequate use of the available radio-electric spectrum. Regulatory bodies in each country are responsible for efficient spectrum utilization.

In the first few years of mobile communications, the increase in demand was fulfilled by opening new frequency bands. For example, in Europe the 450- and 900-MHz bands were opened up for cellular telephony, and the VHF-Band III (174–225 MHz) was opened for trunked systems. In the larger metropolitan areas, however, the saturation phenomenon on the various frequency bands is already a fact. This means that there is a real need to resort to other, more spectrally efficient techniques, such as the deployment of microcells and picocells with limited coverage ranges and/or the use of other access/modulation techniques such as TDMA and CDMA (rather than classical FDMA), which require lower interference protection ratios.

The need to accommodate larger numbers of users and, at the same time, improve spectral efficiency has led to the evolution of cellular systems. First-generation analog systems (AMPS, TACS, NMT, or C) have been replaced by more sophisticated, second-generation digital systems. The dominant systems in world markets are the GSM system developed in Europe and the IS136 TDMA and IS95 CDMA systems developed in the United States. As far as

PMR systems are concerned, the first extremely simple systems (described throughout this chapter) were initially replaced by analog trunked systems and, more recently, by digital systems as, for example, the TETRA standard from the European Telecommunications Standardization Institute (ETSI). This same trend has also been noted in other wireless systems such as cordless and paging.

The most widely used frequencies are found in the VHF and UHF bands where coverage ranges do not exceed a few tens of kilometers (< 30 km). This feature makes frequency reuse possible at relatively short distances thus allowing an efficient use of the radio spectrum.

This chapter concentrates on the conventional PMR systems, which are simple in comparison with the sophisticated cellular or trunked systems. PMR is also known as closed user group radiotelephony as it is used for internal communications by government agencies and companies for applications such as delivery, security, and maintenance. Typical PMR users include electric companies, gas suppliers, police, ambulance services, water suppliers, highway pick-up services, and truck fleets.

As this chapter unfolds, a series of important basic concepts in PMR systems are defined. These concepts are still valid for the more sophisticated systems such as analog and digital cellular, trunked, and cordless, which will be reviewed in some detail in Part III of this book.

Initially, closed user group radio systems may be classified as follows:

- Private systems (PMR) proper;
- Public systems (public access mobile radio (PAMR)).

Private systems usually serve relatively small areas and, generally, use fairly simple configurations. Private systems are related to the concept of self-provision of the communications service, meaning that a given company or agency requiring a PMR system will take care of its operation, actually owning the system.

On the other hand, operators of public systems offer their communication system to the public, freeing users from having to set up their own radio communication systems. Public systems generally have a regional or national coverage and often present a greater technical sophistication.

The required coverage in public systems is achieved using several radio station sites throughout the service area. Multifrequency configurations are used in this case, meaning that each cell, defined by a radio station, is allotted one or a set of radio channels different from those in the adjoining cells, which may be repeated in cells located at a sufficient distance. This is, in part, the well-known cellular concept which is dealt with in Section 2.5.

Table 2.1
Bands Used in Mobile Communications in Europe

Band	Frequency Ranges (MHz)
Low VHF	30.005–47; 68–74.8; 75.2–87.5
High VHF	146–149.9; 150.05–156.7625; 156.8375–174
VHF-Band III	223–230.4875; 273–322; 335.4–399.9
Low UHF	406.1–430; 440–470
High UHF	862–960
Other bands: L, S	1429–1525; 1670–1990; 1700–2655

Table 2.1 presents the most used frequency bands for mobile communications in Europe. The lower bands are used for PMR applications, down to "low" UHF. The "high" UHF band is used for cellular systems and the L- and S-bands are used for several wireless systems including cellular and cordless. Also, these bands will be used in future third-generation mobile systems. Regular meetings of the World Radio Conference (WRC) are held to allocate new frequency bands to services worldwide and to large geographical regions (e.g., ITU Region I, which includes, among other zones, the whole of Europe).

The channel spacings typically used in the PMR bands are 12.5 kHz and 25 kHz and the most widely used modulation is FM. Currently, there is a trend toward using digital modulations as new standards are introduced.

To determine the bandwidth taken up by each transmission, it is essential to account for the technical limitations of the equipment, including, among others, the frequency drift. Typical values may be on the order of ±2.5 kHz in hand-held terminals. There is, therefore, a given bandwidth available to be shared among the FM-modulated signals and a margin (guard band) between adjacent transmissions to allow for drifting. The occupied band corresponds to a maximum modulating frequency f_m of 3 kHz and the FM peak deviation Δf is ±2.5 kHz for the 12.5-kHz channel spacing. Figure 2.1 illustrates these concepts. FM-modulated signal bandwidths are usually calculated using the Carson rule ($BW = 2\Delta f + 2f_m$ (3 kHz)) [1]. Table 2.2 summarizes the parameters for the two typical channel spacings.

2.2 Channel Types in Mobile Communications

The radio channel types used in mobile communications [2] are listed as follows:

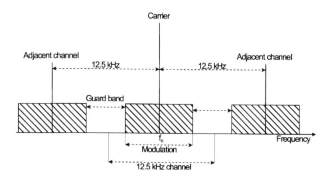

Figure 2.1 Channel spacing in mobile communications [3].

Table 2.2
FM Modulation Characteristics for Channel Spacings at 12.5 and 25 kHz

Channel Spacing (kHz)	Peak Deviation Δf (kHz)	No. of Sub-bands	Bandwidth $BW = 2\Delta f + 2f_m$ (3 kHz)
12.5	2.4	4	10.8
25	4.5	6	15.0

- Single-frequency simplex channels;
- Two-frequency simplex channels;
- Half-duplex channels;
- Full-duplex channels.

Single-Frequency Simplex Channels

Single-frequency simplex channels use the same frequency f_1 for both directions of transmission (mobile-to-base and base-to-mobile). To transmit, the push-to-talk (PTT) button is pressed, and the antenna switch (AS) is activated. This, in turn, connects the transmitter to the antenna and disconnects it from the receiver. See Figure 2.2(a).

Figure 2.3(a) illustrates a possible scenario with two single-frequency simplex systems with their base stations and mobile terminals. The single-frequency simplex configuration allows all the units to listen to all the transmissions within range. This feature is particularly useful for security applications such as the fire or police service, where there is a need to work as a group and where everyone must be aware of what is going on at any given time. For two terminals to be able to communicate with each other, they should both be within their common coverage area.

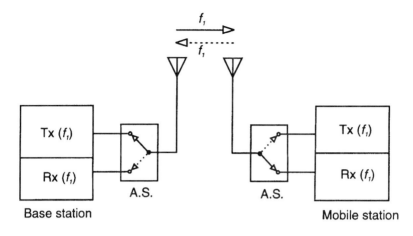

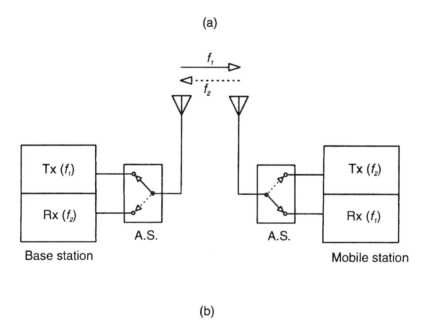

Figure 2.2 Channel types in mobile communications.

The advantages of this channel type include the mutual assistance feature based on the ability of mobiles to talk directly with one another, and an apparent efficient use of the spectrum, as only one frequency is needed for both directions of transmission. One of the drawbacks is that sharing a single-frequency channel with another group of users may lead to problems when a

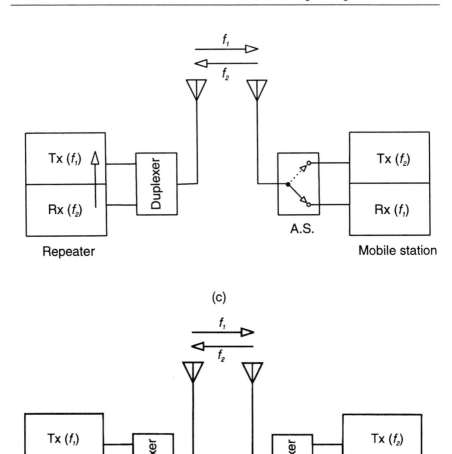

Figure 2.2 (continued).

transmission and a reception occur close to one another. These advantages and disadvantages are dealt with in more detail later in this section.

The mobile-to-base range is usually smaller (due to mobile power limitations) than the base-to-mobile range. The use of the same frequency will, occasionally, make communication possible in cases where one mobile is not

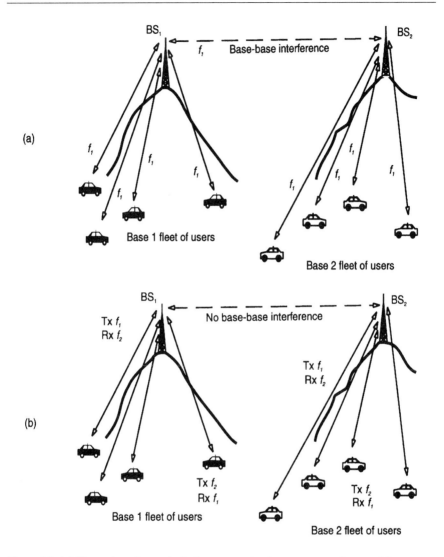

Figure 2.3 (a) Illustration of two single-frequency simplex systems and a possible interference scenario and (b) two-frequency simplex scenario.

able to reach the base directly. This may be done by means of another mobile retransmitting the message back to the base, as illustrated in Figure 2.4(a). Figure 2.4(a) shows how the base station can talk to mobile 1 and mobile 2. Mobile 1 can reach the base, but mobile 3 cannot, nor can the base reach mobiles 3 and 4. Mobiles 1 and 3 can communicate with each other but not with mobiles 2 and 4. If mobile 3 has an urgent message for the base, it can pass it on to mobile 1, which will, in turn, relay it to the base. This advantage

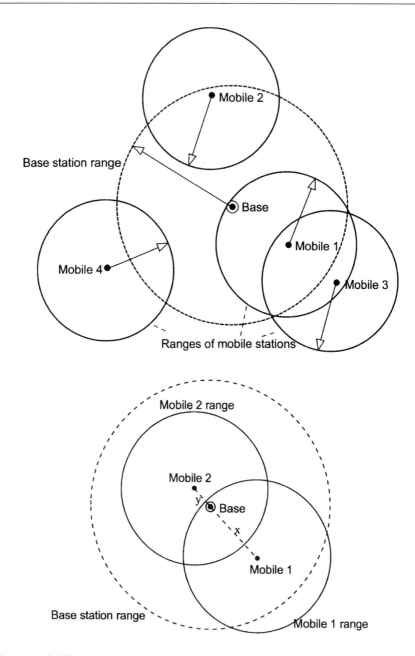

Figure 2.4 (a) Concept of mutual assistance and (b) near-far interference.

of single-frequency simplex systems is known as mutual assistance, as mentioned previously.

One serious drawback to single-frequency simplex systems is the possibility of inadvertently causing interference to a communication in progress as illustrated in Figure 2.4(b). If mobile 1 (M_1) is talking to the base from a distance x—assuming that mobile 2 (M_2) is at a sufficiently large distance from M_1 to make communications between M_1 and M_2 impossible but is at a closer distance from the base (distance y) than M_1—then M_2 may try to reach the base and transmit (assuming that nobody is on the air). As M_2 is so close to the base, the transmitted signal reaching the base will be much stronger than that from M_1. The signal from M_2, therefore, would "capture" the base station receiver, suppressing the signal from M_1 (near-far effect).

Another interference scenario is that shown in Figure 2.5(a). If there are several base stations in the same site, similar capture phenomena may occur if the frequencies used there are close to each other. Assume there are three simplex base stations using frequencies f_1, f_2, and f_3; if BS_2 transmits, this may capture or even block the reception of M_1-BS_1 transmissions and hinder possible M_3-BS_3 communications.

To reduce the influence of a transmitter on a nearby receiver and avoid blocking or the generation of intermodulation products, it is necessary to provide an adequate isolation. This can be achieved by physically separating the stations appropriately. In some cases, in a given site it is not possible to achieve the required spacing of several tens of meters or more, so it is necessary to resort to frequency separation. The required separations are on the order of 4–5 MHz (Figure 2.5(a)). This separation will provide the required isolation thanks to the selectivity of the receivers. It will be noted that within a frequency range of 8–10 MHz, only three frequencies may be used in the same site. This is a clear indication that single-frequency simplex channels use the radio spectrum in an extremely inefficient way.

It is also noted that, with this operation mode, frequency reuse, which is essential in these highly congested bands, is limited considerably (Figure 2.3(a)). If a system requires a wide area coverage, its base station should be set on an elevated site. If at a certain distance another single frequency simplex system operating on the same radio channel is installed on another elevated site, cochannel interference problems will appear, since transmissions from the interfering base will disturb transmissions from the mobiles [3].

To avoid problems like those described above, single-frequency simplex systems use reduced coverages to maximize channel reuse. This is achieved by restricting the maximum effective radiated power (ERP) defined as ERP = $P_{Tx} - L_{Tx} + G_d$ to small values such as 5W [3]. This type of channel is mainly used with low-power hand-helds.

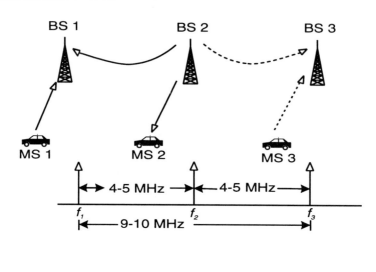

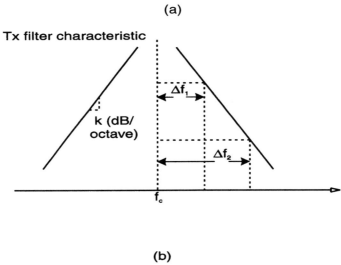

Figure 2.5 (a) Interference phenomenon between single-frequency simplex stations. (b) Characteristic of a transmitter filter.

It may be concluded that, although it appears that the use of a single frequency per station is convenient from the viewpoint of frequency economy, this is by no means the case, especially when several stations must share the same site.

Calculations Taking Into Account Transmitter Filter Effects

The procedure to calculate a carrier-to-interference ratio in which the effects of a transmitter filter are taken into account is as follows. The transmitter filter slope, k, is usually expressed in dB/octave or dB/decade, where

$$k \text{ (dB/octave)} \equiv k' \text{ (dB/decade)} = k/0.3 \quad (2.1)$$

The losses due to the transmitter filter are calculated as follows

$$L_{\text{TxFilter}} \text{ (dB)} = k \log_2 \frac{\Delta f_2}{\Delta f_1} = \frac{k}{0.3} \log_{10} \frac{\Delta f_2}{\Delta f_1} \quad (2.2)$$

where Δf_1 and Δf_2 are frequency separations from the wanted carrier. Figure 2.5(b) illustrates the Tx filter characteristic around the channel center frequency f_c.

Example 2.1. In this example, calculations are shown related to the isolation achieved both by spatial separation and Tx filter attenuation. Assume two transmitters, one with a frequency f_c of 451.225 MHz and the other with a frequency of 200 kHz above f_c. A receiver tuned on f_c is located at 20 km from Tx_1 (wanted signal) and at 5 km from Tx_2 (interfering signal). Assuming free-space propagation conditions, the carrier-to-interference ratio (c/i) may be calculated as shown below. It is assumed that the Tx filter slope is $k = 18$ dB/octave, that the channel spacing is 12.5 kHz, and that $G_{Tx_1} = G_{Tx_2} = 10\text{dB}_i$ and $P_{Tx_1} = P_{Tx_2} = 5\text{W}$. The received power of the wanted signal C is given by the expression

$$C = P_{Tx_1} + G_{Tx_1} - L_{fs}(Tx_1 - Rx) \quad (2.3)$$

and the interfering signal power I is given by

$$I = P_{Tx_2} + G_{Tx_2} - L_{fs}(Tx_2 - Rx) - L_{\text{TxFilter}} \quad (2.4)$$

then, the carrier to interference ratio c/i expressed in decibels ($C - I$) is given by

$$\begin{aligned} C - I &= P_{Tx_1} + G_{Tx_2} - L_{fs}(Tx_1 - Rx) \\ &\quad - (P_{Tx_2} + G_{Tx_2} - L_{fs}(Tx_2 - Rx) - L_{\text{TxFilter}}) \\ &= L_{fs.}(Tx_2 - Rx) + L_{\text{TxFilter}} - L_{fs}(Tx_1 - Rx) \end{aligned} \quad (2.5)$$

$$\begin{aligned} L_{\text{Txfilter}} &= k \text{ (dB/octave) } \log_2 (\Delta f_2 / \Delta f_1) = k/0.3 \log 10 (\Delta f_2 / \Delta f_1) \quad (2.6) \\ &= 18/0.3 \log_{10} (200/6.25) = 90.3 \text{ dB} \end{aligned}$$

$$\begin{aligned} L_{fs}(Tx_1 - Rx) &= 32.4 + 20 \log f(\text{MHz}) + 20 \log d_1 \text{ (km)} \quad (2.7) \\ &= 85.6 + 20 \log d_1 \end{aligned}$$

$$L_{fs}(Tx_2 - Rx) = 32.4 + 20 \log f(\text{MHz}) + 20 \log d_2 \text{ (km)} \quad (2.8)$$
$$= 85.6 + 20 \log d_2$$

finally, the carrier to interference ratio is

$$C - I = 90.3 + 20 \log d_2 - 20 \log d_1 = 78.3 \text{ dB} \equiv 67,608,298 \text{ times} \quad (2.9)$$

Two-Frequency Simplex Channels

The transmit-receive isolation necessary to avoid receiver blocking problems may be achieved in frequency instead of space (Figure 2.2(b)). In this manner, radio channels are established using two frequencies, f_1 and f_2 with a given separation between them. Table 2.3 shows typically used frequency spacings. These values may not be constant for all other networks; certain tolerances, as shown in Table 2.3, are allowed. The transmit-receive spacing shall be wide enough so that intermodulation products in the transmit sub-band do not "fall" on the receive sub-band. In this case, even though two stations are in close proximity to each other—as both transmit, for example, on the "high" frequency sub-band and receive on the "low" sub-band—the receivers will not be affected. Consequently, these systems make a more efficient use of the available spectrum as is illustrated in Figure 2.6(a), which shows four channels for four transceivers (transmitter-receivers) on the same site.

Using two-frequency simplex channels, it is possible to achieve a high rate of frequency reuse. However, two bases using the same radio channel (two frequencies) will not be able to hear each other, thus drastically reducing interference (Figure 2.3(b)). One possible drawback to this is that the mobiles will be unable to listen to each other and this may lead to collisions (i.e., two mobiles calling the base at the same time). In this case, it is possible that two or more mobile terminals will make several call attempts to set up a

Table 2.3
Typical Tx-Rx Separations

Band	Transmit-Receive Frequency Separation (Tolerance)
80 MHz	4,6 ± 1 MHz
160 MHz	4,6 ± 1 MHz
230 MHz	6 MHz
450 MHz	10 ± 2 MHz
900 MHz	45 MHz

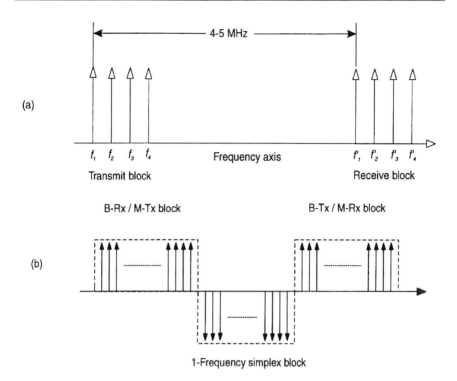

Figure 2.6 (a) Organization of the spectrum for two-frequency simplex channels. (b) Typical PMR frequency plan.

communication, thus producing an unnecessary load on the radio channel. In more complex systems, "bips" are broadcast by the base when this is involved in a communication with one mobile terminal in order to advise other mobiles that they cannot gain access as the channel is engaged. Two-frequency simplex is the most widely used mode of operation in PMR for dispatch applications where one base controls a variable number of mobiles. A typical ERP for bases and mobiles is 25W [3].

Half-Duplex Channels

To make communications between mobiles possible, base stations using two-frequency channels may be configured so that they repeat the signals received from the mobiles. This mode of operation is known as "talk-through." In this case, the base station operates in duplex mode (simultaneous transmit-receive) and the mobiles in simplex mode. This type of circuit, which is simplex at one end and duplex at the other, is known as half-duplex and is used to set up repeaters.

These talk-through systems operate as follows: The base station transmits on f_1' and receives on f_1, and the mobiles transmit on f_1 and receive on f_1'. The base station is equipped with a duplexer and the mobile is equipped with an antenna switch (Figure 2.2(c)). The base station/repeater is able to retransmit automatically on f_1' any signal received from the mobiles on f_1.

This type of system is also widely used in dispatch operations due to its capacity for mobile-to-mobile communication. For an efficient operation, the base station duplexer requires a minimum frequency separation between f_1 and f_1' of at least 600 kHz (antenna repeaters in the 144 MHz band), although the frequency separations authorized between f_1 and f_1' are usually higher, on the order of 5–10 MHz (see Table 2.3).

With this type of configuration, direct mobile-to-mobile communications are not possible without passing through the repeater, unless the mobile stations are also equipped with single-frequency simplex channels. This option is known as link bypassing the repeater. Figure 2.6(b) illustrates a typical frequency plan for the PMR bands. The two-frequency channels use one frequency from the transmit block and the other from the receive block. The simplex frequency is taken from the simplex block. In some cases, the simplex frequency may even belong to a different band, for example, the two-frequency channels may be on VHF and the single-frequency channel on UHF. The single-frequency simplex channel is typically used for mobile-to-mobile communications when away from the base coverage area or for mobile-to-portable operation.

For coverage reasons, repeater stations are set up on elevated sites and are usually left to function unattended and controlled by the incoming signal. Simple carrier activated (free-run) repeaters are not allowed in PMR applications when several organizations share the same radio channel.

Occasionally, when a station is used both as a base as well as a repeater, the mode of operation may be controlled via PTSN (telephone) leased lines or via fixed radio links. The system may be set up in such a manner that it automatically reverts to talk-through mode in the event of link failure or as a decision of the controller.

The following are some of the advantages of half-duplex operation:

- It is easy to set up and operate; this makes it possible for the repeater to work unattended;
- There is a minimum antenna requirement since at the repeater station, only one antenna is needed if a duplexer is used;
- Due to its simplicity, it may be used on mobiles (mobile repeaters).

The disadvantages include the following:

- In free-running mode the repeater may be activated by unwanted signals or interference, thus causing further interference;
- The problem may be solved by using access subaudio tones (see Section 2.4). This complicates the repeater equipment and the mobiles, although it is a widely used technique;
- The duty cycle of the repeater station is usually high. This imposes certain endurance requirements on the equipment;
- The operation of several co-sited talk-through stations may give rise to intermodulation products that make it essential to carefully select the radio channels to be used.

Full-Duplex Systems

In these systems, the base station transmits on f_1 and receives on f_2 and the mobile transmits on f_2 and receives on f_1. Both the base station and the mobiles are equipped with duplexers to allow for simultaneous transmission and reception. See Figure 2.2(d).

Cellular systems use full-duplex channels. Systems using full-duplex channels require a different radio channel (pair of frequencies) to link each mobile with the base (Figure 2.7). Base stations shall then be equipped with several radio channels (i.e., be fitted with as many transceivers as required depending on the traffic generated by the mobiles). As in the two-frequency simplex channel systems, it is not possible to have direct mobile-to-mobile communication without going through the base station.

2.3 Types of Radio Station Control

The two possible types of radio station control are local and remote control. Local control implies that the equipment controlling the operation of a base station (e.g., start-up, shutdown, PTT, channel selector, and volume or squelch controls) is a part of the station itself or lies in very close proximity to it (within about 30m). The remote control configuration means that the equipment controlling the operation of the radio station is located at a certain distance from it (from a few meters or tens of meters to several kilometers). To implement a remote control system, dedicated links using physical lines or radio must be set up (Figure 2.8).

Line-based remote control systems are set up on lines leased from a telephone network operator. Two options are available: two-wire and four-wire lines. For distances of a few kilometers, control signals may be transmitted by continuous current signaling based on the 48V telephone line voltage (Figure 2.9(a)).

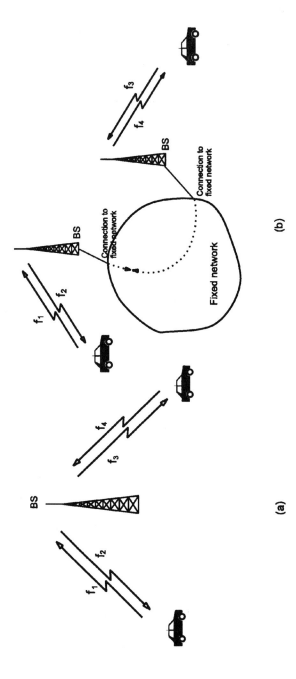

Figure 2.7 (a) Full-duplex communication using the same BS. (b) Full-duplex communication using two BSs and the infrastructure (fixed network) of a cellular system.

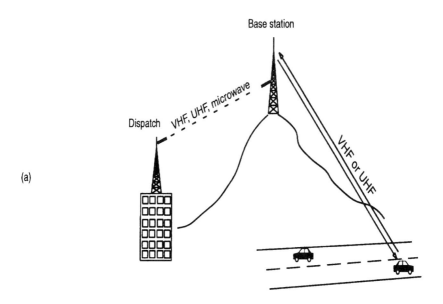

Figure 2.8 (a) Remote control using a fixed radio link [4]. (b) Detailed schematic diagram of a radio link controlled base station [2].

Generally, however, audio lines, which require audio tone signaling, are used. There is no standardized method for this purpose, although techniques such as sending one or two continuous audio tones generated at the control point and detected at the remote control point to activate the transmitter and other switch functions such as start-up/shutdown of talk-through and indicating the state of the line (e.g., loss of continuous tone means failure in the line) are used. A further possibility is to send FSK continuous tones, e.g., 2,898 Hz and 3,011 Hz with specific proportions of ones to zeros (mark/space). It is also possible to use more sophisticated systems by sending data bursts using FFSK modulation. See Figure 2.9(b).

Control may be achieved via single-channel radio links—Figure 2.8(b)—using the same or different bands: VHF, UHF, etc. The radio-links transport the information transmitted and received as well as activation/deactivation signals within the audio band using similar procedures as those described for landlines.

Alternatively, a base station may be operated in talk-through mode from a *stationary mobile* located at the dispatch office. This layout is known as frequency inversion operation.

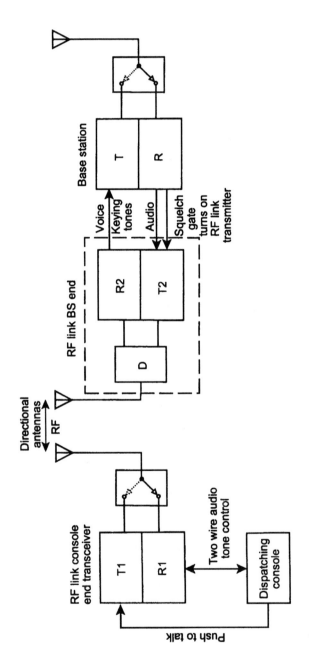

Figure 2.8 (continued).

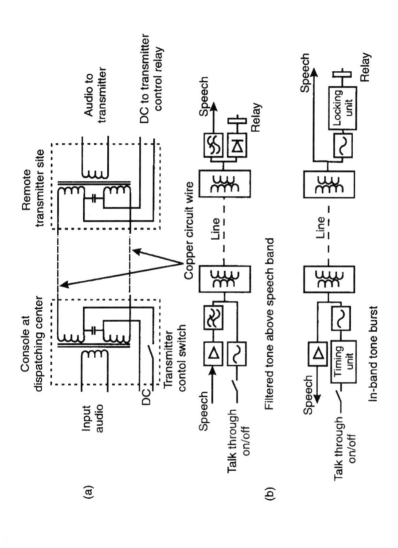

Figure 2.9 Line-based remote control systems. (a) Continuous current system [2]. (b) Audio tone signalling system [5].

2.4 Signaling in PMR Systems

Spectrum overload requires channel sharing between users in the same area. In special cases, there are licenses granting an exclusive use of a channel to a given organization. This is not normally the case, and it means that channels must be shared between several organizations. This sharing causes problems that require some sort of control, which is provided by signaling systems.

Also, even within the same organization, groups are established that have to share the same channel but operate in a totally independent way as, for example, in a large factory, there are maintenance, assembly, security, or storage groups. Signaling techniques make it possible to functionally isolate working groups by coding their transmissions in such a manner that each group is only able to listen to the traffic corresponding to its own group/fleet.

Also, there are signaling schemes that make it possible for individual mobiles to be addressed so that they can receive a communication aimed exclusively at them.

Three possible call types can be distinguished according to the mobile or mobiles addressed:

- Global calls;
- Group calls;
- Individual or selective calls.

Similarly, several signaling techniques may be used:

- Continuous tones;
- Sequential or bursts of tones;
- Combinations of tones;
- Digital signaling.

In PMR systems, digital signaling is used more and more for sending messages back and forth from mobile to base (and vice versa); this makes it possible to exchange a wide variety of possible instructions and responses. The most widely spread systems using digital signaling in PMR and PAMR are trunked systems. In Europe, the most common of these meets the UK DTI MPT 1327 standard.

Squelching Systems

Before describing squelching systems using both subaudio tones and digital codes, it is interesting to summarize the basic concepts related to this type of equipment. There are three types of squelching circuits:

- Carrier-activated;
- Subaudio tone-activated (continuous tone controlled signaling system (CTCSS));
- Digital signal-activated (digitally controlled squelch (DCS)).

Originally, squelching systems were used in mobile communications to avoid the noise generated by high-gain receivers being audible and disturbing the operator. Figure 2.10 illustrates the basic circuit using the carrier activation method. A high-pass filter at the FM detector output is used. The filter output is essentially noise, which is then amplified and rectified producing a continuous voltage. This continuous voltage is applied to the audio stages muting their output. By means of a variable resistance the continuous voltage level and, therefore, the squelch level is controlled. When a modulated carrier is received, the receiver gain is reduced by the limiting stages. This reduces the noise, which, in turn, reduces the squelch voltage, and, thus, the audio circuit in the receiver is activated.

When a repeater is carrier-activated, any transmission will be heard at the dispatch station, even those that do not belong to the same fleet of mobiles. In order to avoid this, squelch systems can be set up with a certain amount of coding, such as CTCSS.

Signaling by Subaudio Tones (CTCSS)

The CTCSS, also known as private line or tone lock, is the most widely used signaling method in PMR applications in which several user groups may share a single channel preventing each group from hearing the others. For this purpose, the transceivers are equipped with a coder/decoder, which adds a specific low-frequency tone to the normal audio signal in the transmitted signal (Figure 2.11(a)). Each user group will use its own tone. The receiver output is monitored by the tone decoder, which will open the receiver only when the incoming transmissions are accompanied by the required tone. There are 32 standardized tones ranging up to 250.3 Hz (Table 2.4).

To avoid accidentally transmitting over a communication that is under way, ways of monitoring channel activity may be put in place. One is to set up a lockout system in the PTT circuitry to make it impossible to transmit if the channel is already engaged with transmissions related to another tone (i.e.,

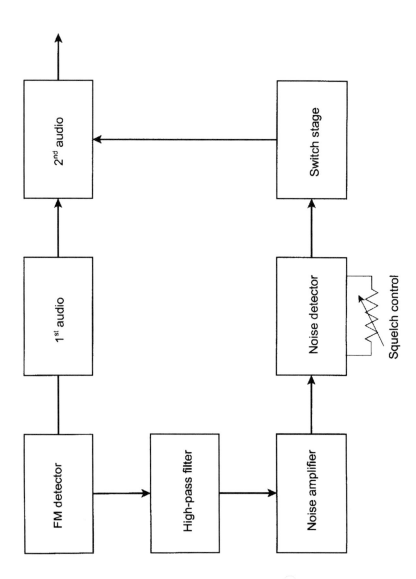

Figure 2.10 Carrier-activated squelch system [2].

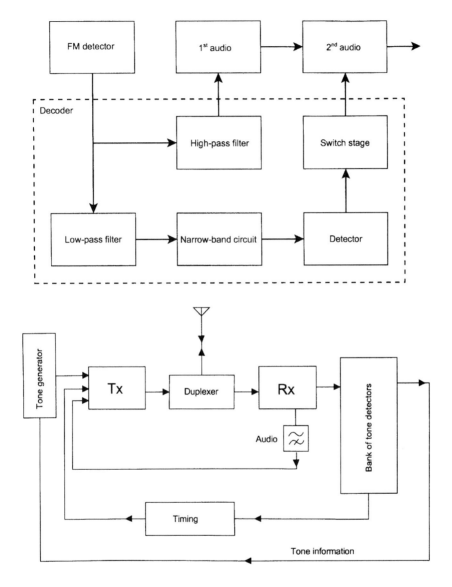

Figure 2.11 (a) CTCSS circuitry [2]. (b) Repeater shared by several fleets or groups [3].

from another user group). With the use of CTCSS, there will be a delay of approximately a third of a second between pressing the PTT and receiving the voice signal.

Shared Repeaters

Shared repeaters may be set up if talk-through bases are fitted with CTCSS signaling. Shared repeaters are very efficient in terms of spectrum economy.

Table 2.4
Standardized Tones for CTCSS Systems [3]

Number	EIA Group A Frequency (Hz)	EIA Group B Frequency (Hz)
1	67.0	71.9
2	77.0	82.5
3	88.5	94.8
4	107.2	103.5
5	114.8	110.9
6	123.0	118.8
7	131.8	127.3
8	141.3	135.5
9	151.4	146.2
10	162.2	156.7
11	173.8	167.9
12	186.2	179.9
13	203.5	192.8
14	218.1	210.7
15	233.6	225.7
16	250.3	241.8

A shared repeater will be used by several user groups, each group having its own CTCSS tone. Figure 2.11(b) shows a system where a CTCSS tone decoder bank detects the tone of an incoming transmission and retransmits both the audio and the subaudio tone.

Decoders ensure that transmissions passing through the repeater are only heard by the addressed group. Activity monitoring facilities may be implemented in the mobile terminals together with PTT blocking circuits. It is also advisable to include transmission timing circuits to limit channel occupation. The dispatch or "trigger base" units, located at the premises of the different organizations, would typically be stationary mobile units. The telecommunications administrations make it compulsory for these units to operate with directional antennas (yagi, as a general rule) pointing toward the repeater and using a limited power level determined by the signal received at the repeater.

Digitally Controlled Squelch (DCS)

A further possibility is a digital private line in which transmissions are accompanied by a low-rate, low-level, continuous binary digital signal that basically consists of a code transmitted repeatedly representing a particular three-digit number (Figure 2.12(a)). The decoder in the receiver (Figure 2.12(b)) will only respond to a unique incoming preprogrammed code. The advantage of

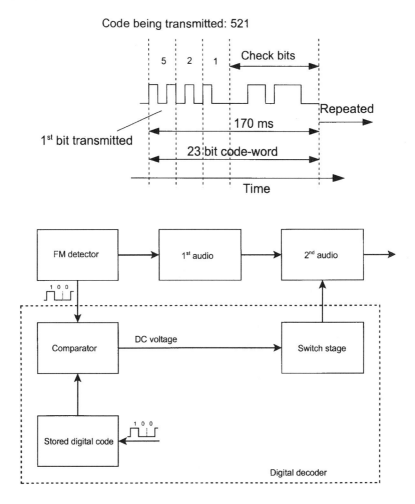

Figure 2.12 (a) Data structure of the DCS system [3]. (b) Receiver circuitry associated with DCS [2].

this technique over CTCSS is that far more combinations are available (up to 104 are recommended) and that the system offers a faster response time. The DCS decoder may also send an "end-of-transmission" code to close the receiver, thus eliminating delays.

Tone Sequence Signaling

In many PMR dispatch systems, it is necessary to call a particular mobile [3] instead of broadcasting general calls for all mobiles as would be done in a fleet of taxis where the messages must be heard by all the mobiles. To address a single mobile, selective call systems should be set up. These are generally based on sending a tone or a sequence of tones within the audio pass band of the transceiver. The factors that determine the signaling system capacity are the number of tones available for each time slot and the number of time slots used.

There are some standardized systems that define the transmission of tone sequences with fixed durations where each tone either stands for a digit or an instruction. Other signaling methods use two tones, like the multifrequency signaling used in conventional telephone networks. The most widespread system is the five-tone system. Table 2.5 shows the values for the tones defined in three different standards.

Each mobile will carry a preprogrammed decoder with a specific code. The dispatch uses a coder at the base station to select and transmit the five-tone code for the mobile called up. The mobile will normally be muted until it receives the correct selective calling code. The receiver may also return an automatic acknowledgment transmission by sending either a single tone or its own five-tone address code.

An alternative is to set up group calls using an additional tone, G. Due to synchronization problems, it is not advisable to send the same tone on consecutive time intervals, so that another tone, R, must be used to replace the second repeated tone. Table 2.6 shows examples of group calls and the use of the repeat tone.

Table 2.5
Frequencies in the Five-Tone Signaling System [3]

Digit	Frequencies (Hz)		
	ZVEI	EEA/CCIR	EIA
1	970	1124	741
2	1060	1197	882
3	1160	1275	1023
4	1270	1358	1164
5	1400	1446	1305
6	1530	1540	1446
7	1670	1640	1587
8	1830	1747	1728
9	2000	1860	1869
0	2200	1981	600
R	2400	2110	459
G	825	1055	2151
Tone duration	70 ms	100 ms	33 ms

Table 2.6
Examples of Calls [4]

Code	Activated Addresses	Number of Terminals
25784	25784	1
2578G	25780–25789	10
257GR	25700–25799	100
25GRG	25000–25999	1,000

The mobile may also use the selective call option to alert the dispatch that it requires attention [3]. This may be done by simply sending its own five-tone code. With such a system, known as vehicle identification (VI) or automatic number indication (ANI), the decoder at the dispatch displays the numbers of the selective calls received. This approach makes it possible to establish a moderately sophisticated fleet control system.

A further possibility provided by five-tone signaling is to set up a message-sending system which is used to report the status of the mobiles back to the dispatch [3]. This simply implies inserting additional digits to the address code sent by the mobile. The additional information may be keyed in by the user. The dispatch will poll all mobiles, which will, in turn, send back their own address codes plus status messages, indicating some information on the progress of their assigned tasks. Also, if the driver is not in the vehicle, the radio would automatically answer, transmitting its "status" message.

Other services may be included, such as a simple emergency call system (e.g., for vehicle security). For example, on single-manned buses, it is possible to install a "panic button" that would start up a specific sequence that would override the requests queued at the dispatch unit [3].

Other widely used selective call applications involve the activation of paging receivers, either directly (overlay paging) or by an automatic reaction of the mobile terminal (revertive or secondary paging) to alert the driver when not inside the vehicle (Figure 2.13) [3].

A further application of selective calling with tone sequences is for obtaining access to a talk-through station or to order the base to switch to talk-through mode. It is also possible to set up systems that give access to the company's own private switchboard (PABX) and make calls to office extensions using the base station as an interface. One or two digits would be added to address the required switchboard extension. There are, however, regulations in force that limit the time "on air," since telephone calls are usually longer than radio-to-radio transmissions [3].

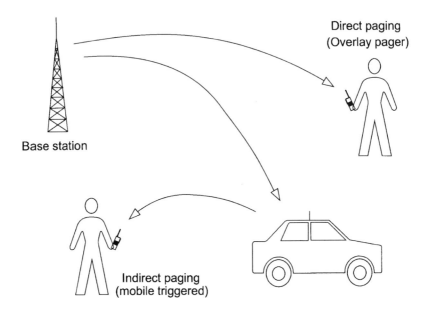

Figure 2.13 Possible types of paging with five-tone signaling [3].

Dual-Tone Multifrequency (DTMF) Signaling

The dual-tone multifrequency (DTMF) system, also known as *touch tone*, uses a frequency matrix to represent the conventional 12-key telephone dial (Figure 2.14). Each digit is sent by the simultaneous transmission of two frequencies, one in the low frequency group and the other in the high frequency group. The frequencies used are in the range between 697 Hz and 1,477 Hz. As it can be observed, these are within the audio pass band. A typical application of the DTMF system is accessing PABX extensions. Usually, the keyboard and the microphone would be part of a single unit.

2.5 Area Coverage Techniques

Obtaining suitable coverages from a single site is limited by factors such as terrain features and buildings, the frequency band being used, and the height of the base station antenna. There are many users that require broader operation areas, and this is only made possible by using two or more sites. Examples of this type of user include the railways, regional authorities, and emergency services. Another reason for using multiple sites would be the need to give adequate coverage to relevant shadow zones within the service area, these being mainly due to terrain irregularity. The most straightforward way to set up

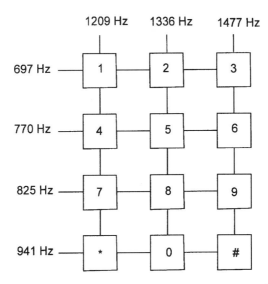

Figure 2.14 Frequencies in the DTMF system.

multiple site systems is to use different channels at each base station. The mobiles would select the appropriate channel at each location, depending on their positions. There are, however, other interesting alternatives that may be adopted for wide area coverage according to the specific requirements of the users.

Multichannel Operation

Most users avoid the complexities and the reception problems associated with the quasi-synchronous systems (dealt with later in this section) by using a set of different channels, each one covering a specific area of interest. The sites are linked to one or more control points and may also be interconnected to link mobiles in different areas. Mobiles would be fitted with multichannel transceivers and should select the appropriate channel for a given operation area. This is, needless to say, a burden on the spectrum. A certain degree of skill is required by the mobile operator to select the appropriate channel at each location. Despite this, it is the most widely used system for wide area coverage. The channel search may be made automatically by means of a frequency scanning system.

Voting Systems

Voting or signal selection systems have several operational advantages over repeater systems. They are oriented towards medium- and low-power mobiles and hand-held units in urban areas or areas where the transmitter units move within certain, clearly defined limits (a motorway, for example). Figure 2.15 shows the structure of a system of this type. These systems operate as follows: A signal selector in the control center receives, via leased telephone lines (or other means), the audio signals from a number of auxiliary or "satellite" receivers evenly distributed throughout the desired service area. The selector chooses the signal with the best quality amongst signals from different receivers.

When a mobile goes through the service area, its transmitted signal may be received by one or several satellite receivers in its vicinity. The received signal in each receiver is conditioned to the appropriate level and sent to the selector unit. The selector connects the line with the best quality to the audio stage. Once the selection has been made, the monitoring of each of the lines continues, and, if required, a new selection is made. Signal transmission from the base station to the mobiles usually takes place via a single, high-powered station. The system coverage range is thus limited by the base-to-mobile link.

Example 2.2. To illustrate the concepts outlined above, note the following calculation of the theoretical improvement to be expected in the return link (mobile-to-base) range of mobile terminals when operating in a voting system [2]. The following formula may be used to calculate the overall coverage probability:

$$p(\%) = 100 - (100 - p_B(\%))(100 - p_{s_1}(\%))(100 - p_{s_2}(\%))$$
$$\ldots (100 - p_{s_n}(\%))$$

(2.10)

where

$p(\%)$ is the probability that a transmission from a mobile is adequately received by the system.

$p_B(\%)$ is the probability that the mobile will be correctly received at the base.

$p_{s_i}(\%)$ is the probability that the mobile will be correctly received at auxiliary receiver i.

Assume that a mobile or hand-held terminal is in a position within the service area where the probability of being adequately received at the base is 50%. Further assume the availability of six "satellite" receivers are arranged in a circle around the base (Figure 2.16) and that their correct reception probabilities are p_{s_1} = 80%, p_{s_2} = 70%, p_{s_3} = 40%. (The other receivers fail to receive a signal level able to intervene in the voting.) The overall coverage probability $p(\%)$ is

Introduction to Mobile and Private Mobile Radio (PMR) Systems 55

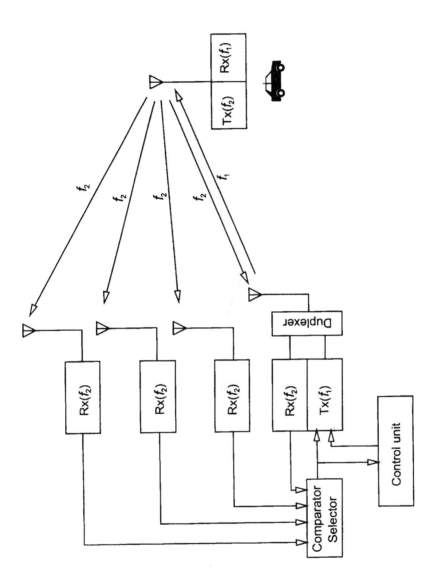

Figure 2.15 Diagram of a voting system.

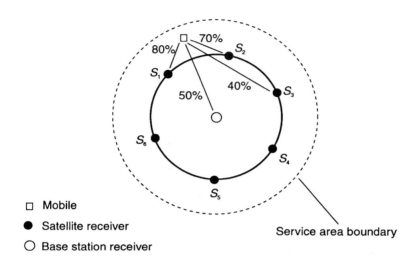

Figure 2.16 Voting system with six auxiliary receivers [2].

$$p(\%) = 100 - (100 - 50)(100 - 80)(100 - 70)(100 - 40) = 98.2\% \tag{2.11}$$

In other words, the probability for an adequate connection in the mobile-to-system (base or "satellite" receivers) direction has increased from 50% to 98.2% by setting up a voting system.

When hand-held terminals are used, the number of satellite receivers should be increased considerably. This is due to the limited range of this type of transceiver due to power and antenna efficiency limitations (see Chapter 5).

Synchronous and Quasi-Synchronous Operation (Simulcast)

For systems with extended service areas, or in smaller areas where there is a need to fill in shadow zones, it may be possible to set up base stations, with all of them operating on the same channel. If all the transmitters are locked on the same frequency, this is known as synchronous mode operation.

This setup, however, presents some drawbacks, especially in areas where the transmissions are received with similar powers (overlap area), i.e., when their difference is less than 10 dB, and, therefore, there is no capture effect. In this case, a standing wave pattern will be generated throughout the overlap region with signal enhancements and deep signal cancellations in much the same way as in the case of multipath propagation. The further away the sites are from each other, the greater the coverage overlap area (Figure 2.17). Also, Doppler shifts due to the movement of the mobiles may cause beat notes of

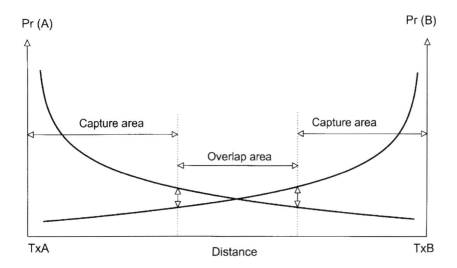

Figure 2.17 Overlap and capture areas.

up to 80–100 Hz. Synchronous operation is achieved at the expense of specialized, costly equipment with which identical frequency transmissions are possible in all the bases.

Quasi-synchronous operation is more practical from the implementation point of view. In this operation mode, the transmitters at each site are offset by a few tens of hertz (5–40 Hz). Although this gives rise to beat notes when the receiver is in the overlap area, the beat notes generated will be well below the audio pass band of the mobile receiver. The received signal enhancements and cancellations change their positions with time. These effects are perceived as a certain degree of fluctuation in transmissions that do not disturb the communication quality significantly.

In quasi-synchronous operation, independent clocks are used at each site; these must be adjusted with care and on a regular basis. An important requirement for both synchronous and quasi-synchronous installations is that the paths from the control point to each transmitter site must be equalized (delays, phases, and bandwidth response). The objective is to avoid distortion in the received signals. Modulation should also be adjusted as regards gain/loss, phase, and frequency response. Such strict requirements may make the use of leased landlines unfeasible (since they are not under the full control of the radio system operator), while favoring microwave lines that will be a part of the user infrastructure.

Since the sites are linked via a control point, it is possible to operate in talk-through mode in all the bases, and all the mobiles may listen to each other. For data applications, system adjustment should be even more thorough in order to avoid large error rates in the overlap areas. Finally, it is obvious that, in the mobile-to-base direction, the system will operate in the same way as a voting system.

2.6 Typical Example of a Dispatch System

This section describes a conventional dispatch system in which coverage is achieved with a multifrequency setup. Tone signaling is used for both selective and group calls. Three base stations BS_1, BS_2, and BS_3 define three coverage areas; these stations are managed from a common control position. This system is comprised of the following basic elements (Figure 2.18):

- Three base stations;
- Several mobile stations fitted with a manual channel selection switch to select the available channel within each area;
- One operating console fitted with one loudspeaker for each base station and a common microphone that can be switched to address one or several areas;

The system has four channels:

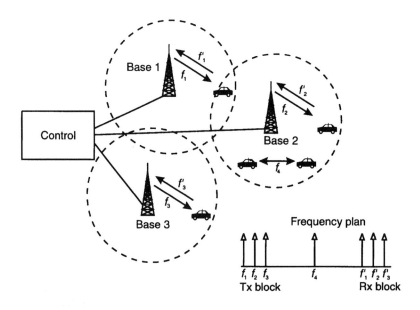

Figure 2.18 Example of a simple dispatch system.

- Three two-frequency channels for simplex or half-duplex operation. Each channel serves one coverage area;
- One single-frequency simplex channel for local communications between mobiles (see Figure 2.18).

Transmit (Tx) f_1, f_2, and f_3 and receive (Rx) frequencies f_1', f_2', f_3' are allocated with a separation of 4–5 MHz. The simplex frequency f_4 is taken from the simplex frequency block between the transmit and receive blocks (Figure 2.18). Typical operation characteristics are described as follows:

1. Selectable two-frequency simplex or half-duplex operation. Each station may be configured as simplex or talk-through from the control station or from an authorized mobile. For example, by day, the system is controlled from the operator console while, by night, the BS may be configured as talk-through to enable mobile-to-mobile communications.
2. Selective calling with or without automatic response. A selective calling system makes it possible to address a particular mobile, a group of mobiles or, even, an entire fleet (general call). This facility avoids the trouble for the mobiles to be listening to all the activity on the channel.
3. Interzone, mobile-to-mobile communications through their corresponding bases, and interbase links.

2.7 Data Transmission in PMR

The rate of information transmission using digital techniques is potentially far greater than using voice [3]. Data transmission also allows direct access to computers and services such as databases, thus reducing the time required for handling messages while improving the reliability of the received messages. Also, since shorter transmission times are needed, increased efficiency in the use of the radio channel is possible, so that a greater number of mobiles may operate in the system without the risk of overloading it. Five to 20 times [3] operation time reductions may be expected if data transmission is used instead of voice. A further advantage is that data transmission grants the privacy that voice communications cannot provide.

The signaling systems described above (five-tone, DTMF) may be considered for data transmission. Although these systems are frequently used, their transmission rates and reliability are low. Typically, telephone system CCITT V-series modems are used for data over radio transmissions. A common data transmission rate is 1,200 bps using FFSK modulation. Designers, however, are free to use any transmission method on the condition that the signal does not exceed the limits of the allocated channel and on the condition that no spurious lateral bands are produced. In practice, for 12.5-kHz channels, the maximum binary rates are on the order of 2,400–4,800 bps.

Data interchange protocols must take into account the specific characteristics of the radio channel as opposed to the telephone line channel: fast fading due to Rayleigh multipath and shadowing effect of such objects as buildings and overpasses. This channel "hardness" should be counteracted using techniques such as error detection together with retransmission requests (ARQ) or correction at the receiver (forward error correction (FEC)).

For low binary rates of up to 2,400 bps, conventional PMR radios can be used. In this case, the data stream modulates an audio subcarrier that is afterward used to FM modulate the RF carrier. For higher binary rates, e.g., 4,800 bps, direct modulation of the carrier (FSK) will be required to fit the modulated signal within the channel limits (Figure 2.19). Peak deviations must be reduced to around 1.2 kHz [3] to avoid exceeding the limits of the channel.

Mobile terminals may transmit data any time they have data available. If two mobiles happen to transmit at the same time, collisions will occur, and the transmitted data will be lost. Several approaches, including the following, may be adopted to handle multiple access systems.

- Pure ALOHA;
- Polling;
- CSMA;
- Slotted ALOHA.

Pure ALOHA systems allow mobiles to transmit at any time in an uncontrolled way, while in slotted ALOHA systems, the base indicates the time intervals or "slots" when the mobiles are invited to transmit. In carrier sense multiple access (CSMA) systems, the base transmits a busy signal during the periods of time when it is receiving data from a mobile, thus advising the others to wait. Once the channel is seen to be free, the mobiles may start to contend for the channel. In the event of a collision, the mobiles will wait randomly before a further attempt is made. These are all contention systems in which collisions may occur. In polling systems, however, each mobile terminal is interrogated in turn to check whether it has available data; it is invited to transmit if this is the case. Although no collisions are possible in polling systems, this setup wastes time unnecessarily, and the maximum possible net data flow (throughput) is reduced.

Introduction to Mobile and Private Mobile Radio (PMR) Systems 61

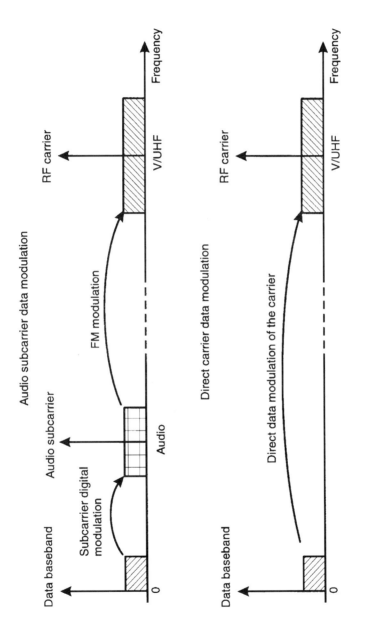

Figure 2.19 Direct and subcarrier data modulations.

References

[1] Carlson, A. B., *Communications Systems: An Introduction to Signals and Noise in Electrical Communication,* 3rd Edition, Singapore: McGraw-Hill International Editions, 1986.

[2] Singer, E., *Land Mobile Radio Systems,* Englewood Cliffs, New Jersey: Prentice Hall, Inc., 1989.

[3] Hanson, D. A., Chapter 5, "Conventional Private Mobile Radio," in *Personal and Mobile Radio Systems,* edited by R. C. V. Macario, London: Peter Peregrinus LTD, IEE Telecommunications Series 25, 1991.

[4] Parsons, J. D., and J. G. Gardiner, *Mobile Communication Systems,* London: Blackie (USA: Halsted Press, a division of John Wiley & Sons, Inc., New York), 1989.

[5] Panell, W. M., *Frequency Engineering in Mobile Radio Bands,* Cambridge, U.K.: Granta Technical Editors in association with Pye Telecomms Ltd., 1979.

II

Propagation Modeling in Mobile Communications

3

Multipath Propagation

3.1 Introduction

This chapter takes a qualitative look at several aspects of the transmission channel generated by multipath. Among other things, it reviews statistical distributions normally used to describe the signal envelope variations; these distributions are used to evaluate the fade margins required for both the uplink and downlink budgets. In addition, this chapter reviews the two types of channels—frequency- and time-dispersive—and the phenomena that cause them. Finally, the chapter summarizes some of the techniques used to measure channel characteristics under both narrowband and wideband conditions.

3.2 Angle of Incidence and Spectrum of the Received Signal

This chapter starts by reviewing the mechanisms that give rise to the received signal time/location variations presented in Chapter 1. Time/location variations are mainly related to the motion of the mobile receiver as it traverses a standing wave pattern created by a number of multipath echoes. Time (or space) selectivity of the channel is closely related to frequency dispersion (Doppler shift) caused by the time variation, due to the mobile displacement, of the propagation channel. The velocity of change in the signal phase translates into a Doppler shift in the received RF carrier.

To illustrate this phenomenon, consider the situation illustrated in Figure 3.1(a) where the mobile runs at a constant speed V along a path between A and A'. Considering only the direct signal (similar arguments may be applied to the different multipath echoes), the increment in traveled distance Δd is

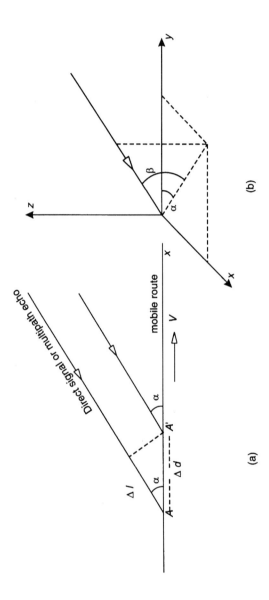

Figure 3.1 (a) Direct signal received by a mobile at two points in its path (2D case). (b) Angles of arrival of multipath echoes (3D case).

given by $\Delta d = V\Delta t$. If the geometry of the scenario is taken into account, the corresponding increment/decrement in the propagation path length is $\Delta l = \Delta d \cos(\alpha)$. This variation in the radio path length, in turn, produces a change in the phase of the received direct signal that is given by the expression

$$\Delta\phi = \frac{2\pi}{\lambda}\Delta l = \frac{2\pi V\Delta t}{\lambda}\cos(\alpha) \qquad (3.1)$$

This phase change produces an offset in the instantaneous received frequency (Doppler shift) given by

$$\Delta f = \frac{1}{2\pi}\frac{\Delta\phi}{\Delta t} = \frac{V}{\lambda}\cos(\alpha) \qquad (3.2)$$

As it can be seen, the change in radio path length depends on the angle α between the direction of arrival of the wave and the direction of movement of the mobile receiver. The waves arriving headlong cause positive Doppler shifts, whereas those arriving from behind generate negative shifts. The waves arriving directly headlong or directly from behind will give rise to the maximum/minimum change in frequency $\pm f_m$

$$\pm f_m = \pm\frac{V}{\lambda} \qquad (3.3)$$

Consequently, if an unmodulated carrier (CW) with frequency f_c is transmitted, the received signal will have a different frequency, $f_c + \Delta f$.

Assuming now an environment where a number of echoes is received from several directions on the horizontal plane and assuming that the vehicle moves along the x-axis at a constant speed V, the Doppler shift of the n-th plane wave reaching the mobile located at point A (Figure 3.1(a)) is

$$\Delta f_n = \frac{V}{\lambda}\cos(\alpha_n) \qquad (3.4)$$

If N plane waves reach the receiver, the fraction of the total incident power corresponding to the infinitesimal angle interval $(\alpha, \alpha + d\alpha)$ will depend on the distribution $p(\alpha)$ of angles of arrival of the waves, where $p(\alpha)$ is a probability density function. Such power fraction is

$$p(\alpha)d\alpha \rightarrow \text{fraction of incident power reaching} \qquad (3.5)$$
$$\text{the receiver between } \alpha \text{ and } \alpha + d\alpha$$

If the receiving antenna has a directivity $g(\alpha)$, the value of the received power fraction from the contributions between α and $\alpha + d\alpha$ is

$$g(\alpha)p(\alpha)d\alpha \tag{3.6}$$

If $p(\alpha)$ follows a uniform distribution in the interval $[0, 2\pi]$ (i.e., multipath echoes arrive from all angles (in the horizontal plane) with equal probabilities), and the receiving antenna is omni-directional (i.e., $g(\alpha)$ = constant), following a statistical development that lies beyond the scope of this text [1], it can be found that, for the vertical component of the electric field, e_z, the following expression is obtained for its corresponding RF spectrum

$$S_{e_z}(f) = A\left[1 - \left(\frac{f-f_c}{f_m}\right)^2\right]^{-1/2} \quad \text{for } |f - f_c| \leq f_m \tag{3.7}$$

$$S_{e_z}(f) = 0 \quad \text{for } |f - f_c| > f_m$$

where A is a constant.

In other words, if a continuous wave of frequency f_c is transmitted through the mobile multipath channel, frequency dispersion will occur. New spectral components will be received in the range ($f_c - f_m$, $f_c + f_m$).

The above expression is illustrated in Figure 3.2 (RF spectrum). This figure also shows the baseband spectrum (i.e., after demodulation by an envelope detector). The situation described corresponds to the *Rayleigh case* where it is assumed that the direct signal is totally obstructed and only multiple echoes with approximately the same amplitudes are received with a uniform angle of arrival distribution in the horizontal plane.

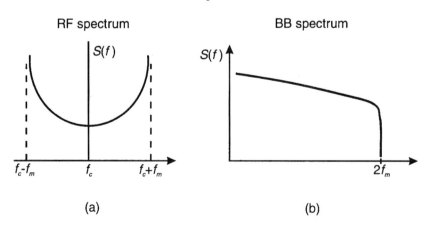

Figure 3.2 RF and BB Doppler spectra for the Rayleigh case.

The results presented in the previous paragraphs were obtained for a theoretical model with scatterers located on the horizontal plane. Now, a more realistic situation is described by introducing an elevation angle of arrival β as shown in Figure 3.1(b). In [2], models are presented that still assume that $p(\alpha) \equiv Uniform\ [0, 2\pi]$, as in [1] but also include angle β and assume different possible distributions of angles of arrival in elevation. These distributions are schematically shown in Figure 3.3(a). The received RF spectra corresponding to these distributions $p(\beta)$ are bounded as in the earlier case but present finite spectral values around $\pm f_m$ (Figure 3.3(b)) instead of an infinite (unrealistic) value, as in the case where $\beta = 0$ degrees.

So far, it has been assumed that multipath echoes originate at static scatterers and that only the mobile is running. If the scatterers were also in motion (e.g., other cars), the received RF Doppler spectrum would no longer be restricted to the $\pm f_m$ interval. In this situation, there would be, in fact, an energy overflow beyond the theoretical limits, as schematically illustrated in Figure 3.4.

Returning now to the case of static scatterers and $\beta = 0°$ (all echoes on the horizontal plane), when there is a predominant signal (*Rice environment*), substantial modifications occur in the received signal RF spectrum. If the main component arrives with an angle α_o, this will give rise to a strong spectral line at $f_c + f_m \cos(\alpha_o)$ in the RF spectrum. In baseband, two peaks will be observed at $f_m(1 \pm \cos(\alpha_o))$, as illustrated in Figure 3.5.

Coming back to the Rayleigh case, it is also interesting to observe the effect of antenna directivity. If, for example, a directive antenna with a *beamwidth*, γ, is used (as illustrated in Figure 3.6), and if the antenna is pointing laterally, a signal with a Doppler spectrum, like that shown in Figure 3.6(a), will be received. If the antenna is pointing forward, the spectrum shown in Figure 3.6(b) will be obtained. The spectral limitation caused by the antenna pattern will also affect the maximum fading frequency experienced by the received signal envelope.

3.3 Received Signal Envelope Statistics

First, the case with no direct or dominant component (Rayleigh case) will be reviewed. Assuming that a large number of plane waves reach the receiver, the vertical component of the received field will be

$$e_z = e_o \sum_{n=1}^{N} e^{j(2\pi f_c t + \phi_n)} = i(t) e^{j\omega_o t} + jq(t) e^{j\omega_o t} \qquad (3.8)$$

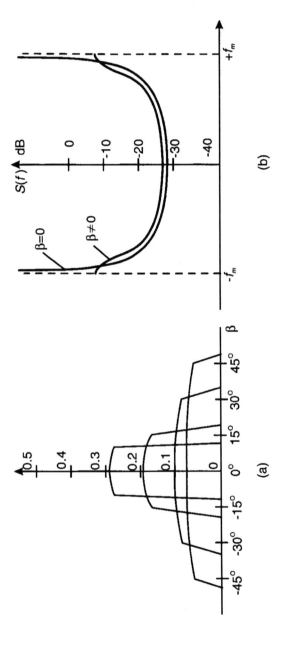

Figure 3.3 (a) Possible distributions of angle β. (b) Doppler spectra for different distributions of β [2].

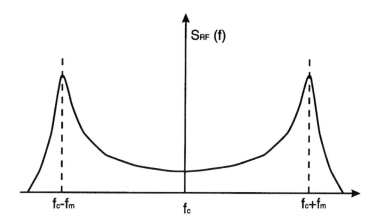

Figure 3.4 Doppler spectrum if the scatterers are also in movement.

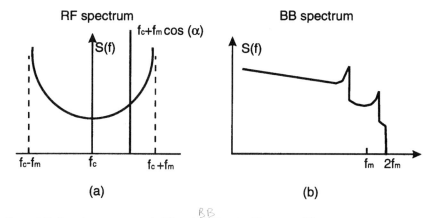

Figure 3.5 Doppler spectrums in RF and BF for the Rice case [2].

If a receiver measures the in-phase, $i(t)$, and quadrature, $q(t)$, components of the received signal, it is possible to observe both the variations in the signal envelope and in the phase as well as the associated Doppler shift. If an unmodulated carrier is transmitted (CW), each received echo, n, is given by

$$e_n = c_n \exp\left[j\left(\omega_c t + 2\pi \frac{V}{\lambda}\cos(\alpha_n)t + \phi_n \right) \right] \quad (3.9)$$

where the amplitudes, c_n, will be approximately equal and the phases, ϕ_n, follow a uniform distribution in the $[0, 2\pi]$ interval. The envelope, $r(t)$, of the total received signal at the mobile will be

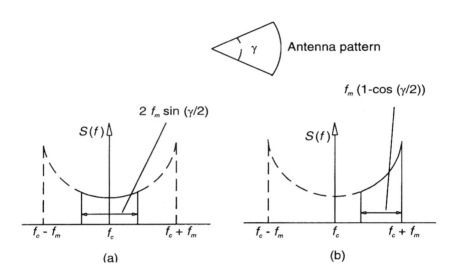

Figure 3.6 RF Doppler spectra for a directive receiving antenna.

$$r(t) = |e_z| = \sqrt{i^2(t) + q^2(t)} \qquad (3.10)$$

The received signal envelope $r(t)$ will follow a Rayleigh distribution whose probability density function (PDF), $p(r)$, (Figure 3.7) is

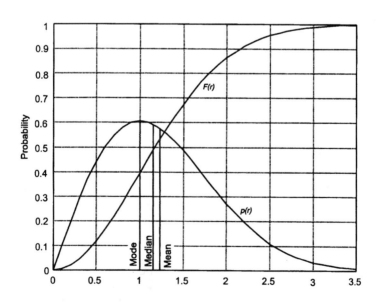

Figure 3.7 Rayleigh probability density function.

$$p(r) = \frac{r}{\sigma^2} \exp\left(-\frac{r^2}{2\sigma^2}\right) \text{ for } r \geq 0 \qquad (3.11)$$

It should be borne in mind that $r(t)$ represents the amplitude variations of a sinusoidal RF carrier. The instantaneous normalized power of the received RF signal is then

$$p_{\text{instantaneous}} = \frac{1}{2} r^2 \qquad (3.12)$$

The cumulative distribution function (CDF) of $r(t)$ is given by the expression

$$\text{CDF} \equiv F(R) = \text{Prob}(r \leq R) = \int_0^R p(r)\,dr = 1 - \exp\left(-\frac{R^2}{2\sigma^2}\right) \qquad (3.13)$$

Other relevant distribution parameters are the *median*, the *variance*, the *mean*, the *mode*, and the *mean-square value*. The mode, σ, is the most probable value and coincides with the maximum in the probability density function curve. Next, the various distribution parameters are given as a function of the mode σ:

$$\text{Mean value } \bar{r} = m_r = E[r] = \int_0^\infty r p(r)\,dr = \sigma\sqrt{\frac{\pi}{2}} = 1.2533\,\sigma \qquad (3.14)$$

$$\text{Mean square value } r_{\text{rms}}^2 = E[(r^2)] = \int_0^\infty r^2 p(r)\,dr = 2\sigma^2 \qquad (3.15)$$

$$\text{Variance } \sigma_r^2 = E[(r^2)] - (E[r])^2 = \sigma^2\left(\frac{4-\pi}{2}\right) = 0.4292\sigma^2 \qquad (3.16)$$

$$\text{Median value } \tilde{r}, \text{Prob}(r \leq \tilde{r}) = 0.5 = 1 - \exp\left(-\frac{\tilde{r}^2}{2\sigma^2}\right) \Rightarrow \tilde{r} \qquad (3.17)$$

$$= \sigma\sqrt{2\ln(2)} = 1.1774\sigma$$

The probability density function is sometimes expressed in terms of any of these parameters instead of being referred to the mode. If the pdf and the CDF are written in terms of the mean square value, r_{rms}^2, the following expressions are obtained

$$p(r) = \frac{2r}{r_{rms}^2}\exp\left(-\frac{r^2}{r_{rms}^2}\right) \quad \text{and} \quad F(R) = 1 - \exp\left(-\frac{R^2}{r_{rms}^2}\right) \quad (3.18)$$

in terms of the mean, m_r

$$p(r) = \frac{\pi r}{2m_r^2}\exp\left(-\frac{\pi r^2}{4m_r^2}\right) \quad \text{and} \quad F(R) = 1 - \exp\left(-\frac{\pi R^2}{4m_r^2}\right) \quad (3.19)$$

and in terms of the median value, $\tilde{r}$

$$p(r) = \frac{2r\ln(2)}{\tilde{r}^2}\exp\left(-\frac{r^2\ln(2)}{\tilde{r}^2}\right) \quad \text{and} \quad F(R) = 1 - 2^{\left(-\frac{R}{\tilde{r}}\right)^2} \quad (3.20)$$

As indicated in Chapter 1, to characterize the total received signal variations, these can be fast and slow. The fast variations are studied for short lengths of the mobile run (small area) of some tens of wavelengths. To study the slow variations, they are separated from the fast variations by averaging (local mean, $m_r(i)$). The slow variations are usually studied for larger sections of the mobile run (larger area) and are statistically characterized by means of a log-normal distribution when expressed in linear units or by a normal distribution if expressed in logarithmic units. The local mean values correspond to the Rayleigh distribution parameter, $m_r(i)$, computed for each point i of the route.

Global Signal Variations

Generally, the study of the total received signal variations is carried out by first separating the fast variations from the slow. However, it is possible to define global *distributions* that characterize the fast and slow variations together. The reader is again reminded of the assumption that the direct signal is totally blocked.

The slow variations are characterized by the evolution of the local mean, $m_r(i)$, which follows a log-normal distribution (normal in logarithmic units). The fast variations occur around or are superposed on the local mean value at each point. According to (3.19), the Rayleigh distribution followed by the fast variations expressed in terms of their corresponding local mean is

$$p(r|m_r) = \frac{\pi r}{2m_r^2} \exp\left(-\frac{\pi r^2}{4m_r^2}\right) \quad (3.21)$$

where m_r is the local mean at the route point of study and $p(r|m_r)$ is the pdf of r conditioned by a mean value m_r. In its turn, the *local mean* follows a normal distribution if expressed in logarithmic units

$$M_r = 20 \log_{10}(m_r) \text{ or, alternatively } m_r = 10^{M_r/20} \quad (3.22)$$

The distribution of M_r is thus

$$p(M_r) = \frac{1}{\sqrt{2\pi}\sigma_L} \exp\left(-\frac{(M_r - \bar{M}_r)^2}{2\sigma_L^2}\right) \quad (3.23)$$

The distribution of m_r, represented in logarithmic units by M_r, follows a normal distribution with a median value $\bar{M}_r$ and a standard deviation σ_L (locations variability).

The *global distribution*, $p(r)$, therefore, is calculated by evaluating the integral of the distribution $p(r|m_r)$ for all possible values of m_r

$$p(r) = \int_{-\infty}^{\infty} p(r|M_r) p(M_r) dM_r \quad (3.24)$$

The following expression (Suzuki distribution) is obtained for the distribution of the received signal envelope (including fast and slow variations)

$$p(r) = \sqrt{\frac{\pi}{8\sigma_L^2}} \int_{-\infty}^{\infty} \frac{r}{10^{M_r/10}} \exp\left[-\frac{\pi r^2}{4 \cdot 10^{M_r/10}}\right] \exp\left[-\frac{(M_r - \bar{M}_r)^2}{2\sigma_L^2}\right] dM_r \quad (3.25)$$

3.4 Received Signal Phase

The phase, θ, may also be expressed in terms of the in-phase and quadrature components of the received signal, $i(t)$ and $q(t)$:

$$\theta(t) = \tan^{-1}\left(\frac{q(t)}{i(t)}\right) \quad (3.26)$$

The phase distribution for the Rayleigh case, i.e., multiple scattered signals with approximately the same amplitude and a random phase, is a uniform in the $[0, 2\pi]$ interval, with a pdf is given by

$$p(\theta) = \frac{1}{2\pi} \quad 0 \leq \theta \leq 2\pi \qquad (3.27)$$

3.5 Second-Order Statistics

First-order statistics are those for which the speed of the mobile exerts no influence. The Rayleigh distribution used to describe the signal variations due to multipath provides information on the percentage of time (or locations) for which the signal envelope is above or below a specified value (e.g., the operation threshold level). This information is relevant to calculate *link budgets* and to specify *fading margins*. Probability density functions and cumulative distributions are first-order statistics.

Systems engineers, however, need to study the received signal variations in more detail—for example, for a quantitative description of the fading rate and the distribution and duration of fades. This kind of information is of great importance when deciding what binary rates, word lengths, or coding schemes should be used for optimum transmission through the mobile channel (i.e., to better protect the system from possible error bursts). The required information is provided by parameters such as the following:

- The *level crossing rate* (lcr);
- The *average fading duration* (afd).

These parameters depend on the speed of the mobile and are known as second-order statistics.

Level Crossing Rate (lcr)

To calculate the *lcr*, the number of crossings through a given signal level, R, with a positive (or negative) slope is computed. If the total number of crossings through level R is N for an observation time of T seconds, the crossing rate is

$$lcr \equiv n(R) = \frac{N}{T} \text{ [crossings/second]} \qquad (3.28)$$

Figure 3.8 shows the procedure for the calculation of this parameter.

The mathematical expression for the *lcr* is [2]

$$n(R) = \int_0^\infty \dot{r} p(R, \dot{r}) d\dot{r} \qquad (3.29)$$

where $p(r = R, \dot{r})$ is a joint probability density function particularized for level R, and $\dot{r} = \dfrac{dr}{dt}$ is the envelope derivative (i.e., the signal variation slope). It has been worked out theoretically [2] that, for the Rayleigh case, assuming an omni-directional vertical monopole, the *lcr* follows the expression [2]

$$n(R) = \sqrt{\frac{\pi}{\sigma^2}} R f_m \exp\left(-\frac{R^2}{2\sigma^2}\right) \qquad (3.30)$$

Normalizing R with respect to the *rms* value, $r_{rms} = \sqrt{2}\sigma$, yields

$$n(R) = \sqrt{2\pi} \rho f_m \exp(-\rho^2) \qquad (3.31)$$

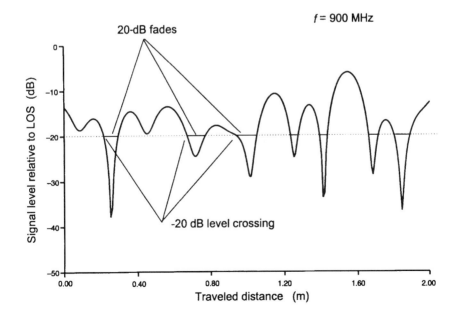

Figure 3.8 Level crossings and fading durations.

where $\rho = \dfrac{R}{\sqrt{2}\sigma} = \dfrac{R}{r_{\text{rms}}}$. $n(R)$ is the number of crossings per second. This is a parameter which depends on the speed of the mobile through $f_m = V/\lambda$. To express the *lcr* independently of the mobile speed, $n(R)$ must be divided by f_m (*normalized lcr*), thus providing the number of crossings per wavelength

$$n'(R) = \frac{n(R)}{f_m} \text{ [crossings/wavelength]} \qquad (3.32)$$

Figure 3.9(a) illustrates the values of the *normalized lcr* for different signal levels when a vertical monopole antenna is used and scatterers with a uniform azimuth angle of arrival distribution are assumed (Rayleigh case). The maximum *lcr* occurs at $\rho = -3$ dB (i.e., 3 dB below the *rms* level) and decreases for higher and lower values of ρ.

The *lcr* is sometimes written in terms of the median signal level $\tilde{r}$ using the expression [2]

$$n'(R) = \frac{n(R)}{f_m} = \sqrt{2\pi\ln(2)}\left(\frac{R}{\tilde{r}}\right)2^{-\left(\frac{R}{\tilde{r}}\right)^2} \qquad (3.33)$$

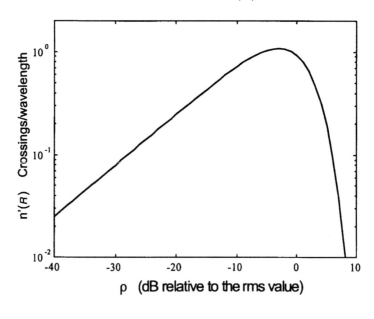

(a)

Figure 3.9 The *lcr* and *afd* for the Rayleigh case.

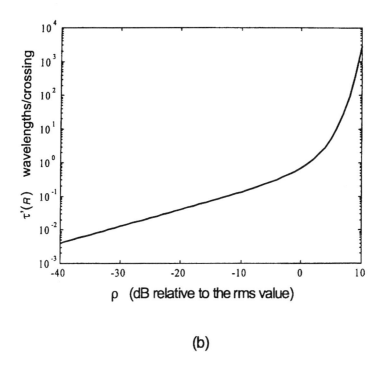

(b)

Figure 3.9 (continued).

Example 3.1. Assuming a carrier frequency f_c = 900 MHz (wavelength 0.33m) and a mobile traveling at a speed of 48 km/h ≡ 13.333 m/s [3], the normalized *lcr* value n' = 1 (maximum value) corresponds to signal level ρ = −3 dB. The absolute value of the maximum crossing rate will be

$$n(R) = n'(R) f_m = \frac{V}{\lambda} = 39 \text{ crossings/second} \qquad (3.34)$$

Average fading duration (afd)

The *afd* is computed by adding up the durations, t_i, of the N fades (below level R) and dividing by N (Figure 3.8),

$$afd = \tau(R) = \frac{\sum_{i=1}^{N} t_i}{N} \text{ [seconds/fade]} \qquad (3.35)$$

If the *lcr* is multiplied by the *afd*, the CDF is obtained

$$n(R) \cdot \tau(R) = \frac{N}{T} \frac{\sum_{i=1}^{N} t_i}{N} = \frac{\sum_{i=1}^{N} t_i}{T} = \text{Prob}(r \leq R) = \text{CDF} \qquad (3.36)$$

or, equivalently

$$lcr \cdot afd = \text{CDF} \qquad (3.37)$$

That is, the time or mobile speed dependency disappears.

The *lcr* and the *afd* are second-order statistics as they depend on speed, whereas the CDF is a first-order statistic, as it is independent of the speed. The theoretical *afd* expression [2] for the Rayleigh case is

$$\tau(R) = \frac{\text{Prob}(r \leq R)}{n(R)} = \sqrt{\frac{\sigma^2}{\pi}} \frac{\exp\left(\frac{R^2}{2\sigma^2}\right) - 1}{R f_m} \quad [\text{seconds/fade}] \qquad (3.38)$$

To express the *afd* in a manner independent from the speed (i.e., in terms of wavelengths per fade) the *afd* must be multiplied by f_m (Figure 3.9(b)), yielding

$$\tau'(R) = \tau(R) f_m = \sqrt{\frac{\sigma^2}{\pi}} \frac{\exp\left(\frac{R^2}{2\sigma^2}\right) - 1}{R} \quad [\text{wavelengths/fade}] \qquad (3.39)$$

The *afd* is sometimes expressed in terms of the received signal's rms and median values

$$\tau(R) = \frac{\exp(\rho^2) - 1}{\rho f_m \sqrt{2\pi}} \qquad (3.40)$$

$$\tau(R) = \frac{1}{\sqrt{2\pi \ln(2)}} \frac{2^{\left(\frac{R}{\tilde{r}}\right)^2} - 1}{f_m \frac{R}{\tilde{r}}} \qquad (3.41)$$

Table 3.1 [2] gives *afd* and *lcr* values for various fade depths referred to the median received signal value, $\tilde{r}$. From Table 3.1, it can be read that, for

Table 3.1
The *afd* and *lcr* for Different Fade Depths [2]

Fade Depth (dB)	afd (λ/fade)	lcr (Crossings/λ)
0	0.479	1.043
−10	0.108	0.615
−20	0.033	0.207
−30	0.010	0.066

example, to detect approximately 50% of the fades with a depth of 30 dB, the received signal should be sampled every 0.01λ, which at 900 MHz ($\lambda = 0.33$m) means every 0.33 cm.

3.6 Random Frequency Modulation

The phase, θ, of the received signal varies with time in a random manner due to the receiver motion. This is equivalent to a random modulation of the phase. This phenomenon is usually termed *random FM*. These random phase variations may be detected by any phase-sensitive detector such as an FM discriminator. The result is the appearance of noise in the receiver. Random FM may be expressed as

$$\dot{\theta} = \frac{d\theta}{dt} = \frac{d}{dt}\left[\tan^{-1}\frac{q(t)}{i(t)}\right] \qquad (3.42)$$

Figure 3.10(a)(b) [2] shows the probability density function and the cumulative distribution of $\dot{\theta}$ (phase derivative with respect to time). Figure 3.10(c) shows the shape of the random FM spectrum. This spectrum is unlimited in frequency (unlike the Doppler spectrum). The energy, however, is mainly confined between 0 and $2f_m$ H. From $2f_m$ upwards, the spectrum decreases at a rate of $1/f$ and is negligible from $5f_m$ onwards. Most of the energy will be confined to the audio frequency band. The greatest frequency excursions are associated with the deepest fades, which rarely occur. Figure 3.11 illustrates a simulated random FM series [4] corresponding to case 5 in Section 1.5.

Example 3.2. For a mobile speed of 105 km/h and a carrier frequency of 1 GHz ($\lambda = 0.3$m), $2f_m$ is 194 Hz and $5f_m$ is 486 Hz [3].

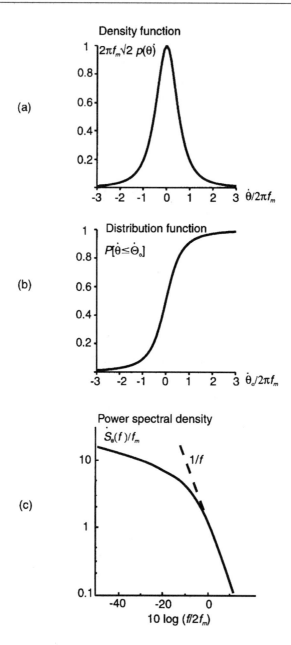

Figure 3.10 (a) and (b) random FM distribution and (c) random FM spectrum [2].

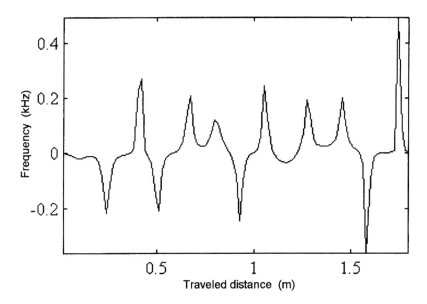

Figure 3.11 Simulated random FM series [4].

In voice communications, the lower limit of the audio passband (limited by filtering) is 300 Hz. This means that the random FM will have a negligible effect on voice communications. In digital signaling transmissions, it is important to use an appropriate binary rate. It is essential to avoid concentrating the transmitted energy on the spectral regions where the random FM is highest (near 0 Hz). To this aim it is essential to use an appropriate line code (e.g., in the cellular AMPS system, a Manchester code is used) [3].

3.7 Fading in the Ricean Case

It has been assumed that the signal reaching the mobile comprises a number of echoes (multipath), all of a similar level. This is the typical situation found in dense urban areas where the direct signal is completely blocked. It may happen, however, that a direct signal reaches the receiver with small attenuation as, for example, at crossroads, in open areas, etc. In such circumstances, the results described above corresponding to the Rayleigh case are no longer valid.

When a direct signal or a contribution originated by a specular reflection with a significant level is present, the received signal variations will follow a *Rice* distribution. The *Rice* probability density function (Figure 3.12) has the expression

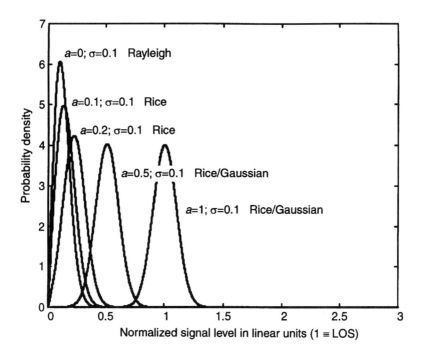

Figure 3.12 Rice probability density function for different (α, σ) values.

$$p(r) = \frac{r}{\sigma^2} \exp\left[-\frac{r^2 + a^2}{2\sigma^2}\right] I_0\left(\frac{ra}{\sigma^2}\right) \text{ for } r \geq 0 \quad (3.43)$$

with I_0 being the modified Bessel function of zero-th order [3]

$$I_0(z) = \sum_{n=0}^{\infty} \frac{z^{2n}}{2^{2n} n! n!} \quad \text{for } z \gg 1 \quad I_0(z) = \frac{e^z}{(2\pi z)^{1/2}}\left(1 + \frac{1}{8z} + \frac{9}{128 z^3} + \cdots\right) \quad (3.44)$$

This distribution presents two extreme cases (Figure 3.12):

- If $a = 0$ (absence of a dominant signal), $p(r)$ becomes a Rayleigh distribution;
- If a is large (dominant signal), $p(r)$ becomes a Gaussian distribution.

The Rice distribution is usually expressed in terms of the carrier-to-multipath ratio or k-factor, $k = c/m$, which may be defined mathematically as follows:

$$k = \frac{c}{m} = \frac{a^2}{2\sigma^2} \quad \text{or, in dB,} \quad K = 10 \log(k) \qquad (3.45)$$

The shape of the Doppler spectrum for the Rice case was already analyzed in Section 3.3. Figure 3.12 shows Rice distributions for various carrier-to-multipath ratios.

3.8 Spatial Correlation of the Field Components

The effects of fast fading may be counteracted by means of spatial diversity techniques. To achieve this goal, the receive antennas should be at a sufficient distance from each other so that the envelopes received by each antenna present a low mutual correlation.

Figure 3.13(a) shows the situation experienced in the mobile [2]. A Rayleigh environment is assumed with echoes of similar magnitudes and uniform distribution of angles of arrival in azimuth. A point P' is considered at a distance ξ along the x-axis (direction of movement) with from point P. P and P' represent the positions of the two antennas located on the mobile.

Figure 3.13(b) shows the correlation in the mobile for the vertical component of the electric field in terms of the antenna spacing expressed in wavelengths, ξ/λ. A marked decorrelation is observed for antenna separations on

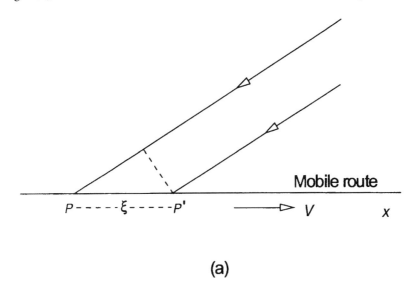

(a)

Figure 3.13 Geometry and spatial correlation values at the mobile.

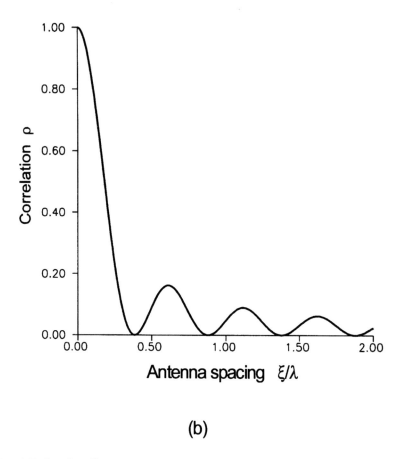

(b)

Figure 3.13 (continued).

the order of 0.3λ. For the frequency bands used in mobile communications, VHF and higher, these distances correspond to very small separations. For example, at 900 MHz ($\lambda = 0.33$), $\xi/\lambda = 0.3$. This means a separation of $\xi = 10$ cm.

It is much more interesting to study the possibility of using *diversity at base stations*. Figure 3.14 illustrates this case. The *principle of reciprocity*, despite being valid for the received field strength for the uplink and downlink, has no reason to be fulfilled for the correlation distance.

The mobile is surrounded by scatterers with an approximately uniform azimuth angle distribution for the Rayleigh case. The base station sees these scatterers only within a small angular sector (Figure 3.14(a)). From Figure 3.14, it is clear that the spatial correlation at a base station also depends on the angle θ between the line joining the antennas and on the configuration of the scenario where the mobile is located (apart from the dependence with

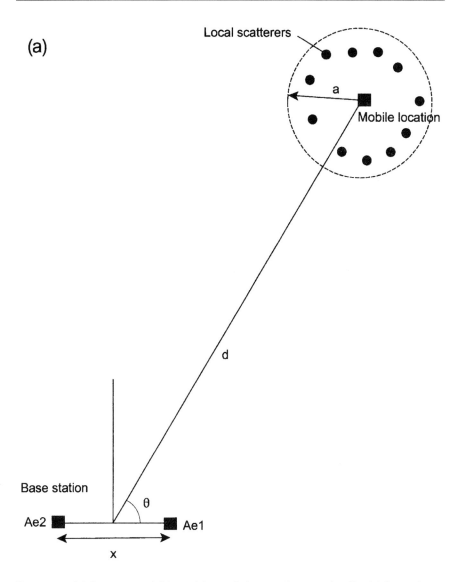

Figure 3.14 (a) Geometry and (b) spatial correlation at a base station (Rayleigh case).

the horizontal separation between antennas). From Figure 3.14(b), it is clear that the required antenna separations at base stations will be much larger than at mobile stations.

3.9 Frequency Selective Fading

So far it has been assumed that an unmodulated carrier (CW) was transmitted. The effects of the channel on the received signal were studied for these condi-

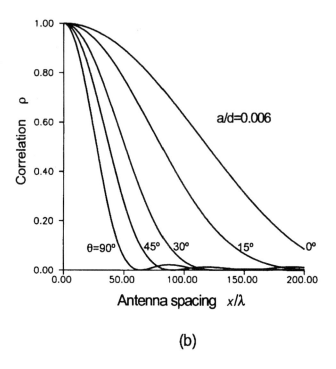

(b)

Figure 3.14 (continued).

tions. It is also important to know the effect of the channel on real signals (i.e., signals that occupy a given bandwidth). Assume two frequencies belonging to a transmitted signal with a given bandwidth. If these two frequencies are close, the different propagation paths (multipath) will have approximately the same electric lengths, $2\pi d/\lambda_1$ and $2\pi d/\lambda_2$. This means that their received amplitudes and phases will vary in approximately the same way. These are called *flat fading* conditions.

If the frequency separation increases, the fading behavior at one frequency tends to be uncorrelated with respect to the fading conditions at the other frequency. This is due to the fact that the phase effects will be significantly different for both frequencies. The degree of correlation will depend on the delay spread caused by the environment (dispersion of path lengths). For large delay spreads, the phase differences between the two received components (f_1, f_2) will differ by several radians.

Signals occupying larger bandwidths will be distorted since the amplitudes and phases of their spectral components will not be the same as those in the original signal. This phenomenon is known as *frequency-selective fading*, and it is caused by a nonuniform frequency response of the multipath propagation channel. The minimum bandwidth for the onset of selective fading effects is known as the *coherence bandwidth*.

The delayed replicas of the transmitted signal reaching the mobile may be associated with specific scatterers in the traversed environment. In order to characterize the propagation channel at a given point on the mobile route, it is not sufficient to know the delay of the various echoes, it is also necessary to measure their angles of arrival. This may be achieved by keeping track of the received Doppler spectrum.

Assuming a simple model where only single interactions are taken into account, the echoes with a given delay value will be located on an ellipse whose focal points are the transmitter and the receiver positions (Figure 3.15) [2]. Considering three scatterers *A, B,* and *C,* the paths *TAR* and *TBR* may be resolved (although having the same angle of arrival) by their different delays. Furthermore, it is possible to distinguish between *TAR* and *TCR* (which have the same delay) by their angles of arrival (AoAs). The AoAs may be obtained by measuring the Doppler shift, $f_m \cos(\alpha_i)$. Doppler measurements present, however, a left-right ambiguity.

The influence of a given propagation scenario may be predicted if the simple channel model presented in the previous paragraph is used. If a template with cofocal ellipses whose focal points correspond to the Tx and Rx terminals is superimposed on a map of the terrain, the possible scattering elements, which may intervene in the transmitted signal propagation in the area studied, may be identified.

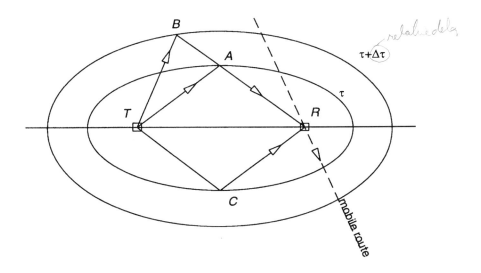

Figure 3.15 Ellipses defining scattering points with equal delays [2].

In the case of a digital transmission, delayed replicas of the signal will be received thus causing intersymbol interference (ISI) (Figure 3.16). In this situation, even though the carrier-to-noise ratio can be improved by increasing the transmitted power, the bit error rate (BER) will not be improved (irreducible BER). In order to improve transmission behavior, it is essential to use channel equalization systems. The effect of ISI is a further manifestation of the selective frequency fading phenomenon.

Two-Ray Model

To illustrate the frequency selectivity phenomenon in channels with time dispersion, the most simple case may be considered [5] (i.e., a situation with only two rays reaching the receiver). It will be assumed that the signal propagates between the transmitter and the receiver via a direct ray and a second ray that runs a greater distance. This is a typical situation in fixed radio links where, due to irregularities in the refraction index, additional echoes may arrive. This phenomenon also appears in systems where a ground reflection occurs.

At the receiver, the received signal will be the sum of two phasors. Assuming a transmitted frequency, f, two path lengths, d_1 and d_2, and two amplitudes, a_1 and a_2, at the receiver, the complex sum of both rays will be

$$s(t) = a_1 \exp(-jkd_1) + a_2 \exp(-jkd_2) \qquad (3.46)$$

To simplify things, the phasor $\exp(j\omega t)$ was omitted. The phase changes proportionally to the traveled distance. The proportionality factor is $k = 2\pi/\lambda$ (wave number). Taking the first (direct) ray as a reference, the previous expression may be rewritten as

$$s(t) = \left\{1 + \frac{a_2}{a_1} \exp[-jk(d_2 - d_1)]\right\} a_1 \exp(-jkd_1) \qquad (3.47)$$

The expression between braces may be considered as an amplitude and phase modulation of the first (direct) signal. Now, more detail is given about the variations in amplitude and phase, that is, the term between braces which is usually called complex envelope $r(t)$.

$$r(t) = 1 + A \exp(-jk\Delta d) = 1 + A \exp(-j2\pi f \Delta \tau) \qquad (3.48)$$

where $A = a_2/a_1$, $\Delta d = d_2 - d_1$ and $k\Delta d = 2\pi f \Delta \tau$.

$\Delta \tau$ is the relative delay between the two components in the received signal. Carrying out a phasor representation (Figure 3.17), if $A < 1$, the second component is smaller than the direct ray; when $\Delta \tau$ changes, the second phasor

Multipath Propagation

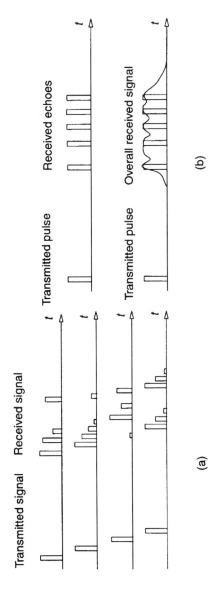

Figure 3.16 Effects of the dispersive and time-variant channel.

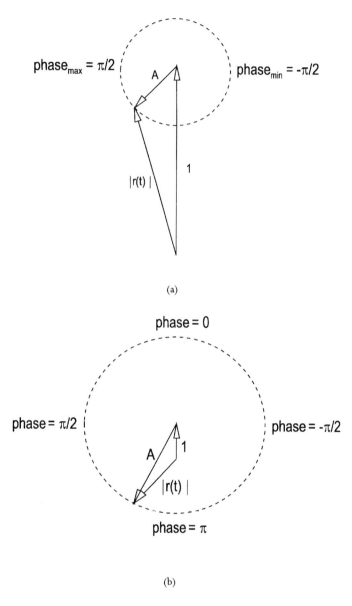

Figure 3.17 Two-ray model, with case 1, $A < 1$, and case 2, $A > 1$ [5].

rotates. As a consequence, the resulting signal amplitude varies between the limits $1 \pm A$ and the phase will do it between $\pm \pi/2$ as a maximum (Figure 3.17(a)). If $A > 1$, the amplitude will again vary between $A \pm 1$ while the phase will rotate continuously (Figure 3.17(b)).

Studying now the amplitude and phase variations for a fixed relative delay $\Delta \tau = 1 \mu s$ with $A = 0.9$ and $A = 1.1$, by carrying out a frequency scan between 900 and 910 MHz, the diagrams shown in Figure 3.18 are obtained. In the amplitude diagrams, channel frequency selectivity is observed (nonflat response). As regards the phase, in the first case it oscillates around a constant value, while in the second case the phase decreases continuously.

Concerning the incidence of a frequency selective channel on a digital (or analog) transmission, it must be remembered that the nondistortion condition occurs when the following is true:

- The *amplitude response* is constant: $|H(f)| = constant$;
- The *phase response* is linear with frequency: $\arg[H(f)] \alpha f$.

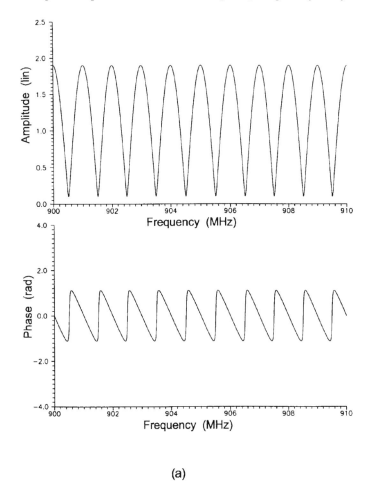

(a)

Figure 3.18 (a) Case 1, with $A = 0.9$ and variations in amplitude and phase with frequency. (b) Case 2, with $A = 1.1$ and variations in amplitude and phase with frequency [5].

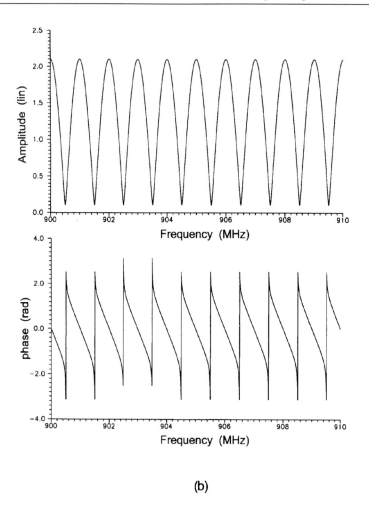

Figure 3.18 (continued).

$$h(t) = A\delta(t - \tau) \leftarrow F \rightarrow H(f) = Ae^{-j\omega\tau} \tag{3.49}$$

$$|H(f)| = \text{constant} \quad \text{and} \quad \arg(H(f)) \alpha f \tag{3.50}$$

As shown earlier, in a channel with a direct ray and one single echo, the nondistortion condition is not met, except for the case of very narrow transmitted signal bandwidths. It may also be observed that the worst conditions occur in the minima of the received amplitude, where both fast changes in amplitude and phase are experienced. Conversely, near the maxima, the responses both in amplitude and in phase are relatively flat and linear, respectively, with very little distortion (small bands with flat fading).

Reality is not, however, as simple as in the model just presented for the following reasons:

- The delay is variable with time (i.e., $|H(f)|$ and $arg\,(H(f))$ are functions of time or position along the route). Channel distortion, therefore, is also time-varying;
- The phase response also varies with time and, thus, Doppler shifts will appear.

As a conclusion, even in the simplest multipath scenario of a propagation channel with a direct signal and one single echo, severe distortions may occur.

3.10 Wideband Channel Characterization

The radio channel created by multipath can be modeled as a linear filter whose characteristics are time-varying. The relation between the inputs and outputs of this filter may be described either in the time and/or frequency domains. This means that several system functions are possible. The variables of these functions are those indicated in Table 3.2.

The time-variant impulse response, $h(\tau; t)$, is one of these system functions. If there were no time variations for the scenario shown in Figure 3.19, this would give rise to the following system function (impulse response).

$$h(\tau) = a_0\delta(\tau - \tau_0) + a_1\delta(\tau - \tau_1) + a_2\delta(\tau - \tau_2) + \ldots \quad (3.51)$$

By the simple fact that the receiver is on the move, the path lengths change and, consequently, $h(\tau)$ will be time-variant, $h(\tau; t)$. In this case, the time-varying system function would be

Table 3.2
Parameters of the Functions That Characterize the Multipath Channel

τ	delay
ν	Doppler shift
f	frequency
t	time

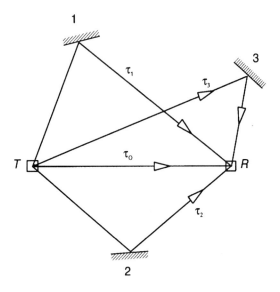

Figure 3.19 Schematic representation of a multipath channel.

$$h(\tau; t) = a_o(t)\delta(\tau - \tau_o(t)) + a_1(t)\delta(\tau - \tau_1(t)) + a_2(t)\delta(\tau - \tau_2(t)) + \ldots \tag{3.52}$$

When analyzing the influence of the channel on transmission systems, the consideration of the channel as a linear system makes it possible to model it as a time-varying linear filter in simulations. Examples of this time-varying linear representation are shown in Figure 3.20(a)(c) [2, 5].

Other system functions are: $T(f; t)$, $H(f; \nu)$, $S(\tau; \nu)$. One of these, $H(f; \nu)$, is illustrated in circuit form in Figure 3.20(b). All these functions are inter-related by means of Fourier transforms (F):

$$t \leftarrow F \rightarrow \nu \quad \text{and} \quad \tau \leftarrow F \rightarrow f$$

The previous channel representations are deterministic. A deterministic study of the mobile channel is difficult even though attempts are being carried out using ray-tracing techniques (see Chapter 4). Given its complexity, the mobile channel is normally described statistically. The statistical characterization of the channel is highly involved since it is necessary to know the joint probability density functions of the different intervening variables (τ, t, f, ν). In practice, three simple functions are used:

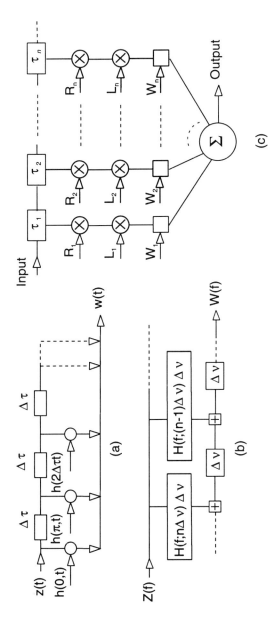

Figure 3.20 Circuit representation of system functions (a) $h(\tau; t)$ and (b) $H(f; \nu)$ [2]. (c) Diagram of a channel simulator (COST 207) [5].

- The power delay profile (PDP);
- The frequency correlation function;
- The scattering function.

The PDP $P_h(\tau)$ provides an indication of the distribution of the received echoes, their magnitudes and delays. A PDP measures the width of the channel impulse response $h(\tau; t)$. Figure 3.21(a) illustrates a measured PDP for a suburban environment. A parameter closely related to the PDP is the *frequency correlation function*, also known as *coherence function*. Both parameters are related via the Fourier transform

$$R_H(\Delta f) = \int_{-\infty}^{\infty} P_h(\tau) e^{-j2\pi\Delta f \tau} d\tau \qquad (3.53)$$

This relation is shown graphically in Figure 3.22. Figure 3.21(b) also illustrates the corresponding coherence function.

The interpretation of the frequency-separation correlation function is as follows: The abscissa indicates the frequency separation, Δf, and the ordinate gives a measure of the correlation between two separate frequencies of the transmitted signal spectrum that are Δf Hz apart. The correlation/decorrelation criteria depends on the modulation scheme used, among other factors.

The value of B_c (coherence bandwidth) is usually specified for correlation values of 0.9 or 0.5. As shown for the channel in Figure 3.21, $B_c(0.5) = 2$ MHz. This value indicates that, if the transmitted signal has a bandwidth greater than B_c, it will be subject to *frequency selective fading* (i.e., the different spectral components of the transmitted signal will fade independently, and the signal received will be distorted). When the transmitted signal has a bandwidth of less than B_c, *flat fading* will occur.

Finally, the third function is the *scattering function/matrix*, $S(\tau; \nu)$, which provides a complete "radiography" of the channel. The scattering function gives the received echoes classified by delays and Doppler shifts (or, in other words, angles of arrival). Figure 3.23 shows a scattering function corresponding to a highly dispersive environment.

Last, the effects of multipath propagation on signal transmission will be reviewed in a simplified form. Propagation channels may be classified according to two criteria:

- Frequency dispersion (time variability): slow/fast varying channels;
- Time dispersion (frequency selectivity): frequency selective/non-selective (flat) channels.

Multipath Propagation

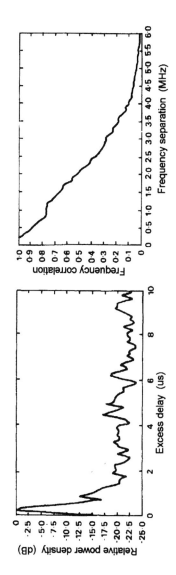

Figure 3.21 (a) Measured PDP and (b) coherence function for a suburban environment.

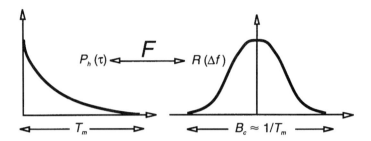

Figure 3.22 Relation between the PDP and the coherence function.

A general mobile channel presents both characteristics: time- and frequency-dispersion. Figure 3.24(a, b) [3] schematically shows the effect of channel dispersion on the transmitted signal.

Deterministic Scattering Model

It is interesting to attempt to model the behavior of the channel in a deterministic manner [2, 6]. The function $S(\tau; \nu)$ may be expressed in terms of the radar cross section values, $\sigma_i (m^2)$, of the elements located in the scenario. Similarly, other propagation mechanisms like specular reflections may be characterized using reflection coefficients. In this way, a physical insight into the main features of the channel can be obtained.

It will be assumed that the radar cross section or reflection coefficients for all the elements in the surroundings intervening in the multipath phenomenon are known. Besides, to simplify, it will also be assumed that only single interactions occur. The following contributions that comprise the multipath structure may be considered [6]:

- Direct signal (with or without obstruction);
- Specular reflected signals (characterized by the reflection coefficient of the surface);
- Scattered (diffuse) signals (characterized by the radar cross section of the scatterer).

The effects of *absorption losses* (as the transmitted signal passes through trees or walls) may also be included, as well as *diffraction losses*. The *direct signal*, except for losses due to absorption or diffraction, is characterized (assuming isotropic antennas) by the expression

$$\frac{p_r(\text{direct signal})}{p_t} = \left(\frac{1}{4\pi d^2}\right)\left(\frac{\lambda^2}{4\pi}\right) \quad (3.54)$$

Multipath Propagation

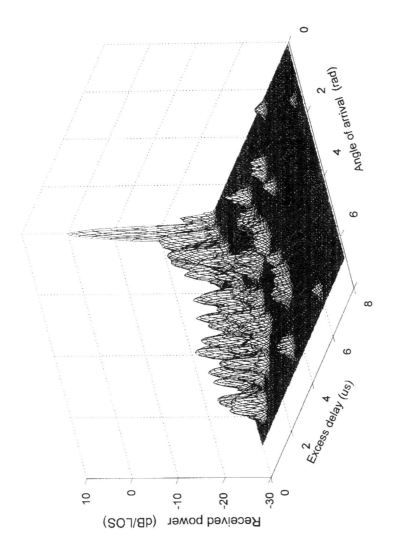

Figure 3.23 Scattering function for a highly dispersive environment.

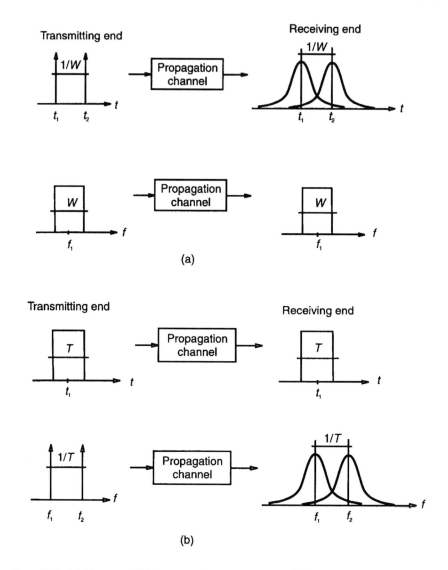

Figure 3.24 (a) Time- and (b) frequency-dispersive channel [3].

The *reflected components* are characterized by the expression

$$\frac{p_r(\text{reflected signal})}{p_t} = \left(\frac{|R|^2}{4\pi(d_1+d_2)^2}\right)\left(\frac{\lambda^2}{4\pi}\right) \quad (3.55)$$

and the *diffuse scattered components* are characterized by the *bistatic radar equation*

$$\frac{p_r(\text{scattered signal})}{p_t} = \left(\frac{1}{4\pi d_1^2}\right)\sigma\left(\frac{1}{4\pi d_2^2}\right)\left(\frac{\lambda^2}{4\pi}\right) \qquad (3.56)$$

Example 3.3. Assume an environment sketched in Figure 3.25(a) through which the mobile travels. A scattering function as the one illustrated in Figure 3.25(b) will describe the channel characteristics. It is noted that the direct signal produces a strong peak with a relative delay zero and negative Doppler equal to $-f_m$. A strong echo from the hills is also observed in the direction of the route giving rise to a positive Doppler equal to $+f_m$. There is also a further echo with zero Doppler from a group of trees perpendicular to the route, and, finally, one last echo comes from another group of trees with negative Doppler.

Example 3.4. A study will now be made on a given environment with a mobile traveling at 40 km/h. It is observed that there are several specular reflectors and diffuse scatterers. Assume that the elements described in Figure 3.26(a) are the relevant factors of multipath conditions. The Doppler spectrum at the point on the route shown in the figure will be computed and illustrated in a diagrammatic form, assuming an omni-directional receiving antenna (case A). The Doppler spectrum will also be computed for the case where an antenna with a cosine shaped radiation pattern with its maximum orientated in the direction of travel is used (case B). The working frequency is 900 MHz, and the distance from the transmitter to the scenario is 10 km, whereas the distance from the point scatterers to the mobile is 100m and from the reflectors to the mobile is 1 km.

Since the mobile velocity is $V = 40$ km/h ≈ 11.11 m/s and the wavelength is $\lambda = 0.33$m, the maximum Doppler shift will be $f_m = V/\lambda = 33.33$ Hz. Each echo will have a Doppler shift that will depend on the angle of incidence with respect to the direction of the route, according to the expression

$$f_i = f_m \cos(\alpha_i) \qquad (3.57)$$

To graphically represent all the multipath components, their magnitudes will be referred to the direct signal and then expressed in decibels (20 log). The total received field strength, $|e_r|$, will be proportional to

$$|e_e| \propto (\text{dir. signal}) \frac{1}{\sqrt{4\pi}} \frac{1}{d_{\text{dir}}}$$
$$+ (\text{refl. signals}) \frac{1}{\sqrt{4\pi}} \sum \frac{R}{d_1 + d_2} \qquad (3.58)$$
$$+ (\text{scatt. signals}) \frac{1}{\sqrt{(4\pi)^2}} \sum \frac{\sqrt{\sigma}}{d_1 \cdot d_2}$$

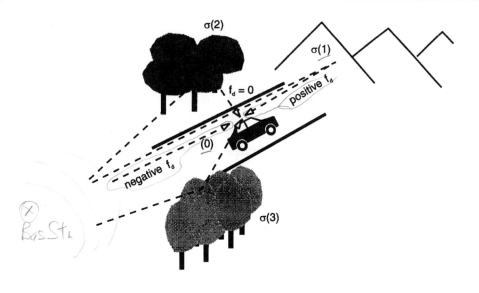

(a)

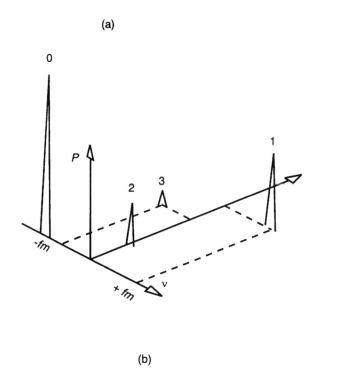

(b)

Figure 3.25 (a) Example mobile scenario and (b) corresponding scattering function.

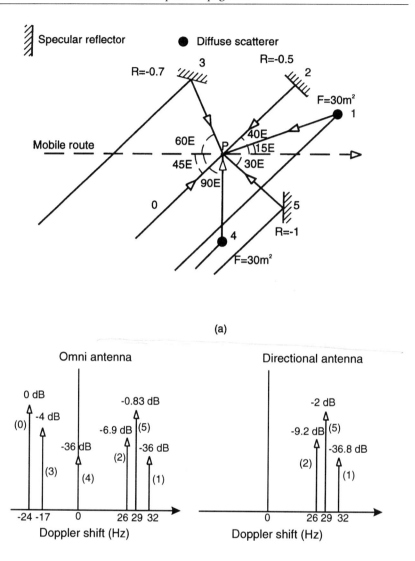

Figure 3.26 (a) Environment with diffuse scatterers and specular reflectors. (b) Doppler spectra (cases A and B).

The Doppler spectra for cases (A) and (B) will be those indicated in Tables 3.3 and 3.4 and in Figure 3.26(b).

3.11 Global Wideband Characterization of the Land Mobile Channel

This section briefly examines the manner in which the wideband characterization of the channel for land mobile communications is made. The basic parameter for its characterization in the time domain is the PDP, $P_h(\tau)$. The study of PDPs is made in two stages, as it was already done for the narrowband case: ← Power Delay Profile

- *Small scale* (for short runs of approx. *20* to *40* λ, small area);
- *Large scale* (for larger areas).

For the *small scale* characterization, a number of instantaneous PDPs are averaged over short route sections where there are no significant variations in multipath conditions. Generally, two parameters are computed from averaged PDPs:

Table 3.3
Amplitudes of Echoes for Figure 3.26(a).

Direct S. (Reference)	d_{Dir}	10 km	1 (lin)	0.0 dB	$1/(\sqrt{4\pi}\ 10000)$
Reflected S.	$d_1 + d_2$	11 km	0.91 (lin)	−0.83 dB	$1/(\sqrt{4\pi}\ 11000)$
Scattered S.	$d_1 \times d_2$	10 × 0.1	0.0028 (lin)	−31.0 dB	$1/(4\pi\ 1000)$

Table 3.4
Doppler Spectra for the Environment in Figure 3.26(a).

Ray No.	Amplitude (lin)	Level (A) (dB)	+20 log(cos (θ))	Level (B) (dB)	Doppler (Hz)
0	1	0.0	—	—	−25.6
1	$0.0028 \times \sqrt{30}$	−36.22	−0.54	−36.82	32.2
2	0.91×0.5	−6.84	−2.31	−9.15	25.5
3	0.91×0.7	−3.92	—	—	−16.7
4	$0.0028 \times \sqrt{30}$	−36.28	—	—	0.0
5	0.91×1	−0.83	−1.25	−2.08	28.9

- The average delay, D;
- The delay spread, S.

These parameters are defined in mathematical terms as follows:

$$D = \frac{\int_0^\infty \tau P_h(\tau) d\tau}{\int_0^\infty P_h(\tau) d\tau} \qquad S^2 = \frac{\int_0^\infty (\tau - D)^2 P_h(\tau) d\tau}{\int_0^\infty P_h(\tau) d\tau} \qquad (3.59)$$

In the studies carried out in the framework of the Euro-COST 207 [5] project, which was mostly related to the GSM system, the conclusion was reached that it was necessary to use two additional parameters (Figure 3.27) to provide a suitable characterization of a PDP:

- The delay window, W_q;
- The delay interval, I_p.

The *delay window* is the duration of the intermediate portion of the delay profile containing $q\%$ of the total energy, typically, $q = 90\%$.

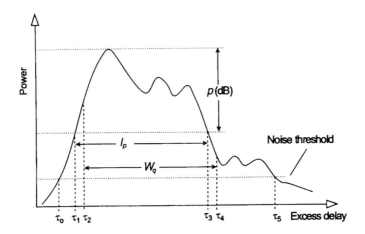

Figure 3.27 Delay profile model (COST 207) [5].

$$W_q = \tau_4 - \tau_2 \quad \text{with} \quad \int_{\tau_2}^{\tau_4} P_h(\tau)d\tau = q \int_{\tau_o}^{\tau_5} P_h(\tau)d\tau \quad (3.60)$$

The *delay interval* is the difference in delay between the points where the profile crosses a given level $p(dB)$ for the first time and the point at which it passes through the same level $p(dB)$ for the last time (see Fig. 3.27).

$$I_p = \tau_3 - \tau_1 \quad (3.61)$$

Description in the Frequency Domain

In this case, the most important parameter is the *coherence bandwidth* that depends on the channel coherence function, which is the Fourier transform of the PDP (Figure 3.22).

$$R_H(\Delta f) = \int_{-\infty}^{\infty} P_h(\tau) \exp(-j2\pi\Delta f\tau)d\tau \quad (3.62)$$

The *coherence bandwidth* or correlation bandwidth, B_c, is taken for values of 0.5 and/or 0.9 of $R_H(\Delta f)$. As P_h and R_H are related via the Fourier transform, the characterization and measurement of the mobile channel may be performed either in the time or frequency domain. Going from one representation to the other is, thus, straightforward.

Large-Scale Characterization

For larger areas, the study is made of the statistical distribution of the parameters measured on a small scale, such as the *delay spread* or the *coherence bandwidth*. The statistical characterization of S or of B_c is made for different types of environments: urban, suburban, open, flat, hilly, etc. Figure 3.28(a, b) [7] shows distributions of these parameters for a suburban environment.

Other Power Delay Profile-Related Models

In the work carried out within COST 207 [5] that studied the effects of propagation on mobile digital systems, especially GSM, the use of *triangular distributions* (log power vs. delay) was recommended for channel simulations (Figure 3.29(a)). These triangular distributions present different spreads in time depending on the type of environment considered. For hilly areas where significant echoes with long delays from quasi-specular reflections may occur nearby mountains, PDPs with two peaks are used (Figure 3.29(a)).

Multipath Propagation 109

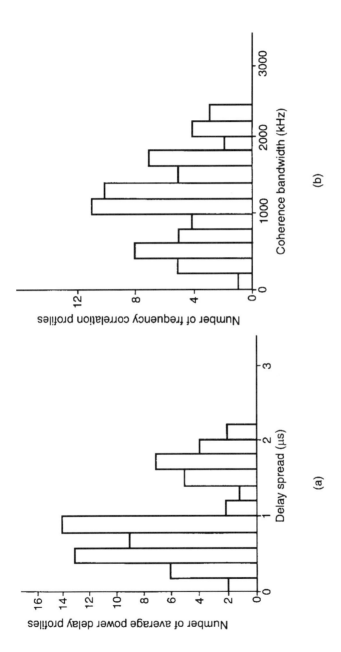

Figure 3.28 Measured distributions of S (a) B_c (b) for a suburban area [7].

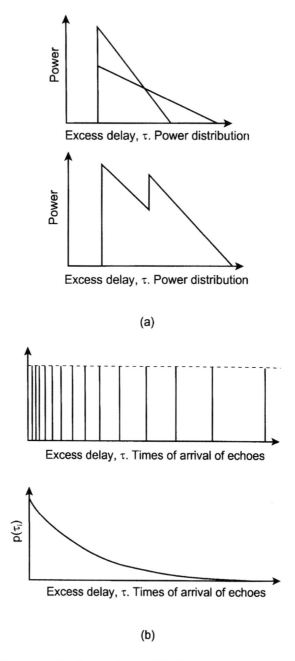

Figure 3.29 (a) Triangular distributions (COST 207). (b) Exponential model of PDP.

In some instances, for the characterization of a PDP, not only the decay rate with delay is important. There are models that provide a statistical description of the times of arrival of multipath echoes relative to the direct signal (whether it exists or not). A widely accepted model is the *negative exponential* model. Figure 3.29(b) illustrates how echoes arriving with smaller delays are more frequent than long delayed echoes. The negative exponential model is given by the following expression

$$p(\tau_i) = \frac{1}{\sigma_{av}} \exp\left(-\frac{\tau_i}{\sigma_{av}}\right), \quad \tau_i \geq 0 \tag{3.63}$$

where τ_i is the excess delay (relative to the direct signal) and σ_{av} is the mean (and also the standard deviation) of the negative exponential distribution.

3.12 Land Mobile Channel Measurements

So far, the channel has been characterized by several parameters. It is also important to know how such parameters can be measured. Basically, there are two types of measurements: narrowband and wideband channel sounding.

Narrowband Measurements

For narrowband measurements, continuous wave (CW) transmissions (i.e., unmodulated carriers) are used. Logarithmic receivers are generally used, as they make it possible to record signal variations directly in a logarithmic scale (*dBm* or *dBµV*). Normally, the data acquisition and storage is carried out with a PC connected via a IEEE 488 bus or an A/D card to an RF receiver (field strength meter). It is also important to record the instantaneous mobile speed by using a so-called "vehicle fifth wheel" (odometer). Time information is directly available from the sampling rate used. Figure 3.30 illustrates a measurement system or narrowband channel sounder. If information on the phase (for example, to measure Doppler shifts) is required, two receivers should be used, one in-phase and the other in-quadrature ($i(t) + jq(t)$). In this case both transmitter and receiver local oscillators must be phase-locked.

Signal Sampling

To avoid missing important signal variation features such as deep fades, the received signal envelope must be suitably sampled. Table 3.1 showed that the average duration of 30-dB fades for Rayleigh distributed signals is 0.01λ, which, at 900 MHz, for example, corresponds to a distance on the order of

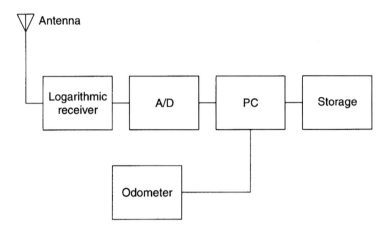

Figure 3.30 Received signal variations measurement system.

0.33 cm. The fading depth measurement requirements will determine the sampling rate to be used.

Also, it is well-known that the fast variations are superimposed on the slow variations. To conduct studies and to compare models and measurements, it is useful to separate the two types of variations (shadowing and fading). The usual method used to separate the slow variations from the fast ones is to apply a discrete low-pass filter. Generally, a simple averaging technique is used (running average)

$$\bar{x}(i) = \frac{1}{2N+1} \sum_{k=-N}^{N} x(i+k) \qquad (3.64)$$

Figure 1.4 illustrates the results of this filtering process. The value of N should be adjusted to separate the fast variations from the slow in a suitable manner. In general, $2N$ should cover samples corresponding to several tens of wavelengths. Normal values are in the 20–40λ range.

Wideband Channel Sounding

In wideband channel studies, the main pieces of information sought are PDPs and related parameters: the delay spread and the coherence bandwidth. Measurements are normally carried out in the time domain (i.e., PDPs are measured) since the different channel functions are related via Fourier transforms. Sounding may also be carried out in the frequency domain by scanning the measurement bandwidth to obtain an estimate of the channel transfer function. The most widely used methods for measuring PDPs are the following:

- Sounding with regular pulses;
- Sounding with pulse compression techniques.

With the first option, short duration pulses (pseudo-impulses) are used. The received signal is actually the convolution of the sounding pulse with the channel impulse response. To record the estimated variations of $h(\tau; t)$, repeated pulses are sent. These pulses must be sufficiently close to each other so as to keep track of the channel variations along the route but sufficiently distant from each other so that all the significant echoes reach the receiver before the following impulse is received (Figure 3.31(a)). Typical values are [2]:

τ_1 = 0.1 μs, which allows for resolutions of 30m (e = Vt = $3 \times 10^8 \times 0.1 \times 10^{-6}$ = 30m);
τ_2 = 100 μs, which allows maximum power delays of 30 km (e = Vt = $3 \times 10^8 \times 100 \times 10^{-6}$ = 30km) to be measured;
P_{Peak} = 10w.

With this method, high peak powers are required to sufficiently illuminate the scatterers in the environment. This produces interference problems with systems operating on the same or adjacent bands. Generally, *pulse compression techniques* are used to overcome these problems. These are based on generating sounding signals with similar characteristics to those of white noise. If white noise, $n(t)$, is applied to the input of a linear system and the cross correlation of the received signal, $w(t)$, with a replica of the transmitted signal shifted at time τ, $n(t - \tau)$, is computed, $h(\tau)$ will be obtained.

One important feature of white noise is that its autocorrelation is a delta function [2]

$$E[n(t)n^*(t - \tau)] = R_n(\tau) = N_o\delta(\tau) \quad (3.65)$$

where $R_n(\tau)$ is the autocorrelation function of $n(t)$ and N_o is the unilateral noise power spectral density. The output of the measurement system (channel sounder) is given by [2]

$$w(t) = \int_{-\infty}^{\infty} h(\xi)n(t - \xi)d\xi \quad (3.66)$$

Computing the cross-correlation between the input and the output yields [2]

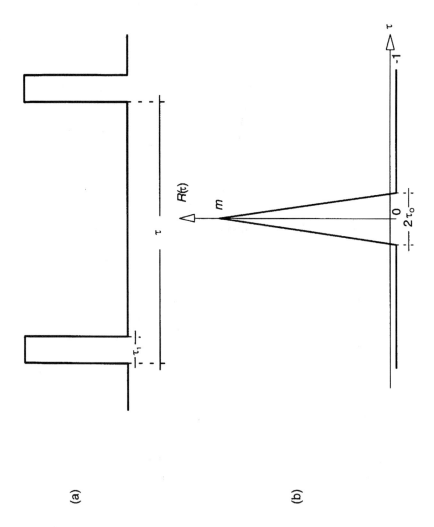

Figure 3.31 (a) Channel sounding by regular pulses. (b) Autocorrelation function of a PN sequence.

$$E[w(t)n^*(t-\tau)] = E\left[\int_{-\infty}^{\infty} h(\xi)n(t-\xi)n^*(t-\tau)d\xi\right] = \ldots \quad (3.67)$$

$$\ldots = \int_{-\infty}^{\infty} h(\xi)R_n(\tau-\xi)d\xi = N_o h(\tau)$$

It can be observed that the result of this operation will be a sample of the channel impulse response for a specific delay τ.

In practice, deterministic signals known as pseudo-noise sequences, PN, of length m are used. These signals have similar characteristics than those of white noise with an autocorrelation function that approximates a delta function (Figure 3.31(b)) where τ_o is the PN sequence clock period.

3.13 Cost 207 Channel Model

As a result of the work conducted by the COST 207 [5] study groups, several wideband propagation models were proposed for the practical realization of both hardware and software simulators. These models are based on the following transversal filter elements (Figure 3.20(c)):

1. A number of taps, each one with its corresponding delay and average power;
2. A Rayleigh or Rice amplitude distribution for each tap with an associated Doppler spectrum $S(\tau_i; \nu)$ where i indicates the tap number (Figure 3.20.c).

    Four types of Doppler spectra have been defined:

1. *Classic spectrum* (CLASS), used for delays not greater than 500 ns;

$$S(\tau_i; f) = \frac{A}{\sqrt{1-\left(\frac{f}{f_m}\right)^2}} \text{ for } f \in (-f_m, f_m) \quad (3.68)$$

with f_m being the maximum Doppler shift.

2. *Gaussian spectrum 1* (GAUS1) is the sum of two Gaussian functions and is used for the delay range between *500 ns* and *2 μs*.

$$S(\tau_i; f) = G(A, -0.8f_m, 0.05f_m) + G(A_1, 0.4f_m, 0.1f_m) \tag{3.69}$$

where A_1 is 10 dB below A, with

$$G(A, f_1, f_2) = A \exp\left(-\frac{(f-f_1)^2}{2f_2^2}\right) \tag{3.70}$$

3. *Gaussian spectrum 2* (GAUS2) is, as in the previous case, the sum of two Gaussian functions. It is used for paths with delays in excess of $2\mu s$

$$S(\tau_i; f) = G(B, 0.7f_m, 0.1f_m) + G(B_1, -0.4f_m, 0.15f_m) \tag{3.71}$$

where B_1 is 15 dB below B.

4. *Rice spectrum* (RICE) is the sum of a classic Doppler spectrum and a delta function representing the direct ray. (This spectrum is used for the shortest path corresponding to the first tap of the channel transversal filter representation in rural areas.)

$$S(\tau_i; f) = \frac{0.41}{2\pi f_m \sqrt{1 - \left(\frac{f}{f_m}\right)^2}} + 0.91\delta(f - 0.7f_m) \text{ for } f \in (-f_m, f_m) \tag{3.72}$$

Figure 3.32(a) shows these types of spectra.

Models

Several models are defined depending on the type of environment to be characterized. For each environment, a PDP was defined

$$P_h(\tau_i) = P_o \int_{-f_m}^{+f_m} S^2(\tau_i; f) df \tag{3.73}$$

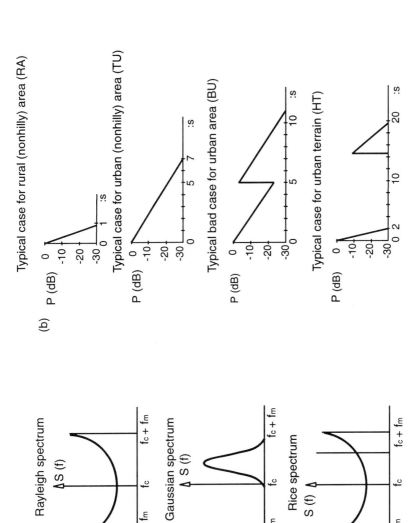

Figure 3.32 (a) Types of Doppler spectra in the COST 207 model. (b) COST 207 PDPs [5].

where $P_h(\tau_i)$ is the power corresponding to filter tap i and P_o is a normalization power. The schematic diagram of the COST 207 simulator is shown in Figure 3.20(c).

$R_1, \ldots, R_n$ are independent zero-mean complex Gaussian generators.
$L_1, \ldots, L_n$ are log-normal modulators with zero dB mean.
$W_1, \ldots, W_n$ are weighing factors.

This scheme is applicable both to hardware and software simulators. The defined PDPs, $P_h(\tau)$, are shown in Figure 3.32(b). The approximation of three of these PDPs by discrete coefficients is given in Tables 3.5, 3.6, and 3.7.

Table 3.5
Typical Case for Urban Areas With Nonmountainous Terrain (TU) [5]

Tap No.	Delay [μs]	Power (lin)	Power [dB]	Doppler Category	SD [μs]
1	0	0.4	−4	CLASS	
2	0.2	0.5	−3	CLASS	
3	0.4	1	0	CLASS	
4	0.6	0.63	−2	GAUS1	
5	0.8	0.5	−5	GAUS1	
6	1.2	0.32	−3	GAUS1	1.0
7	1.4	0.2	−7	GAUS1	
8	1.8	0.32	−5	GAUS1	
9	2.4	0.25	−6	GAUS2	
10	3.0	0.13	−9	GAUS2	
11	3.2	0.08	−11	GAUS2	
12	5.0	0.1	−10	GAUS2	

Table 3.6
Typical Worst Case for Urban Area With Mountainous Terrain (BU) [5]

Tap No.	Delay [μs]	Power (lin)	Power [dB]	Doppler Category	SD [μs]
1	0	0.2	−7	CLASS	
2	0.2	0.5	−3	CLASS	
3	0.4	0.79	−1	CLASS	
4	0.8	1	0	GAUS1	
5	1.6	0.63	−2	GAUS1	
6	2.2	0.25	−6	GAUS2	2.5
7	3.2	0.2	−7	GAUS2	
8	5.0	0.79	−1	GAUS2	
9	6.0	0.63	−2	GAUS2	
10	7.2	0.2	−7	GAUS2	
11	8.2	0.1	−10	GAUS2	
12	10.0	0.03	−15	GAUS2	

Table 3.7
Typical Case of Mountainous Terrain (HT) [5]

Tap No.	Delay [μs]	Power (lin)	Power [dB]	Doppler Category	SD [μs]
1	0	0.1	−10	CLASS	
2	0.2	0.16	−8	CLASS	
3	0.4	0.25	−6	CLASS	
4	0.6	0.4	−4	GAUS1	
5	0.8	1	0	GAUS1	
6	2.0	1	0	GAUS1	5.0
7	2.4	0.4	−4	GAUS2	
8	15.0	0.16	−8	GAUS2	
9	15.2	0.13	−9	GAUS2	
10	15.8	0.1	−10	GAUS2	
11	17.2	0.06	−12	GAUS2	
12	20.0	0.04	−14	GAUS2	

References

[1] Clarke, R. H., "A Statistical Theory of Mobile Radio Reception," *Bell Syst. Tech. J.*, 47 (July–Aug. 1968), pp. 957–1000.

[2] Parsons, J. D., *The Mobile Radio Propagation Channel*, London: Prentech Press, 1992.

[3] Lee, W. C. Y., *Mobile Communications Design Fundamentals,* New York: John Wiley & Sons, Inc., 1993.

[4] Fontan, F. P., A. Seoane, and M. A. V. Castro, "Matlab for Windows Software Aid in a Mobile Communications Course," *Int. Journal of Electrical Engineering Education,* Vol. 32, No. 4, October 1995, pp. 341–349.

[5] *Digital Land Mobile Radio Communications,* Euro Cost 207, Final Report, 1989, Commission of European Communities, Brussels.

[6] Fontan, F. P., M. A. V. Castro, and P. Baptista, "A Simple Numerical Propagation Model for Nonurban Mobile Applications," *IEE Electronic Letters,* Vol. 31, No. 25, December 7, 1995, pp. 2212–2213.

[7] Parsons, J. D., and J. G. Gardiner, *Mobile Communication Systems,* Glasgow: Blackie, 1989.

4

Propagation Path Loss

4.1 Introduction

Several propagation models have been developed over the last decades for mobile communication network planning. This chapter looks at some of the models used in *conventional large cells*. Chapter 6 deals with the models used in *microcell* and *indoor* systems.

It is essential to know precisely what the propagation models are trying to predict. As explained in earlier chapters, the study of signal variations with movement is divided, more or less arbitrarily, into slow or long-term variations and fast or short-term variations. Different route lengths are associated with the concepts of fast and slow variations: "*small areas*" covering tens of wavelengths and "*larger areas*" covering tens to hundreds of meters.

As regards received signal levels, the different route lengths correspond to different signal values, namely, the instantaneous value, associated with a given position on the route, the local mean within a small area and the distribution of local means within a larger area. Propagation models normally try to predict the median value of the distribution of the local means within a given larger area. Reported larger area sizes are variable depending on the authors, with values from 50×50, 100×100, 200×200, ... 1000×1000 m^2 [1] or areas of up to 1–1.5 km in radius [2]. These areas should have homogeneous characteristics (i.e., should not include at the same time two or more land-usage types such as open and built-up areas).

Instantaneous variations in the received signal are usually characterized by means of a Rayleigh or Rice distribution. To study the received signal, the fast variations are usually separated from the slow variations and are characterized separately. The most widely used model for describing the slow variations is the log-normal distribution (normal distribution in logarithmic units).

The standard deviation of the normal distribution (in logarithmic units), σ_L, is known as *locations variability* and depends on the frequency, the type of environment in which the mobile is located (urban, suburban, open area, etc.) and the terrain irregularity (flat, undulating, hilly, etc.). A short-term or fast variability (Rayleigh) is superimposed on this long-term or slow variability.

This chapter basically reviews two models that may be considered representative of two possible approaches used in large-cell mobile system planning, both cellular and PMR/PAMR:

- Okumura-Hata model [2, 3];
- Modified JRC model [4, 5].

In addition to these approaches, propagation models are presented that make use of detailed information of the environment, such as the shape and distribution of buildings in an urban area (urban databases (UDBs)). Examples of these are the Ikegami model [6] and the COST 231 Walfish-Ikegami model. Finally, a description is given of the new trends in propagation modeling, particularly for predictions in small macrocells and microcells in urban areas using UDBs and ray-tracing techniques [7].

The introduction of these new models means a qualitative step forward in the way predictions are made. Prediction detail or spatial resolution is drastically increased. For example, with the Ikegami model [6], there is a change from predicting the median value of the received field in a *larger area* to performing predictions at points much nearer to each other. The predictions obtained in this manner correspond to the values of the local mean at those points. In this way, the median value and the standard deviation of the slow signal variations in, for example, a street, are calculated directly from the series of local mean values obtained using the Ikegami model.

An additional improvement would be to try to predict not only the received signal's slow variations but the overall variations (fast and slow) together with other parameters (e.g., power delay profiles and scattering functions). This would be made possible by the use of detailed UDBs and terrain databases together with ray-tracing techniques to identify all possible paths between the transmitter and the receiver (including direct, reflected, double reflected, diffracted, and reflected-diffracted) and advanced electromagnetic models (geometrical/uniform theory of diffraction (GTD/UTD) [7]) for evaluating the amplitudes, phases, delays, and polarizations of each ray.

Now, coming back to the issue of spatial resolution of predictions, Figure 4.1 schematically illustrates possible prediction points on a digitized map. Each one of these points corresponds to a more or less extensive area with similar morphographic or land-usage characteristics. Each larger area is represented by an elementary surface element or pixel of 50 × 50 up to 500 × 500 m².

To end this introductory section, two concepts used throughout this chapter are now briefly summarized: the *basic loss*, L_b, and the *excess loss*, L_{excess}. The basic loss is defined as the total path loss between two hypothetical isotropic antennas

$$L_b \text{ (dB)} = L_{\text{isotropic}} = 10 \log\left(\frac{p_t}{p_r}\right) \quad (4.1)$$

The excess losses are those in addition to those due to free-space propagation, L_{fs}, and are defined as follows:

$$L_{excess} \text{ (dB)} = 20 \log\left(\frac{e_o}{e_r}\right) \quad (4.2)$$

where e_o is the received field under free space conditions and e_r is the actual received field. Obviously, the following equation holds

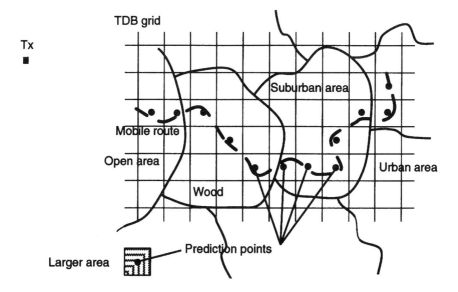

Figure 4.1 Propagation calculations for "larger areas" along the mobile route.

$$L_b \text{ (dB)} = L_{fs} + L_{excess} \qquad (4.3)$$

4.2 The Okumura Model

The Okumura et al. model [2] is a fully empirical method developed from a series of measurements made in Japan at several frequencies representative of mobile communications services (up to 1920 MHz). Curves were fitted to measured values as a function of a number of basic propagation parameters such as the type of environment, the terrain irregularity, and the antenna heights. The model is thus based on the use of a number of *correction factors*. These factors make it possible to associate the results of the model to the type of environment, terrain irregularity and antenna heights corresponding to the actual mobile path being studied. The following elements have been defined:

- Terrain features;
- Environment types.

Urban area propagation conditions are considered as a *reference* and a median field strength value is computed for this *reference environment*. To study other propagation conditions, a number of correction factors are introduced as explained below.

4.2.1 Terrain Features

The method was developed to predict received field strengths without knowing in very much detail the actual radio path between the base and the mobile; instead, certain easily estimated general terrain features are used as inputs to the model. Also, at the time the model was developed (late 1960s), it was common to study a limited number of radials (e.g., every 45 degrees or 30 degrees) around a radio station for coverage studies. The terrain profiles predictions were drawn manually from topographic maps. Nowadays, with the widespread use of terrain databases, this model is still very much used, and it is applied to radial profiles obtained automatically by a computer with any angle separation (e.g., 1 degree, 5 degrees, 10 degrees).

The terrain features defined in the Okumura model are listed below:

1. *BS effective height* (h_{te}). The effective height is defined as

$$h_{te} = h_{ts} - h_{ga} \text{ (m)} \qquad (4.4)$$

where h_{ts} is antenna height above sea level (Figure 4.2(a)), and h_{ga} is the average height of the terrain taken between 3 and 15 km (or less, if the path length is < 15 km).

2. *Terrain undulation* (Δh). This parameter is defined as the "interdecile range" of the terrain heights taken in a 10-km profile segment from the position of the mobile toward the base (Figure 4.2(b)).

   The most evolved models tend not to use these parameters to evaluate terrain irregularity losses. On the contrary, they consider the terrain effects in detail by taking into account the losses caused by each obstacle along the radio path.

3. *Isolated mountain and path parameter* (h). This parameter is used to account for the diffraction effects caused by an isolated ridge in the

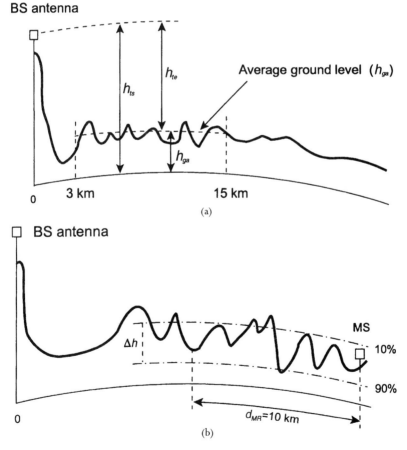

Figure 4.2 (a) Definition of the base station effective antenna height, (b) profile undulation, and (c) isolated mountain model and associated parameters [2].

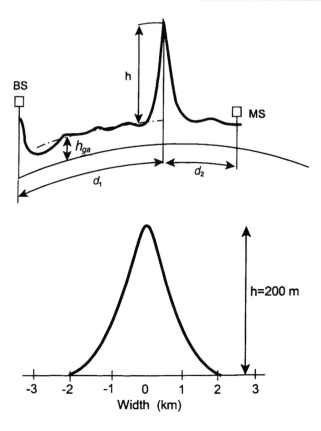

Figure 4.2 (continued).

radio path (Figure 4.2(c)). When there is only one ridge blocking the signal path, this obstacle is said to be isolated. The losses introduced are calculated using the knife edge diffraction model, which has been proved to be valid for frequencies in the VHF and UHF bands. The h parameter is evaluated with respect to the average terrain height, as illustrated in Figure 4.2(c). Figure 4.2(d) shows the theoretical model by which a real mountain is represented, where a standard height of 200m is assumed.

4. *Mean slope of the terrain* (θ_m). In sections of the terrain profile where a certain slope is observed for at least 5–10 km, a slope parameter, θ_m, may be defined as shown in Figure 4.3(a).

5. *Mixed sea-land path parameter* (β). This parameter is used to quantify the effects of propagation on paths that partially traverse water spans, such as lakes and bays. Several cases may be considered, depending

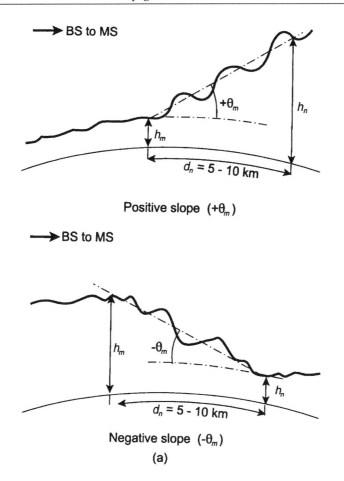

Figure 4.3 (a) Definition of the terrain slope parameter and (b) parameter describing mixed see-land paths [2].

on the order in which the stretches of land and water paths are arranged (Figure 4.3(b)).

Example 4.1. Figure 4.4(a) illustrates a terrain profile between a base station and a mobile. Both the effective antenna height and the profile undulation are calculated.

- *Effective height.* The average terrain height value, h_{ga}, between 3 and 15 km is 1,732m. In the case of an antenna on a 30m mast in a site at 1989m above sea level, this gives a height of $h_{ts} = 30 + 1989 = 2,019$m. The effective height is then $h_{te} = h_{ts} - h_{ga} = 287$m.

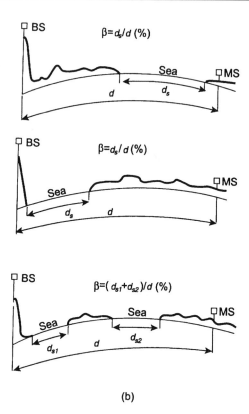

Figure 4.3 (continued).

- *Undulation.* The determination of the undulation parameter is necessary to calculate the distribution function of the profile heights and to calculate the difference between the 90% and the 10% deciles, as illustrated in Figure 4.4(b). The approximate undulation value, in this case, is $\Delta h \approx 100$m.

4.2.2 Environment Types

In order to take the shadowing/blockage effects of the immediate surroundings of the mobile into account (the base station is assumed to be clear of obstructions in its neighborhood), the land-usage type in the vicinity of the mobile must be considered. The Okumura model clearly distinguishes between three environment types:

1. *Open area:* A given environment may be considered as an open area if there are no obstacles over 300–400m in the direction of the base

Propagation Path Loss

129

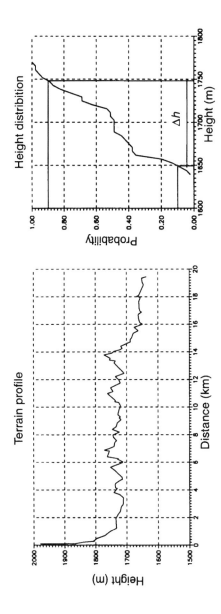

Figure 4.4 (a) Profile in Example 4.1 and (b) calculation of its undulation.

station and, in general, around the position of the mobile. Further, two correction factor curves (presented in Section 4.2.4) are given in the model for the following:
- Open areas;
- Quasi-open areas.

2. *Suburban area:* Suburban areas are those with some obstacles in the vicinity of the mobile, although low in density.
3. *Urban area:* Urban areas are cities with tall buildings and houses with more than two stories. Further, two urban area correction factors are provided in the model for the following:
- Large city;
- Average sized town.

As summarized above, actually a total of five different environment types are considered in the model. Additionally, "water" may be considered here as an environment type (even though it is described within the terrain irregularity parameters: mixed paths).

One necessary morphographic category that is not included in the Okumura model is the "wooded area" class. In some works performed in Germany [8], it has been proven that attenuations in wooded areas are similar in magnitude to those observed for urban areas.

It is noted that the classification of environments is a fairly subjective issue and depends to a large degree on the type of town, country, or region. So, an urban area in the United States is different, for example, from one in Japan or Europe [9]. In order to overcome this, the Okumura model has been complemented with subsequent studies where ways of quantifying building density, as shown in Section 4.2.5, are established. The modified JRC model, described in Section 4.4, also quantifies the effects of building density in urban areas.

4.2.3 Calculating the Received Median Field Strength Value

The following equation gives the median field strength value of the distribution of local means (slow signal variations) in a prediction area (larger area) of 1–1.5 km in radius. (Recall that Rayleigh-distributed fast variations are superimposed on the slow variations predicted by the model)

$$\bar{E} \text{ (dB}\mu\text{V/m)} = E_{fs} - A_{mu}(f, d) + H_{tu}(h_{te}, d) + H_{ru}(h_{re}, f) \quad (4.5)$$
$$+ \sum \text{Correction Factors}$$

where E_{fs} is the received field strength, for a given transmitter EIRP, under free-space propagation conditions. E_{fs} can be calculated using the following expression:

$$e_{fs}(\mu V/m) = \frac{\sqrt{30\ \text{eirp (w)}}}{d(m)} 10^6 \quad (4.6)$$

This is the *rms* value of the received field strength. In practical units, the following expression may be used:

$$E_{fs}(dB\mu V/m) = 74.78 + 20 \log \text{eirp (w)} - 20 \log d\ (km) \quad (4.7)$$

Normally, the curves given by the Okumura model are referred to a 1 kW ERP. The ERP and the EIRP are defined as follows:

$$\text{eirp (dBW)} = P_t(dBW) + G_t(dB_i) - L_t(dB) \text{ and} \quad (4.8)$$
$$ERP = P_t(dBW) + G_t(dB_d) - L_t(dB)$$

where P_t is the transmitter output power, L_t is the loss between the transmitter and the antenna, and G_i and G_d are both antenna gains, one expressed with respect to an isotropic reference antenna and the other with respect to a half-wave dipole. The half-wave dipole gain being 2.15 dB_i, we have

$$G(dB_i) = G(dB_d) + 2.15 \quad (4.9)$$

- $A_{mu}(f, d)$ (*urban area excess loss*) is the median attenuation value relative to free space in an urban area (reference environment), assuming a base station effective height of h_{te} = 200m and a mobile antenna height h_{re} = 3m. Figure 4.5 provides values for this parameter.
- $H_{tu}(h_{te}, d)$ is the *BS height gain factor*. A h_{te} = 200m reference value is taken so that, for this height, the BS height gain parameter is equal to 0 dB. Figure 4.6 shows its dependence with distance and BS effective antenna height.
- $H_{ru}(h_{re}, f)$ is the *MS height gain factor*. A h_{re} = 3m reference value is taken so that, for this height, the MS height gain factor is zero. Figure 4.7 illustrates its dependence with frequency and MS antenna height.

The Okumura model also provides a collection of curves for typical mobile communications frequency bands (150, 450, 900, and 1,500 MHz)

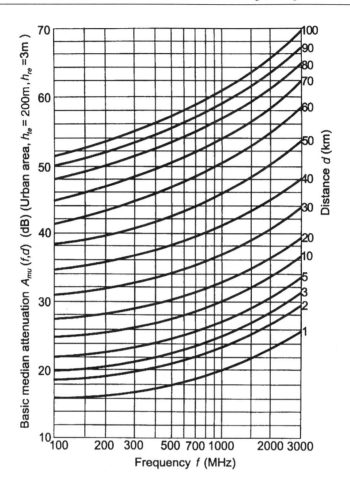

Figure 4.5 Excess loss for an urban area [2].

that give the predicted median field strength value for different BS effective antenna heights (h_{te} = 30, 50, 70, 100, 150, 200, 300, 450, 600, 800 and 1,000m), for an MS antenna height of 1.5m and for a 1-kW ERP (Figures 4.8–4.11). These curves correspond to a flat urban environment. In order to use them in other environments and under other terrain irregularity conditions, the *correction factors* described below must be applied.

4.2.4 Correction Factors

In this section the correction factors defined by Okumura are described in some detail.

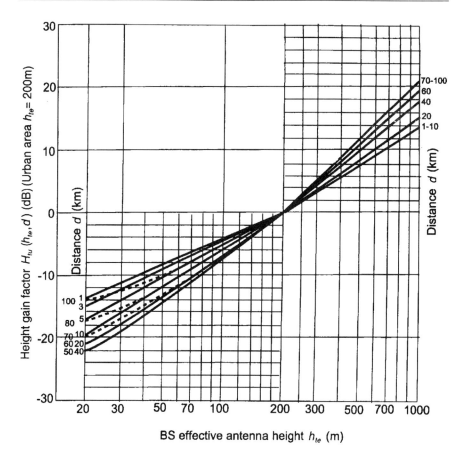

Figure 4.6 BS height gain factor [2].

1. *Urban area street orientation correction factor* (k_{al} and k_{ac}). These factors take into account the wave-guiding effect observed in street canyons when aligned with the base station and the loss increment when the street is perpendicular to the BS-MS path. Figure 4.12 shows the values for this parameter. On the curves it can be read that this correction factor has a range of 7 − (−5) = 15 dB at 5 km from the base station. This range decreases with distance to 3 − (−2) = 5 dB at 100 km from the base. Frequency dependent variations are not considered for this parameter.

2. *Suburban area correction factor* (k_{mr}). An urban reference environment is considered in this model. To tailor the model predictions to fit other environments, correction factors are used. Figure 4.13 provides a curve for the suburban area correction factor. In Figure 4.13, no

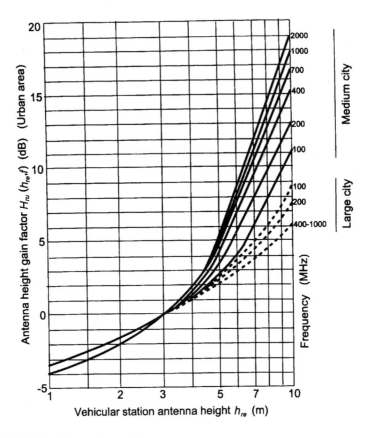

Figure 4.7 MS height gain factor [2].

distance dependence is observed while there is a marked dependence with frequency. Thus, a 8.5-dB value may be read for 450 MHz, 10 dB for 900 MHz, and 12.5 dB for 2 GHz.

3. *Open-area correction factor* (Q_o and Q_r) (Figure 4.14). Two types are distinguished, open area and quasi-open area with their corresponding factors, Q_o and Q_r. There is no dependence on distance, although the parameter is frequency-dependent, and so, in open areas, this correction factor has a value that goes from 26 dB for 450 MHz to 29 dB for 900 MHz and 32 dB for 2 GHz.

4. *Sloping terrain correction factor* (k_{sp}) (Figure 4.15). The correction factor varies with the distance, but in any case on the ascending slopes, greater field strength values will be experienced than on the descending slopes. Correction values for distances other than those given in the curves may be obtained by interpolation.

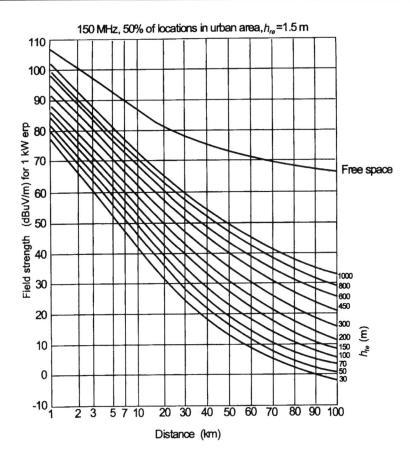

Figure 4.8 Received field strength in an urban area at 150 MHz [2].

5. *Correction factor for mixed sea-land paths* (k_s) (Figure 4.16). For paths traversing water spans, an increase in received field values is observed (recovery effect). For example, in the case of a link with a d_s/d ratio of 50% and a total path distance $d < 30$ km, there will be an increase of $k_s = 7.5$ dB in the received signal with respect to a path wholly over land.

6. *Undulating terrain correction factor* (k_h) (Figure 4.17). The model considers a flat ($\Delta h \approx 15$m) urban area as the reference environment. In order to take into account the decrease in the median received field strength for paths over irregular terrain, this frequency-dependent correction factor is introduced.

7. *Isolated mountain correction factor* (k_{im}, α). When the radio path is found to be obstructed by an isolated ridge obstacle, the isolated

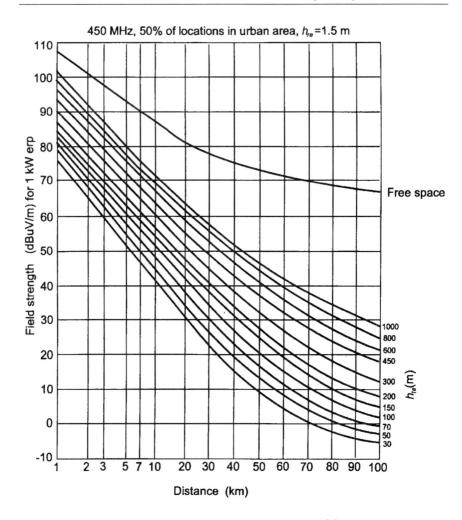

Figure 4.9 Received field strength in an urban area at 450 MHz [2].

mountain correction, k_{im}, will be used. The values given in Figure 4.18(a) are for different distances d_2 from the obstacle to the mobile for a standard obstacle height, h, of 200m. For other heights, the actual correction is obtained by multiplying the correction factor k_{im} by a conversion factor α (Figure 4.18(b)).

4.2.5 Quantification of Built-Up Area Losses

The Okumura model presents a good deal of uncertainty as to the quantification of losses in built-up areas. The configuration of built-up areas varies enormously from country to country. In order to be able to use models derived empirically

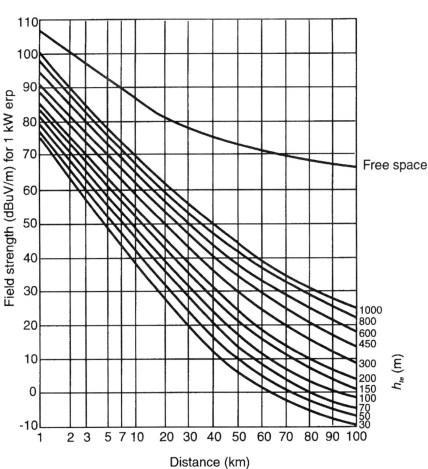

Figure 4.10 Received field strength in an urban area at 900 MHz [2].

in one country and apply them in another, it is necessary to introduce an objective measurement of the building density.

Several works have been carried out in this regard. Here, two curves are presented that quantify the effects of propagation in built-up areas in terms of the percentage of area occupied by buildings. Kozono and Watanabe [10] studied the variation in the received field in terms of the following parameters:

- The percentage of built-up surface area in the test area, 2%;
- The percentage of built-up surface area in the extended area, 2%;

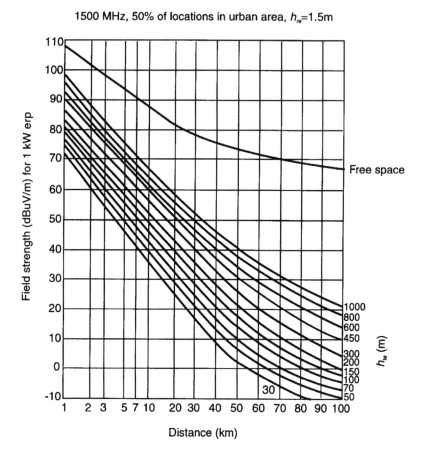

Figure 4.11 Received field strength in an urban area at 1500 MHz [2].

Table 4.1
Corrections of the Median Value in a Gaussian Distribution

x % of Locations	k (x%L)
50	0
75	0.67
90	1.28
95	1.64

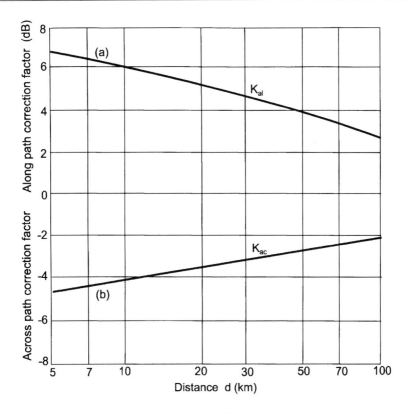

Figure 4.12 Street orientation correction factor [2].

- The volume of construction in the test area, β;
- The volume of construction in the extended area, β.

In the study, the *test area* is a circle with a radius of 250m and the *extended area* is the test area plus an additional area 500m long in the direction of the transmitter (Figure 4.19). The parameter α was found to be the most appropriate for quantifying the losses in built-up areas. The following expression was proposed to correct the values predicted by the Okumura model.

$$S(\text{dB}) = -25 \log \alpha + 30 \text{ with } 3 < \alpha < 50 \qquad (4.10)$$

Figure 4.20 shows an empirical curve provided by the ITU-R produced from the measurements made in Italy [11].

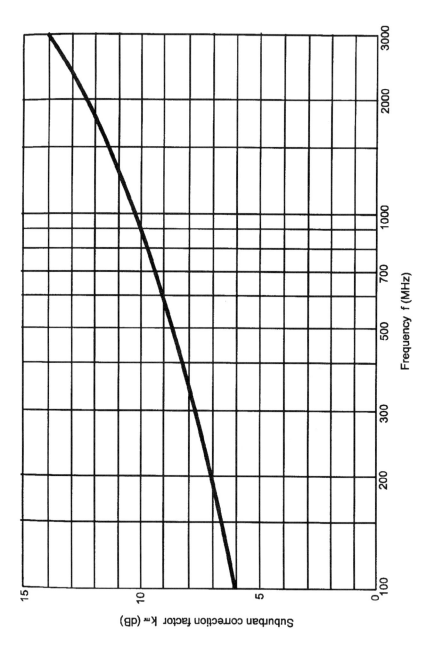

Figure 4.13 Suburban area correction factor [2].

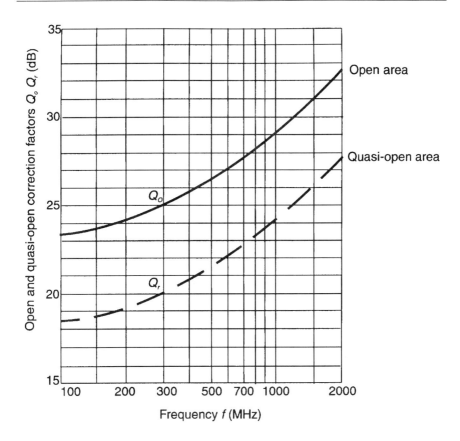

Figure 4.14 Open area correction factor [2].

4.2.6 Locations Variability

The Okumura model predicts the median received field strength value. This value represents the median value of the distribution of the local means of the received signal for "larger areas" of approximately 1–1.5 km in radius. Variations in the local mean (slow variations) approximately follow a log-normal distribution in linear units and, therefore, a normal or Gaussian distribution in logarithmic units (dBμV/m).

The Okumura model, besides giving the median value of the received field, also provides location variability values, σ_L (dB), (standard deviation of the Gaussian distribution of the local means in the study area). Figure 4.21 shows two curves giving location variability values for several environments, as a function of the frequency.

In coverage calculations, both in noise- and interference-limited systems, an attempt is made to guarantee exceeding the operation threshold for high

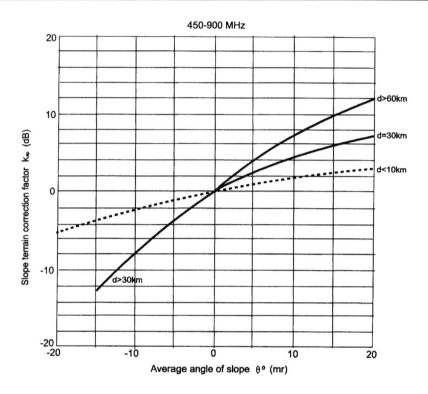

Figure 4.15 Sloping terrain correction factor [2].

percentages of locations (e.g., 90% or 95%). To achieve this, it is necessary to correct the median value of the received field and to convert it to the value exceeded for the desired percentage of locations. To do this, the median value given by the Okumura model shall be corrected according to the following expression:

$$E(x\%L) = \bar{E} - k(x\%L)\sigma_L \tag{4.11}$$

Table 4.1 gives different values of $k(x\%L)$ for various practical values of $x\%$.

Example 4.3. For an urban area, assuming a working frequency in the 500-MHz band, the location variability is 6 dB. If the median value of the received field given by the Okumura model is 45 dBμV/m, the value exceeded for 90% of locations is

$$E(90\%L) = 45 - 1.28 \cdot 6 = 37.32 \text{ dB}\mu\text{V/m} \tag{4.12}$$

Figure 4.16 Mixed path correction factor [2].

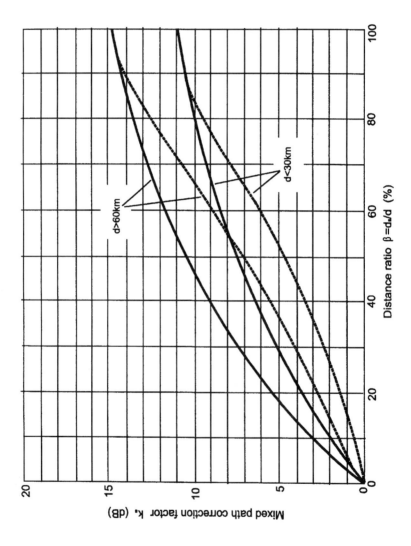

Figure 4.16 (continued).

Propagation Path Loss

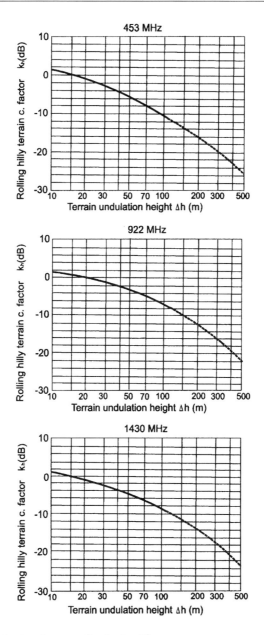

Figure 4.17 Irregular terrain correction factor [2].

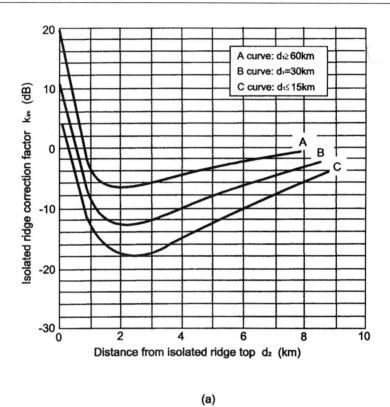

(a)

Figure 4.18 Isolated mountain correction factor and conversion factor for $h \neq 200$ m [2].

4.2.7 Calculations Using the Okumura Model

In this section, illustrative examples of the use of the Okumura model are provided.

Example 4.4. Assume that a mobile travels through a quasi-open rural area and that the BS effective height is 25m and the MS antenna height is 1m. The MS is 10 km away from the base and the working frequency is 450 MHz. The terrain is not too irregular; the undulation parameter, Δh, value is 50m. The MS has an effective gain of 0 dBi (comprising antenna gain and cable losses). An antenna impedance of 50Ω is assumed. The base station has a transmitter power of 5W and a gain of 5 dB_i.

Next, received power values using the free-space, plane Earth, and Okumura models are computed (see in Appendix 4A the discussion of the plane Earth model).

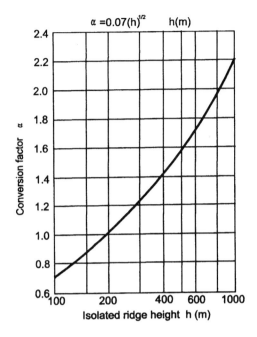

(b)

Figure 4.18 (continued).

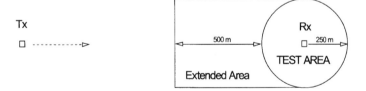

Figure 4.19 Test area and extended area [10].

Free space. If isotropic antennas are considered, the received power is

$$p_r(\text{for isotropic antennas}) = \frac{p_t}{4\pi d^2}\frac{\lambda^2}{4\pi} = \Phi\left(\frac{W}{m^2}\right)A_{\text{eff}}(m^2) \quad (4.13)$$

Free-space losses may be expressed in terms of practical units as

$$L_{fs}(\text{dB}) = 32.4 + 20\log f(\text{MHz}) + 20\log d\ (\text{km}) = 105.6\ \text{dB} \quad (4.14)$$

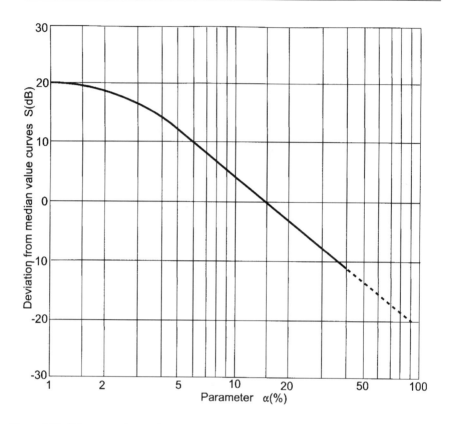

Figure 4.20 Urban area losses in terms of the percentage or area occupied by buildings [11].

The transmitted power is 5W ≡ 7dBW ≡ 37dBm. Thus the received power is

$$P_r = P_t + \sum G - \sum L = 7 + (5 + 0) - 105.6 = -93.6 \text{ dBW} \equiv -63.6 \text{ dBm} \quad (4.15)$$

This is a good received power value for PMR/PAMR applications. Typical receiver sensitivities are on the order of −113 dBm (or, equivalently, 0.5μV for a receiver impedance of 50Ω, see Chapter 5).

Plane Earth model. The path losses in the case of the plane Earth model (in linear units) are

$$l_{pe} = \frac{d^4}{(h_t h_r)^2} \quad (4.16)$$

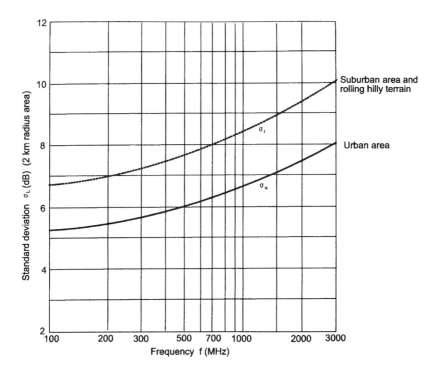

Figure 4.21 Location variability [2].

or, using practical units,

$$L_{pe}(\text{dB}) = 120 + 40 \log d \text{ (km)} - 20 \log (h_t(m)h_r(m)) = -132 \text{ dB} \quad (4.17)$$

The received power in this case is

$$P_r = P_t + \sum G - \sum L = 7 + (5 + 0) - 132 = -120 \text{ dBW} \equiv -90 \text{ dBm} \quad (4.18)$$

Okumura. The median value of the received field strength for a flat urban area, given by the Okumura model, has the following general expression:

$$\bar{E} \text{ dB}\mu\text{V/m} = E_{fs} - A_{mu} + H_{tu} + H_{ru} \quad (4.19)$$

Reading the appropriate values on the model curves and filling in the different parameters, we have

$\bar{E}$ (Urban area, flat terrain, BS effective height 25 m, MS height 1m) (4.20)
$= E_{fs} - 27 - 18 - 4 = E_{fs} - 49 \text{ dB}\mu\text{V/m}$

To this field strength value, correction factors corresponding to the terrain undulation and to a quasi-open area must be added. From the curves, the following correction factor values can be read:

$$k_r \text{ (undulation } \Delta h = 50 \text{ m)} = -5.5 \text{ dB}$$

$$Q_r \text{ (quasi-open area)} = 21 \text{ dB}$$

Hence, introducing these values in the median received field expression,

$\bar{E}$ (Quasi-open area, $\Delta h = 50$, BS effective height 25m, MS height 1m)(4.21)
$= (E_{fs} - 49) + 21 - 5.5 \text{ dB}\mu\text{V/m}$

The excess loss will then be

$$L_{excess} = E_{fs} - \bar{E} = 49 - 21 + 5.5 = 33.5 \text{ dB} \quad (4.22)$$

and the total losses are

$$L = L_{fs} + L_{excess} = 105.6 + 33.5 = 139.1 \text{ dB} \quad (4.23)$$

So, in this case, the received power is

$$P_r = P_t + \sum G - \sum L = 7 + (5 + 0) - 139.1 = -127.1 \text{ dBW} \equiv -97.1 \text{ dBm} \quad (4.24)$$

Appendix 4A: Review of the Plane Earth Model

Figure 4.22 illustrates the geometry of the plane Earth model. This model corresponds to the situation in which low antennas are used and both ends have direct visibility of each other. This model yields a $n = 4$ power propagation law with distance, which is very similar to that observed in mobile communication measurements.

Two rays are assumed to reach the receiver, namely, the direct ray and a reflected ray. These two contributions add up coherently at the receiver. The total received field strength will therefore be

$$e = e_{direct}(1 + Re^{-j\Delta\phi}) \quad (4.25)$$

Appendix 4A: Review of the Plane Earth Model 151

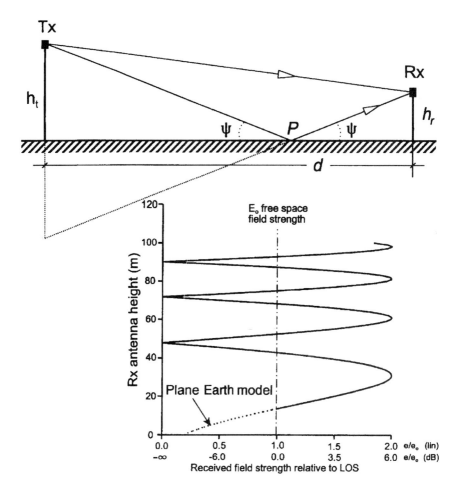

Figure 4.22 Plane Earth model and interference pattern for different receiver antenna heights.

where R is the ground reflection coefficient

$$R = |R|e^{-j\beta} \quad (4.26)$$

The phase difference between the two paths is

$$\Delta\phi = \frac{2\pi}{\lambda}\Delta l \quad (4.27)$$

where Δl is the path length difference between the direct and the reflected rays. Its value is

$$\Delta l = \mathrm{TPR} - \mathrm{TR} = [d^2 + (h_t + h_r)^2]^{1/2} - [d^2 + (h_t - h_r)^2]^{1/2} \quad (4.28)$$

$$= d\left[1 + \left(\frac{h_t + h_r}{d}\right)^2\right]^{1/2} - d\left[1 + \left(\frac{h_t - h_r}{d}\right)^2\right]^{1/2}$$

Using the first two terms of the series expansion for the square roots in the above expression, we get

$$\Delta l \approx d\left[1 + \frac{1}{2}\left(\frac{h_t + h_r}{d}\right)^2\right] - d\left[1 + \frac{1}{2}\left(\frac{h_t - h_r}{d}\right)^2\right] = \frac{2 h_t h_r}{d} \quad (4.29)$$

Returning now to the expression of the total received field strength and introducing the magnitude and phase of the reflection coefficient, we have

$$e = e_{\text{direct}}(1 + |R|e^{-j(\Delta\phi + \beta)}) \quad (4.30)$$

the magnitude of the total received field is

$$|e| = |e_{\text{direct}}|(1 + |R|^2 + 2|R|\cos(\Delta\phi + \beta))^{1/2} \quad (4.31)$$

Taking into account that

$$\sin^2(\alpha) = \frac{1 - \cos(2\alpha)}{2}$$

and that $d \gg h_t\, h_r$ ($\varphi \approx 0$), which implies $|R| \approx 1$ and $\beta \approx \pi$, the magnitude of the total received field is

$$|e| = 2|e_{\text{direct}}|\sin\left(\frac{\Delta\phi}{2}\right) \approx 2|e_{\text{direct}}|\sin\left(\frac{2\pi h_t h_r}{\lambda d}\right) \quad (4.32)$$

From this expression it is clear that the combination of a direct and a reflected ray gives rise to an interference pattern with amplitudes ranging from 0 to $2e_{\text{direct}}$. Figure 4.22 illustrates the received field relative to the direct signal

Appendix 4A: Review of the Plane Earth Model

field when an antenna height scan is conducted on the receiver station side. The transmitter station antenna is assumed to have a height of 30m and the link is 1 km in length.

Going now into the mobile communications case where low antennas are used (i.e., $h_t, h_r \ll d$), the argument within the sin function is small and thus the sin function can be replaced by its argument, yielding

$$|e| \approx |e_{\text{direct}}| \frac{4\pi h_t h_r}{\lambda d} \qquad (4.33)$$

In other words, in the plane Earth model, the receiver will always be moving below the first maximum of the interference pattern shown in Figure 4.22. Losses in excess of those for free space are then

$$l_{\text{excess}} = \left| \frac{e_{\text{direct}}}{e} \right|^2 \approx \left(\frac{\lambda d}{4\pi h_t h_r} \right)^2 \qquad (4.34)$$

The total losses expressed in linear units are the product of the free-space losses and excess losses

$$l_{\text{total}} = l_b = l_{\text{fs}} l_{\text{excess}} = \left(\frac{4\pi d}{\lambda} \right)^2 \left(\frac{\lambda d}{4\pi h_t h_r} \right)^2 = \left(\frac{d^2}{h_t h_r} \right)^2 \qquad (4.35)$$

It should be noted that any reference to the wavelength (or frequency) has disappeared. It is also observed that the path loss distance dependence follows a $n = 4$ power law or a 40-dB/decade decay rate. In practical units, the earth loss is

$$L_{pe} \text{ (dB)} = 120 + 40 \log d \text{ (km)} - 20 \log (h_t(m) h_r(m)) \qquad (4.36)$$

Example 4.5. Assume a mobile radio system [12] with the following characteristics:

- The system operates in the 1-GHz band.
- The power of the base transmitter is 10W.
- The base station antenna gain (omni-directional) is 10 dBi.
- The MS antenna gain is 2 dBi with a height of 2m. The characteristics of the area to be served are shown in Figure 4.23(a).

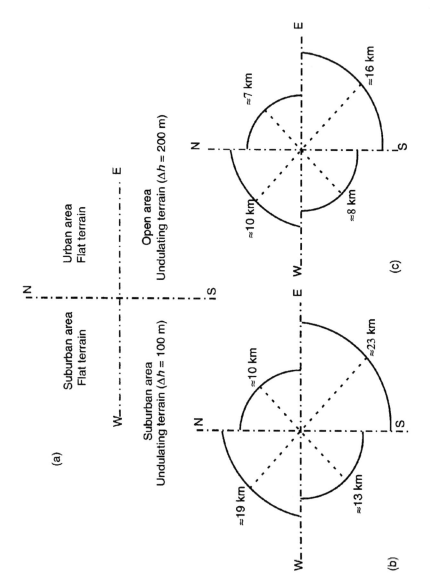

Figure 4.23 Description of the service area and coverage contour for (a) 50% and (b) 90% of locations.

Appendix 4A: Review of the Plane Earth Model

Further assume that the BS effective antenna height is 100m and the transmission bandwidth is 25 kHz and that frequency modulation is used. The mobile receiver has a $F_n = 7$ dB *noise figure*. In order to guarantee an adequate transmission quality, an additional 10-dB margin is added to the standard 10-dB FM margin above the noise level.

The objective is to obtain coverage contours for the four quadrants in the service area (Figure 4.23(a)) for the following cases:

- A 50% locations probability;
- A 90% locations probability.

Only one distance or coverage range is asked for in both case (a) and (b). To carry out these calculations, the use of the spreadsheet in Table 4.2 is recommended.

Calculations. Free-space losses (first column in Table 4.2) are computed for the different distances in the table using the known expression in practical units

$$L_{fs} \text{ (dB)} = 32.4 + 20 \log f \text{(MHz)} + 20 \log f \text{(km)} \qquad (4.37)$$

Then, the received power under free-space conditions (second column in Table 4.2) is computed for all the distances in the table using the link budget expression

Table 4.2
Coverage Calculations Spreadsheet [12]

Distance (km)	L_{fs} (dB)	P_r (dBm) (Free Space)	A_m (dB)	H_b (dB)	H_m (dB)	P_r (dBm) (Urban Area)
2						
5						
10						
15						
20						
25						
30						
40						
50						

Notes: $K = 1.38 \cdot 10^{-20}$ mJ K^{-1} (Boltzman constant); $T = 300°$K; σ_L(suburban) = σ_L(open) = 8 dB; σ_L(urban) = 5 dB; and $Q(k = 1.28) = 0.1$ (Gaussian distribution tail).

$$P_r = P_t + \sum G - \sum L = 40 + 10 + 2 - L_{fs} = 52 - L_{fs} \text{ (dBm)} \quad (4.38)$$

Since 10W = 40dBm, G_{BS} = 10 dB$_i$, and G_{MS} = 2 dB$_i$.
The receiver noise power is

$$N = 10 \log(1 \cdot 3 \cdot 8 \cdot 10^{-20} \cdot 300 \cdot 25{,}000) + F_n \quad (4.39)$$
$$= -129.85 \text{ dBm} + 7 \text{ dB} = -122.85 \text{ dBm}$$

The operation threshold is set up taking into account the two specified margins, i.e.,

$$P_{\text{Threshold}} = N + 10 + 10 = -102.8 \text{ dBm} \quad (4.40)$$

At this point, the table has to be completed using the Okumura curves (rest of columns in Table 4.2). The results are shown in Table 4.3. The coverage ranges found for the four quadrants for case (a) (i.e., a 50% locations percentage) and case (b) (i.e., 90% locations percentage) are worked out below.

The last column in Table 4.3 shows the received signal level for a flat urban area: 1st quadrant. In Table 4.4, median received signal levels are given for the four quadrants taking into account the following corrections:

k_{mr} (suburban area) = 10 dB;
k_h (undulating terrain, Δh = 100m) = −7 dB;
k_h (undulating terrain, Δh = 200m) = −12 dB;
Q_o (open area) = 29 dB.

Now, comparing the results shown in the table with the operation threshold, the coverage map for a 50% locations probability in Figure 4.23(b) is

Table 4.3
Spreadsheet for an Urban Area

Distance (km)	L_{fs} (dB)	P_r (dBm) (Free Space)	A_m (dB)	H_b (dB)	H_m (dB)	P_r (dBm) (Urban Area)
2	98.5	−46.5	23	−4	−2	−75.5
5	106.5	−54.5	27	−5	−2	−88.5
10	112.5	−60.5	30	−6	−2	−98.5
15	116.0	−64.0	31	−6	−2	−103.0
20	118.5	−66.8	33	−6	−2	−107.8
25	120.5	−68.5	35	−7.3	−2	−112.8
30	122.0	−70.0	37	−7.3	−2	−116.3
40	124.5	−72.5	41	−7.5	−2	−123.0
50	126.5	−74.5	44	−7	−2	−127.5

Table 4.4
Spreadsheet for the Four Quadrants for a 50%L Probability

Distance (km)	1st Quadrant (Flat Urban Area)	2nd Quadrant (Flat Suburban Area)	3rd Quadrant (Undulating Suburb. Area)	4th Quadrant (Undulating Open Area)
2	−75.5	−65.5	−72.5	−58.5
5	−88.5	−78.5	−85.5	−71.5
10	−98.5	−88.5	−95.5	−81.5
15	−103.0	−93.0	−100.0	−86.0
20	−107.0	−97.8	−104.8	−90.8
25	−112.8	−102.8	−109.8	−95.8
30	−116.3	−106.3	−113.3	−99.3
40	−123.0	−113.0	−120.0	−106.0
50	−127.5	−117.5	−124.5	−110.5

obtained. For a 90% locations probability, it is necessary to take into account a Gaussian statistical correction, which is different depending on the value of the locations variability parameter, σ_L.

- Quadrant 1: $k\,\sigma_L = 1.28 \times 5 = 6.4$ dB
- Quadrants 2, 3 and 4: $k\,\sigma_L = 1.28 \times 8 = 10.24$ dB

A new table with the received powers exceeded in 90% of locations in each quadrant may be computed (Table 4.5). Comparing the threshold value with the values in Table 4.5, the coverage map shown in Figure 4.23(c) is obtained.

4.3 Hata Model

The Hata model [3] is a version of the Okumura model developed to be used in computerized coverage calculation applications. Hata obtained mathematical expressions fitting the values provided by the Okumura model curves. He also gives expressions for the more commonly used correction factors. Hata obtained a series of expressions to calculate the basic propagation loss, L_b, (loss between isotropic antennas) for urban, suburban, and rural environments. The Hata model expression for the basic loss is the following:

$$L_b \text{ (dB)} = 69.55 + 26.16 \log(f) - 13.82 \log(h_t) \quad (4.41)$$
$$- a(h_m) + (44.9 - 6.55 \log(h_t))\log(d)$$

Table 4.5
Spreadsheet for the Four Quadrants for a 90%L Probability

Distance (km)	1st Quadrant (Flat Urban Area)	2nd Quadrant (Flat Suburban Area)	3rd Quadrant (Undulating Suburb. Area)	4th Quadrant (Undulating Open Area)
2	−81.9	−75.7	−96.9	−68.7
5	−94.9	−88.9	−109.9	−81.7
10	−104.9	−98.7	−105.7	−91.7
15	−109.4	−103.2	−110.2	−96.2
20	−114.2	−108.0	−115.0	−101.0
25	−119.2	−113.0	−120.0	−106.0
30	−122.7	−116.5	−123.5	−109.5
40	−129.4	−123.2	−130.2	−116.2
50	−133.9	−127.7	−134.7	−120.7

where f is in megahertz, h_t and h_m are in meters, and d in kilometers. These losses correspond to a *flat urban area*. For an MS antenna height of 1.5m, $a(h_m) = 0$. Model corrections are given next.

For a medium-small city:

$$a(h_m) = (1.1 \log(f) - 0.7)h_m - (1.56 \log(f) - 0.8) \qquad (4.42)$$

For a large city:

$$a(h_m) = 8.29(\log(1.54 h_m))^2 - 1.1 \quad f \leq 200 \text{ MHz} \qquad (4.43)$$
$$a(h_m) = 3.2(\log(11.75 h_m))^2 - 4.97 \quad f \geq 400 \text{ MHz}$$

For a suburban area:

$$L_{bs} = L_b - 2(\log(f/28))^2 - 5.4 \qquad (4.44)$$

For a rural area:

$$L_{br} = L_b - 4.78(\log(f))^2 + 18.33 \log(f) - 40.94 \qquad (4.45)$$

The validity limits of this model are the following:

$$150 \leq f(\text{MHz}) \leq 1500 \qquad 30 \leq h_t \text{ (m)} \leq 200$$
$$1 \leq h_m \text{ (m)} \leq 10 \qquad 1 \leq d \text{ (km)} \leq 20$$

where h_t is the BS effective antenna height and h_m is the MS height.

The Hata model is applicable for the frequency range given above. Recently, in view of the need to plan new systems, such as the GSM 1800 (1,800-MHz band version of the GSM 900-MHz system), a new revision of the Hata model (COST 231-Hata [1]) has been developed using similar methods to those used by Hata based on the Okumura model. The COST 231-Hata model follows the expression given below

$$L_b = 46.3 + 33.9 \log(f) - 13.82 \log(h_t) - a(h_m) \qquad (4.46)$$
$$+ (44.9 - 6.55 \log(h_t))\log(d) + C_m$$

where $a(h_m)$ has the same expression as in the original model for a medium-small city, and C_m is equal to 0 dB for medium sized cities and suburban centers with an average density of trees and 3 dB for metropolitan centers. The application range for the model is the same as for the original model, except for the frequency range which is $1{,}500 < f(\text{MHz}) < 2{,}000$.

4.4 Modified JRC Model

The modified JRC (UK's Joint Radio Committee of Power Industries) [4, 5] model is described here. It has been selected for presentation in this chapter as it uses a different approach to that of the Okumura-Hata model for dealing with terrain irregularity effects. In this model, detailed terrain profiles are used, and to compute the losses due to terrain obstacles, diffraction models are applied.

In the Okumura model, to describe terrain features and irregularity, the undulation, the effective antenna height and the terrain slope were used. In the case of the JRC model, the approach is completely different since individual obstacles need to be identified in the first place. As regards the determination of the losses due to elements in the neighborhood of the mobile (that cannot be read on a terrain database), the model provides expressions for the excess losses in urban areas.

The JRC model is described in [4], while the modifications due to Ibrahim and Parsons to take into account the excess losses in built-up areas are described in [5].

For this model, a *reference loss* is defined. This is either the free space loss

$$L_{fs} \text{ (dB)} = 32.4 + 20 \log f(\text{MHz}) + 20 \log d \text{ (km)} \qquad (4.47)$$

or the plane Earth loss

$$L_{pc} \text{ (dB)} = 120 - 20 \log(h_t(\text{m})h_r(\text{m})) + 40 \log d \text{ (km)} \quad (4.48)$$

To compute the losses due to the terrain irregularity, several different situations or path types are defined, as listed in Table 4.6.

Having obtained a detailed terrain profile, either manually by reading the heights on a topographic map or by using a terrain database, the profile is then classified into one of the six path types listed in Table 4.6. Prior to carrying out this classification, the profile curvature must be corrected to take into account the radius of the Earth and refraction effects.

Normally, straight line rays are used in propagation studies, and it is assumed that the Earth, instead of having a radius of R_o = 6,370 km, has an *effective radius*, kR_o, where the factor k takes into account refraction effects (coverage ranges larger than optical). The value of k for a standard atmosphere is 4/3.

Figure 4.24(a) illustrates the procedure normally used in the calculation of fixed radio links. There, both ends are considered to be fixed, and the intermediate heights of the profile are corrected using a parabolic approximation of the Earth's curvature.

Obstacle heights are modified to take into account both the Earth curvature and the effects of refraction (ray bending), as shown in Figure 4.24(b). The expression used for this correction is

$$B(x) = \frac{x(d-x)}{2kR_o} \quad (4.49)$$

where $B(x)$ is the height correction for an obstacle at a distance x from the transmitter end.

Table 4.6
Types of Radio Paths Considered in the JRC Model

a	Line-of-sight	With sufficient path clearance
b		With insufficient path clearance
c	Non-line-of-sight	1 obstacle
d		2 obstacles
e		3 obstacles
f		More than 3 obstacles

Appendix 4A: Review of the Plane Earth Model 161

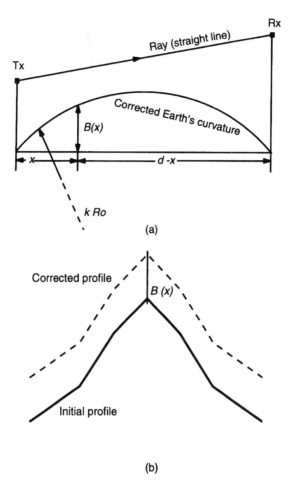

Figure 4.24 Earth model with effective radius kR_o and correction of obstacle heights.

For area coverage studies, only the base station height is considered to be fixed, and the profile is corrected from this point onward with another parabolic formula which approximates the Earth's circumference, as shown in Figure 4.25. The expression used in this case to correct obstacle heights is

$$y_i = h_i - \frac{x_i^2}{2kR_o} \tag{4.50}$$

where

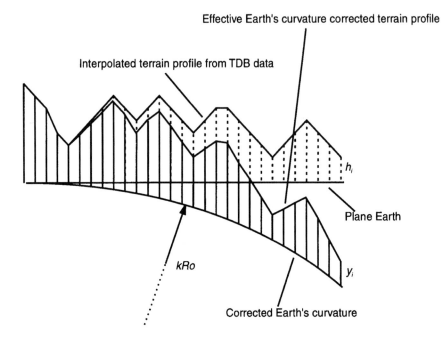

Figure 4.25 Terrain height corrections for coverage studies.

y_i is the corrected height;
h_i is the height read on the map;
x_i is the distance to the base station.

After correcting the profile curvature, the next step is to classify the terrain profile into one of the six categories in Table 4.6. In case of a direct line-of-sight path, a study of the path clearance parameter, c, must be carried out. The path clearance parameter is the opposite of the obstruction parameter, $h(c = -h)$, used in nonline-of-sight paths.

The path clearance shall be compared with the radius of the first Fresnel zone, R_1, to verify whether the direct ray is clear from the most outstanding terrain feature by at least 60% of R_1. If this is the case, diffraction effects will be negligible. Figure 4.26 shows a line-of-sight profile and illustrates the path clearance concept.

Also, in Figure 4.26 the knife-edge diffraction model parameters are illustrated for a radio path with a single obstacle. Instead of using the obstruction-to-first Fresnel zone ratio, h/R_1, a normalized obstruction parameter is used to compute path losses,

$$\nu = \sqrt{2}\frac{h}{R_1} \qquad (4.51)$$

Appendix 4A: Review of the Plane Earth Model 163

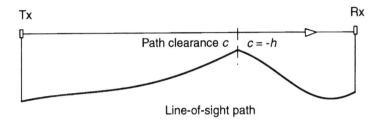

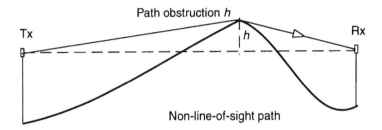

Figure 4.26 Geometry of the knife-edge model. Path clearance in a LOS link and path obstruction in a NLOS link.

Now, expressing ν in terms of R_1,

$$R_1 = \sqrt{\frac{\lambda d_1 d_2}{d_1 + d_2}} \Rightarrow \nu = h\sqrt{\frac{2}{\lambda}\frac{d_1 + d_2}{d_1 d_2}} \quad (4.52)$$

The losses due to diffraction, L_d, must be added to the reference losses, as indicated below. To compute L_d, the curve shown in Figure 4.27 can be used, or, alternatively, the following approximate formulas may be used:

$$L_d \text{ (dB)} = \begin{cases} 0 \\ 6.02 + 9.0\nu + 1.65\nu^2 \\ 6.02 + 9.11\nu - 1.27\nu^2 \\ 13 + 20\log(\nu) \end{cases} \text{ for } \begin{cases} -0.8 > \nu \\ -0.8 \leq \nu < 0 \\ 0 \leq \nu < 2.4 \\ \nu > 2.4 \end{cases} \quad (4.53)$$

Continuing with the line-of-sight path case, if it is observed that there is an insufficient path clearance, i.e., $\nu > -0.8$ ($h \approx 60\% \ R_1$), diffraction losses L_d must be computed.

For case (a), *sufficient path clearance,* the total loss is equal to the free-space loss

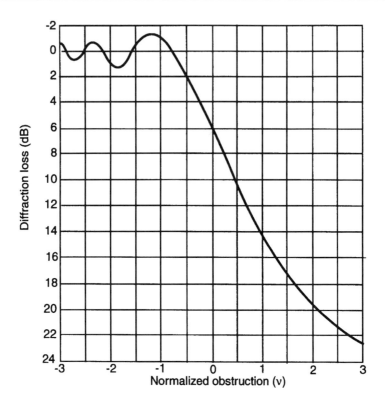

Figure 4.27 Diffraction losses as a function of the normalized obstruction parameter v.

$$L_b = L_{fs} \qquad (4.54)$$

For case (b), *insufficient path clearance,* the total loss is the sum of the reference losses (maximum of free space and plane Earth losses) and the losses due to diffraction

$$L_b = \max(L_{fs}, L_{pe}) + L_d \qquad (4.55)$$

For case (c), *a single obstacle,* total losses will follow the same expression

$$L_b = \max(L_{fs}, L_{pe}) + L_d \qquad (4.56)$$

In case (d), *two obstacles,* the total losses due to diffraction should be computed taking into account both obstacles. The procedure used by the JRC model is due to Epstein and Peterson [13]. Basically, this method assumes that the link comprises two sublinks in tandem: the first between the transmitter and the second obstacle with diffraction on the first obstacle and the second sublink between the first obstacle and the receiver with diffraction on the second obstacle. Figure 4.28(a) gives a sketch of the procedure for the calculation of the obstruction parameters h_1 and h_2. In this case, the total losses are given by

$$L_b = \max(L_{fs}, L_{pe}) + L_{d1} + L_{d2} \tag{4.57}$$

Case (e), *three obstacles* is illustrated in Figure 4.28(b). To compute diffraction losses the Epstein and Peterson method shall be applied successively to the three obstacles considering three sublinks as illustrated in Figure 4.28(b). The total loss is given by

$$L_b = \max(L_{fs}, L_{pe}) + L_{d1} + L_{d2} + L_{d3} \tag{4.58}$$

Finally, case (f), *more than three obstacles,* is shown in Figure 4.28(c). In this case, it is advisable to reduce the number of obstacles to a maximum of three (i.e., reduce the profile to case (e)). In order to do this, the procedure suggested by Bullington [14] is used. This involves describing the total diffraction effects by means of a virtual obstacle, as also shown in Figure 4.28(d).

4.4.1 Additional Losses in Built-Up Areas

Ibrahim and Parsons [5] proposed an empirical correction to the JRC model to take into account the losses in built-up areas. To do this, a factor, β, termed *excess clutter loss* is defined. The total losses are then

$$L_b = L_{JRC} + \beta \tag{4.59}$$

The β parameter is given by the expression

$$\beta \text{ (dB)} = 20 + \frac{f(\text{MHz})}{40} + 0.18L \text{ (m)} - 0.34H + K_4 \tag{4.60}$$

with

$K_4 = 0.094 \, U(\%) - 5.9$ for highly urbanized areas;
$K_4 = 0$ for other areas.

where

L is the *land usage factor* which is defined as the building percentage in an area of 500×500 m^2 where the propagation prediction is performed.

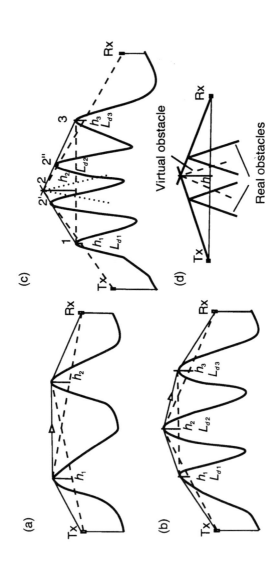

Figure 4.28 Epstein & Peterson method for (a) two and (b) three knife-edges. (c) JRC model for more than three knife-edges and (d) Bullington multi-edge diffraction method.

U is the *urbanization factor*, which is defined as the percentage in an area of 500 × 500 m² with buildings of four or more stories.

H is the *mobile relative spot height*, which is defined as the height of the point where the mobile is located relative to the height of the base station.

Example 4.6. A link between a base and a mobile is studied here using the JRC model. The terrain profile corresponding to the radio path is illustrated schematically in Figure 4.29(a). The base station antenna is 20m high and the mobile antenna is 1m high. The working frequency is 150 MHz. It is assumed that the point where the MS is located corresponds to an open area. The location variability due to terrain undulation effects is assumed to be $\sigma_L = 6$ dB.

The median field strength in the vicinity of the mobile (500 × 500 m² area) will be computed. Also, the field strength value exceeded for 90% of locations will be computed. The base station EIRP is assumed to be 10W.

It can be observed that the terrain profile presents two knife edges. The joint effect of both edges is evaluated applying the Epstein and Peterson method. First, however, it is necessary to correct the heights of the main terrain features to take into account the Earth's radius and the *k* factor. The $B(x)$ correction for both edges is (4.49):

$$B(x) = \frac{10(30 - 10)}{2 \cdot 1.333 \cdot 6370} = 0.01177 \text{ km} = 11.77\text{m}$$

Now, computing the radius of the first Fresnel zone, R_1, which will be equal in both subpaths

$$\lambda = \frac{300}{150} = 2\text{m} \quad R_1 = \sqrt{\frac{\lambda d_1 d_2}{d}} = \sqrt{\frac{2 \cdot 10000 \cdot 10000}{20000}} = 100\text{m}$$

The obstruction parameters, h_1 and h_2, are calculated by means of geometrical considerations shown in Figure 4.29(b). Then, the normalized obstructions and knife-edge diffraction losses are determined

$$\nu_1 = \sqrt{2}\frac{h_1}{R_1} = 0.578 \Rightarrow L_1 \approx 11 \text{ dB}$$

$$\nu_2 = \sqrt{2}\frac{h_2}{R_1} = 0.295 \Rightarrow L_2 \approx 8.6 \text{ dB}$$

The free-space loss is

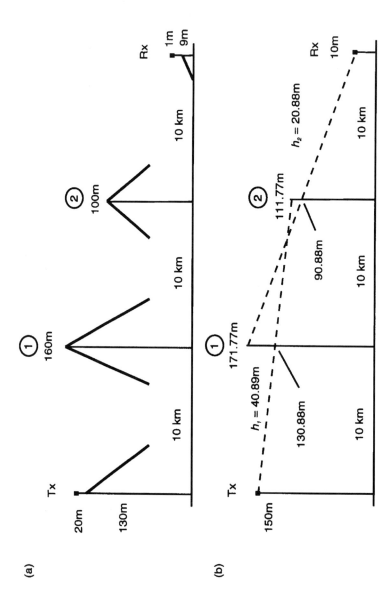

Figure 4.29 Terrain profile in Example 4.6 and study of the obstruction parameters h_1 and h_2.

$$L_{fs} \text{ (dB)} = 32.4 + 20 \log f + 20 \log d$$
$$= 32.4 + 20 \log(150) + 20 \log(30) = 105.6 \text{ dB}$$

and plane Earth loss is

$$L_{pe} \text{ (dB)} = 120 - 20 \log(h_t h_r) + 40 \log d$$
$$= 120 - 20 \log(150 \cdot 10) + 40 \log 30 = 115.6$$

Hence, the total loss is

$$L_b = \max(L_{fs}, L_{pe}) + L_{d1} + L_{d2} = 115.6 + 11 + 8.6 = 135.2 \text{ dB}$$

The excess loss is

$$L_{excess} = L_b - L_{fs} = 135.2 - 105.6 = 29.6 \text{ dB}$$

Assuming a base station eirp of 10W, the field strength at the mobile location in free-space conditions is

$$e_{fs} = \frac{\sqrt{30 \text{ eirp}}}{d} = \frac{\sqrt{30 \cdot 10}}{30000} \cdot 10^6 = 577.4 \ \mu\text{V/m} \equiv 55.2 \text{ dB}\mu\text{V/m}$$

The field strength for the conditions imposed by the terrain profile in Figure 4.29(a) is

$$E = E_{fs} - L_{excess} = 55.2 - 29.6 = 25.6 \text{ dB}\mu\text{V/m}$$

If a locations variability $\sigma_L = 6$ dB is assumed, the field strength exceeded for 90% of locations in the vicinity of the point where the calculations were made, is

$$E(90\%L) = E(50\%L) - k\sigma_L = 25.6 - 1.28 \cdot 6 = 17.9 \text{ dB}\mu\text{V/m}$$
$$\text{where } E(50\%L) = \bar{E}$$

4.5 Physical-Geometrical Models

This section presents an alternative approach (physical-geometric models) to the classical one based on empirical considerations, represented by the Okumura model. These models require a far more detailed representation of the "contour conditions" influencing propagation. These conditions are taken from building layout information contained in UDBs. It is interesting to point out the significant difference as regards the amount of information to be dealt with in the two approaches.

Empirical models use the following:

- Terrain databases with resolutions of 250 × 250 to 25 × 25m^2, for example;
- Land-usage databases with up to 15 morphographic categories.

Physical/deterministic models use the following:

- A detailed description of the geometry of buildings, including heights. This means large amounts of information contained in UDBs: groups of linear segments defining the layout of each building.

As examples of this new approach, which is more of a *deterministic* nature, the Ikegami model is described in greater detail in Section 4.5.1. Also, Section 4.5.2 presents the fundamentals and formulation of the COST 231 Walfish-Ikegami model. Finally, Section 4.5.3 discusses some concepts regarding models using ray-tracing and GTD/UTD techniques [7].

4.5.1 Ikegami Model

It has been concluded, from measurements of direction of arrival of echoes, that part of the rays comprising the multipath structure may be predicted by applying geometric considerations [6]. These contributions may be termed "*theoretical rays*" and correspond to specular reflections (coherent components). Also, the multipath will comprise many other, probably more attenuated components, corresponding to diffuse reflections which may be termed "*non-theoretical rays*" (Figure 4.30). The theoretical rays will be predominant. The received signal will be the coherent sum of all contributions

$$e_{\text{total}} = \left| \sum_{i=1}^{N} e_i e^{-j\frac{2\pi}{\lambda}\cos(\phi_i)x_i} e^{-j\theta_i} \right| \quad (4.61)$$

The received signal power at a point x is $p(x) = 1/2 e_{\text{total}}^2(x)$. As indicated in Chapter 3, the *local mean* of the received signal is computed for short stretches of the traveled route $2L \approx 20$–40λ long.

Appendix 4A: Review of the Plane Earth Model

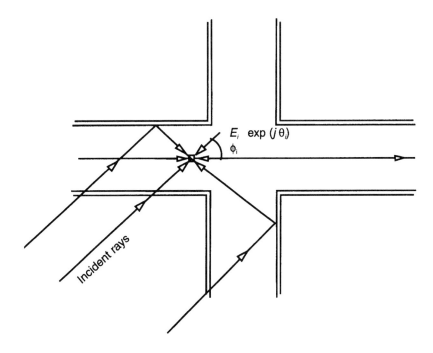

Figure 4.30 Ikegami model geometry.

$$\overline{p}(x) = \frac{1}{2L}\int_{-L}^{L} p(x)\,dx = \frac{1}{2}\frac{1}{2L}\int_{-L}^{L} e_{\text{total}}^2(x)\,dx \quad (4.62)$$

By averaging over a sufficiently long distance, $2L$, $\overline{p}(x)$ will be equal to the power sum of all received multipath components

$$\overline{p}(x) = \frac{1}{2}\sum_{i=1}^{N}\overline{e}_i^2(x) = \sum_{i=1}^{N}\overline{p}_i(x) \quad (4.63)$$

where $\overline{p}_i(x)$ is the average power of the i-th component evaluated over a route length $2L$. Consequently, the average field strength value may be obtained by taking the square root of the previous expression

$$\overline{e}(x) = \sqrt{\sum_{i=1}^{N}\overline{e}_i^2} \quad (4.64)$$

In urban areas, field strength variations are divided into *long-term* variations due to the shadowing effect of buildings and *short-term* variations due to interference between multipath components. The Ikegami model tries to predict the long-term variations, and, therefore, it is assumed that the fast variations have been filtered out from the received signal.

Assuming that the power of each multipath echo is constant over the averaging route of length $2L$. Since the total received power is the sum of the powers of individual echoes, a limited number of which present much higher levels than the majority, if it assumed that the average received power is only due to these stronger components, a certain error will be committed, albeit a small one.

If the contributions considered are Σp_a and the neglected contributions are Σp_d, the error will be

$$\Delta = \log \frac{\Sigma p_a + \Sigma p_d}{\Sigma p_a} = 10 \log(1 + \delta) \text{ dB with } \delta = \frac{\Sigma p_d}{\Sigma p_a} \quad (4.65)$$

$\delta = \frac{\Sigma p_d}{\Sigma p_a}$ is less than 3 dB the prediction may be considered acceptable.

Model Development

The Ikegami model assumes that the most important contribution to the total received power is due to two rays (Figure 4.31):

1. The *direct ray* that may be subject to diffraction on building tops in the neighborhood of the MS.
2. A *reflected ray* on building faces on the side of the road opposite to the transmitter. This may also be subject to diffraction effects.

In order to develop the model, an urban area with a regular structure is assumed with uniform building heights. The angle between the direction of travel and the incident ray is ϕ (street angle), which will be in the $0 \leq \phi \leq 90$-degree range. In order to calculate the losses due to diffraction for both rays, e_1 and e_2, approximate expressions are used based on the knife-edge model. For normalized obstruction values $\nu > 1$, diffraction losses may be expressed in the following way:

$$\frac{e}{e_o} \approx \frac{0.225}{\nu} \quad \text{or equivalently} \quad L_d \approx 13 + 20 \log(\nu) \text{ dB} \quad (4.66)$$

with an error of less than 1 dB.

Appendix 4A: Review of the Plane Earth Model

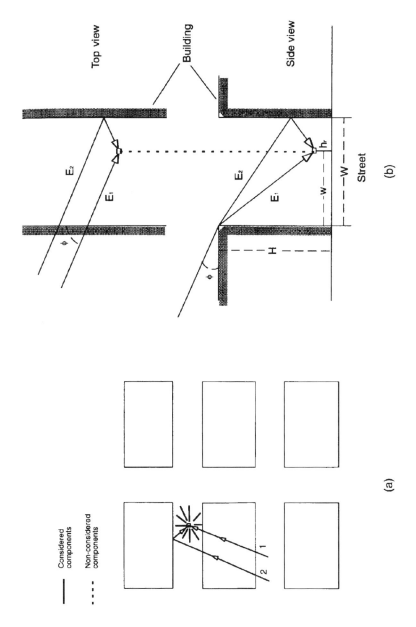

Figure 4.31 Components considered in the Ikegami model and model geometry.

The parameter ν has the following expression in terms of the radius of the first Fresnel zone, R_1, and of the obstruction value, h,

$$\nu = \sqrt{2}\frac{h}{R_1} \text{ with } R_1 = \sqrt{\frac{\lambda d_1 d_2}{d}} \qquad (4.67)$$

For the geometry of the model (Figure 4.31), the normalized obstructions 1 and 2 have the following approximate expressions:

$$\nu_1 \approx \sqrt{2}(H-h_r)\sqrt{\frac{\sin(\phi)}{\lambda w}} \quad \nu_2 \approx \sqrt{2}(H-h_r)\sqrt{\frac{\sin(\phi)}{\lambda(2W-w)}} \qquad (4.68)$$

Since $d_1 \gg d_2$ the value of the Fresnel radius is $R_1 = \sqrt{\lambda d_2}$. Assuming furthermore that the elevation angle θ is 0 and that d_1 (distance from the transmitter to the obstacle) $\gg w, 2W-w, H$, the expressions for fields e_1 and e_2 become

$$\frac{e_1}{e_o} \approx \frac{0.225}{\sqrt{2}} \frac{\sqrt{\lambda w}}{(H-h_r)\sqrt{\sin(\phi)}} \qquad (4.69)$$

$$\frac{e_2}{e_o} \approx \frac{0.225}{\sqrt{2}} \frac{\sqrt{\lambda(2W-w)}}{l_r(H-h_r)\sqrt{\sin(\phi)}} \qquad (4.70)$$

where

e_o is the received field under free-space conditions;
λ is the wavelength;
l_r are the losses due to reflection for ray 2.

The value of l_r depends on the *reflection coefficient* of the building surfaces. Its value in dB is $L_r = 20\log(l_r)$. The field $\bar{e}$ (local mean) is the power sum of both components

$$\bar{e} = \sqrt{e_1^2 + e_2^2} \qquad (4.71)$$

$$\bar{e} \approx e_o \frac{0.225}{\sqrt{2}} \frac{\sqrt{\lambda\left(w + \frac{2W-w}{l_r^2}\right)}}{(H-h_r)\sqrt{\sin(\phi)}} \qquad (4.72)$$

Now, the variations of the received field $\bar{e}/e_o$ are evaluated for different receiver positions across the street. Figure 4.32 shows these variations. It may be observed that, for four different reflection coefficients (reflection losses), the total field level is practically constant across the street. Thus, to simplify the model, the received field will be calculated for a representative point on the center of the street. So, the final expression of the model (for $w = W/2$) is

$$\bar{e} \approx e_o \frac{0.225}{\sqrt{2}} \sqrt{1 + \frac{3}{l_r^2}} \frac{\sqrt{\lambda w}}{(H - h_r)\sqrt{\sin(\phi)}} \quad (4.73)$$

Putting this final expression in decibels and using practical units

$$\bar{E} = E_o + 10 \log\left(1 + \frac{3}{l_r^2}\right) + 10 \log(W) - 20 \log(H - h_r) \quad (4.74)$$
$$- 10 \log(\sin(\phi)) - 10 \log(f) + 5.8$$

where W, H, h_r are in meters and f in MHz. Thus, an expression has been developed to describe propagation effects in urban areas in terms of a limited number of independent parameters W, H, h_r, f:

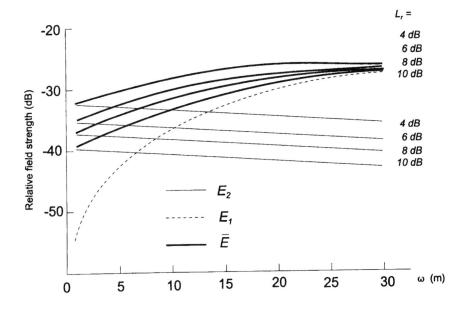

Figure 4.32 Fields E_1, E_2, and the total E across a street.

As regards L_r, a value of 6 dB is recommended (i.e., reflections on building faces are assumed to produce a loss of half the amplitude. That is, the reflection coefficient magnitude is assumed to be equal to 0.5 to take into consideration the roughness effect of building faces.

4.5.2 COST 231 Model Walfish-Ikegami

The COST 231 model [14], also known as the Walfish-Ikegami model, is an evolution of the Ikegami model. This model is applicable to urban areas where no direct visibility of the transmitter is available due to obstructing buildings. This model, in addition to considering the influence of the street where the mobile is located, includes the contribution to the total path loss due to the fact that the signal illuminating the street where the mobile propagates above numerous buildings (multiedge diffraction). This contribution was evaluated by Walfish and Bertoni [15], thereby yielding the name of the model. The geometry of the model is shown in Figure 4.33. The following street and path parameters intervene in the model:

- h_B: Base station antenna height above the ground (m);
- h_m: MS antenna height above the ground (m);
- h_R: Mean building height (m) ($h_R > h_m$);
- w: Width of the street where the MS is located (m);
- b: Distance between building centers (m);
- d: BS-to-MS distance (km);

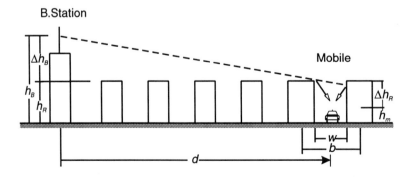

Figure 4.33 Geometry of the Walfish-Ikegami-COST 231 model.

- φ: Angle of the radio path with respect to the street axis (degrees);
- $\Delta h_B = h_B - h_R$: BS height over the mean building height (m);
- $\Delta h_R = h_R - h_m$: Mean height of the buildings over the MS antenna height.

The general expression for the Walfish-Ikegami-COST 231 model is

$$L_b = L_{fs} + L_{rts} + L_{msd} \qquad (4.75)$$

where L_{fs} are the free space losses

$$L_{fs} \text{ (dB)} = 32.4 + 20 \log f \text{(MHz)} + 20 \log d \text{ (km)} \qquad (4.76)$$

L_{rts} is the loss due to the "rooftop-to-street" diffraction. It has the following expression:

$$L_{fs} \text{ (dB)} = -16.9 - 10 \log(w) + 10 \log(f) + 20 \log(\Delta h_R) + L_{ori} \qquad (4.77)$$

if $L_{rts} \leq 0$ dB, a value of $L_{rts} = 0$ dB will be taken. L_{ori} considers the orientation of the street relative to the transmitter

$$L_{ori} = \begin{cases} -10 + 0.3574\varphi & 0 < \varphi < 35° \\ 2.5 + 0.075(\varphi - 35°) & 35° \leq \varphi < 55° \\ 4 - 0.114(\varphi - 55°) & 55° \leq \varphi \leq 90° \end{cases} \qquad (4.78)$$

where φ is the angle between the direct radio path and the axis of the street.

L_{msd} is an estimate of the multi-obstacle diffraction effects the ray experiences between the transmitting antenna and the building closest to the receiver, due to intermediate buildings. This follows the expression

$$L_{msd} = L_{bsh} + k_a + k_d \log(d) + k_f \log(f) - 9 \log(b) \qquad (4.79)$$

The parameters involved in this expression are calculated as follows:

$$L_{bsh} = -18 \log(1 + \Delta h_B) \qquad (4.80)$$

If $h_B < 0$, $L_{bsh} = 0$

$$k_a = \begin{cases} 54 & \Delta h_B \geq 0 \\ 54 - 0.8\Delta h_B & \Delta h_B < 0 \text{ and } d \geq 0.5 \\ 54 - 0.8\Delta h_B \, (d/0.5) & \Delta h_B < 0 \text{ and } d < 0.5 \end{cases} \quad (4.81)$$

$$k_d = \begin{cases} 18 & \Delta h_B \geq 0 \\ 18 - 15\Delta h_B/h_R & \Delta h_B < 0 \end{cases} \quad (4.82)$$

$$k_f = \begin{cases} -4 + 0.7\left(\dfrac{f}{925} - 1\right) & \text{for medium size cities and suburban areas with moderate vegetation density} \\ -4 + 1.5\left(\dfrac{f}{925} - 1\right) & \text{for large metropolitan centers} \end{cases}$$

$$(4.83)$$

The applicability of the COST 231 model corresponds to the following parameter ranges:

- $800 \leq f\,(\text{MHz}) \leq 2000$;
- $4 \leq h_B\,(\text{m}) \leq 50$;
- $1 \leq h_m\,(\text{m}) \leq 3$;
- $0.02 \leq d\,(\text{km}) \leq 5$.

If data on the urban environment are unknown, the following default values are recommended:

- b: 20–50m;
- w: $b/2$;
- φ: 90 degrees;
- h_R: 3m × (N. of floors) + roof-height (m) (with attic: 3m (pitched)), 0 m (flat).

Example 4.7. Assume a cellular mobile communications scenario in an urban environment with high built-up density. The following data/parameters are used in this example:

$f = 900$ MHz;
$d = 1.5$ km;

h_B = 30m;
φ = 37 degrees;
h_R = 20m;
w = 20m;
h_m = 1.5m;
b = 40m.

The calculations are as follows:

- Δh_B = 10m;
- Δh_R = 18.5m;
- L_{ori} = 2.5 + 0.075(37 − 35) = 2.65;
- L_{rts} = −16.9 − 10 log 20 + 10 log 900 + 20 log 18.5 + 2.65 = 27.63 dB;
- k_f = −4 + 1.5(900/925 − 1) = −4.04;
- k_d = 18;
- k_a = 54;
- L_{bsh} = −18 log(1 + 10) = −18.75;
- L_{msd} = −18.75 + 54 + 18 log 1.5 − 4.04 log 900 − 9 log 40 = 12.08 dB;
- L_{bf} = 32.45 + 20 log 900 + 20 log 1.5 = 95.06 dB.

Accordingly, the total losses are L_b = 95.06 + 27.63 + 12.08 = 134.77 dB.

4.5.3 Ray-Tracing Based Models

As UDBs have become widely available, specially for urban development/planning applications, it is also possible to use them for applications in radio system calculations. It is a well-known fact that the nature of the propagation channel in mobile communications is conditioned by shadowing and multipath phenomena. One of the current trends in propagation modeling is to use *ray-tracing techniques* on UDBs to identify the paths of the relevant contributions reaching the receiver.

Ray-tracing techniques must be accompanied by electromagnetic models, which make it possible to calculate the magnitude, phase, and polarization of the different rays found in the tracing stage. A very frequently used high-frequency electromagnetic technique is the Uniform Theory of Diffraction (UTD) [7]. Figure 4.34 shows a ray-tracing example on an UDB.

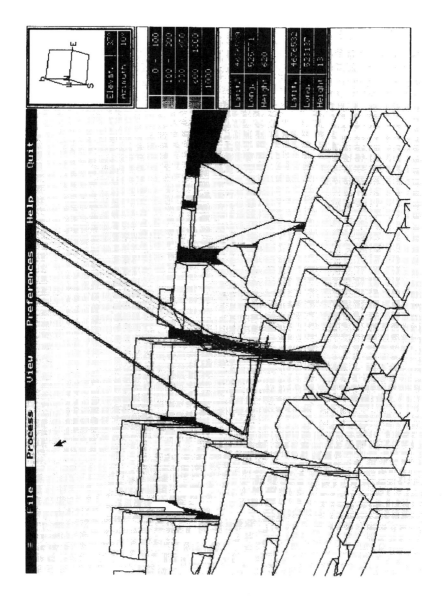

Figure 4.34 Example of ray-tracing using an UDB.

Appendix 4A: Review of the Plane Earth Model

In the case of ray-tracing, it is necessary to follow a series of rules which characterize the behavior of the different ray types considered, namely the following:

- Direct (Dir);
- Reflected (R);
- Diffracted (D).

Depending on the working frequency bands, rays penetrating building walls may also be considered. These rays would be traced following the same rules as for direct rays but without taking the obstructions into account, except to introduce absorption losses.

Reflected rays will follow the *Snell law* of reflection law (i.e., to trace them, all planes in the UDB (the ground and building faces) would be considered and must be checked to see if, at any point on those surfaces, the angle of incidence and the angle of reflection are equal and that neither of the subpaths are obstructed).

In order to trace diffracted rays, the *Keller law* is used (i.e., it will be checked that the angle of incidence on a (vertical or horizontal) edge of a building is equal to the angle of diffraction). Figure 4.35 shows this angle, normally termed β. Also in the figure the so-called Keller cone of possible diffracted rays directions is illustrated.

Figure 4.36 shows a study of a simple urban area with one reflection on the ground, a reflection on a building face, and a diffraction on an edge of the same building. Figure 4.36 also schematically illustrates the fact that, for each subpath, the possibility of obstruction by any building must be checked to verify the existence of each traced ray.

For all possible rays, including diffracted and reflected rays, general expressions such as those given below are used to characterize their electromagnetic behavior

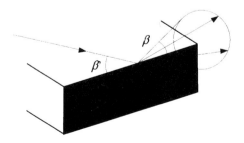

Figure 4.35 Keller law for diffracted rays.

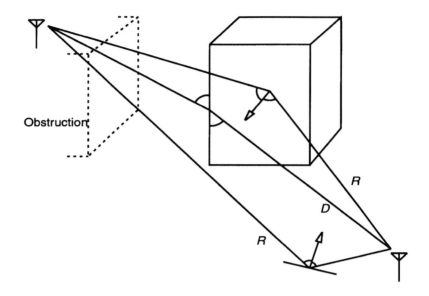

Figure 4.36 Diagram showing different contributions reaching the receiver in a simple urban scenario.

$$e(Q_x) = e(s_1)e^{-jks_1}A(s_1) \tag{4.84}$$

$$e(p) = e(Q_x)\text{Coeff}\, e^{-jks}A(s) \tag{4.85}$$

where $e(Q_x)$ is the field of a ray reaching point Q_x which may represent the receiver or be on a plane or an edge and give rise to reflected or diffracted rays. $e(p)$ is the field at an observation point (receiver) after a reflection or diffraction. s_1 is the distance from the source of the ray to Q_x, and s is the distance from Q_x to the observation point P. The second element in the second equation is a *coefficient* that may be either a reflection, R, or diffraction, D, coefficient depending on the ray type. The third term is the *phase* due to the traveled path by the ray from Q_x to P. Finally, A is the *spread factor*, which takes into account the losses with distance and the type of wave (spherical for reflections, cylindrical for diffractions).

The reflected on diffracted rays are characterized by the *reflection* and *diffraction coefficient*. In every case, there will be a coefficient for the field component perpendicular to the plane of propagation of the ray $\perp$ and another one for the field component contained in the propagation plane $\parallel$ ($R_\perp$, $R_\parallel$, $D_\perp$, $D_\parallel$). Figure 4.37(a) shows the geometry of reflection and Figure 4.37(b) shows the geometry of diffraction.

Appendix 4A: Review of the Plane Earth Model

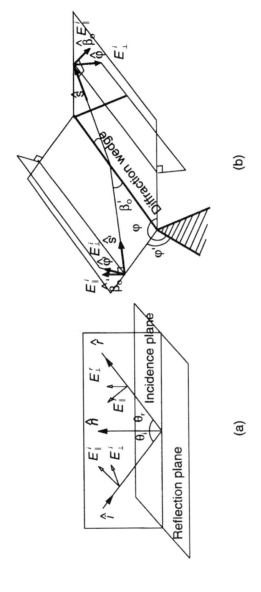

Figure 4.37 (a) Components $\perp$ and $\parallel$ of the reflected field. (b) Components of the edge diffracted field.

In order to illustrate the concept of diffracted ray, a simple layout will be studied with one edge and one ray impinging it perpendicularly, with the receiver located on a circumference of radius s on the same plane (Figure 4.38(a)). *Three regions* may be defined in this figure, the *shadow region* where only the diffracted ray exists, another region where both the direct and the diffracted rays reach the receiver and finally, a region where also reflected rays may reach the receiver. (Figure 4.38(b)).

Using the ray-tracing + UTD approach, multipath propagation in urban areas may be studied. A selectable number of interactions may be considered depending on an accuracy-computation time tradeoff. Thus, *multiple interactions* of the following types: *DD, DDD, RR, RRR, RRRR, DR, RD, RDR, DRD, DDR*, where R and D denotes reflection on diffraction, respectively. (Figure 4.39) may be considered.

One fundamental feature of this type of propagation modeling approach is that not only received field time series along a given route can be produced (including fast and slow variations), as shown in Figure 4.40. It is also possible to obtain wideband parameter predictions, particularly PDPs since, when tracing rays, these may be classified by their times of arrival. Figure 4.41 presents a delay profile calculated in this manner, where the different echoes are represented by delta functions.

Appendix 4A: Review of the Plane Earth Model 185

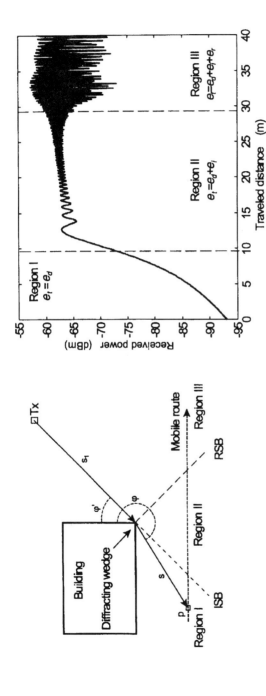

ISB: Incident shadow boundary
RSB: Reflection shadow boundary

Figure 4.38 Simple diffraction geometry and received fields in regions I, II, and III.

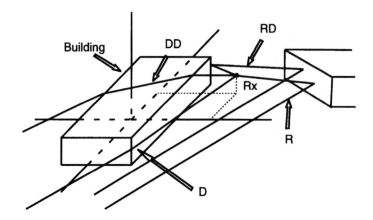

Figure 4.39 Multiple interactions in an urban environment.

Appendix 4A: Review of the Plane Earth Model 187

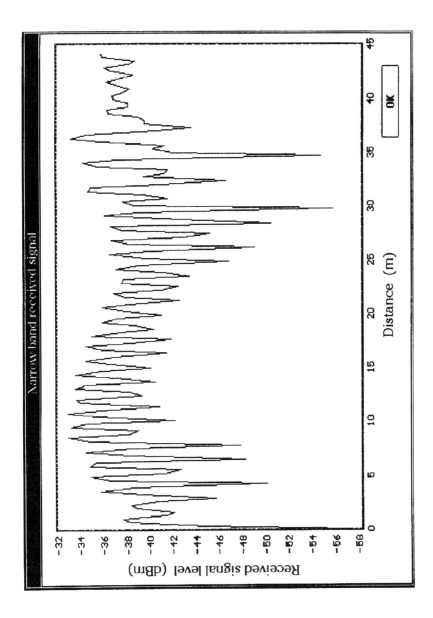

Figure 4.40 Calculation of the signal variations along a mobile route.

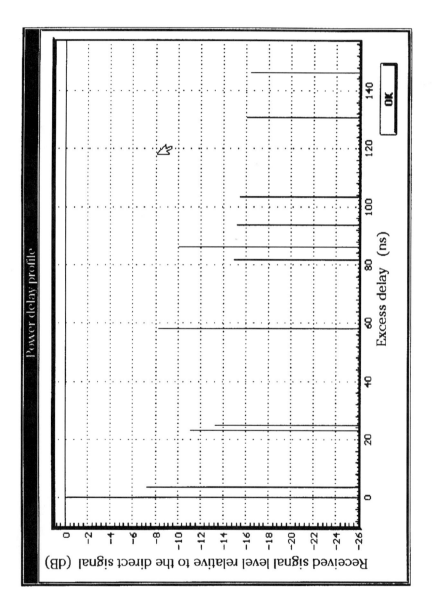

Figure 4.41 Calculated power delay profile.

References

[1] Evolution of land mobile radio (including personal) communications, EURO-COST 231 Project, Final Report, Draft, 1996.

[2] Okumura, Y., et al., "Field Strength Variability in VHF and UHF Land Mobile Service," *Rev. Elec. Comm. Lab.,* Vol. 16, No. 9–10, Sep–Oct 1968, pp. 825–873.

[3] Hata, M., "Empirical Formula for Propagation Loss in Land Mobile Radio Services," *IEEE Trans. Veh. Tech.,* VT-29, No. 3, August 1980, pp. 317–325.

[4] Edwards, R., and J. Durkin, Computer Prediction of Service Areas for VHF Mobile Radio Network, *IEE Proc.,* Vol. 116, No. 9, September 1969, pp. 1483–1500.

[5] Ibrahim, M. F. A., J. D. Parsons, and D. E. Dadson, "Signal Strength Prediction in Urban Areas Using Topographical and Environmental Databases," IEEE Conference, ICC 83.

[6] Ikegami, F., S. Yoshida, T. Takeuchi, and M. Umehira, "Propagation Factors Controlling Mean Field Strength on Urban Streets," *IEEE Trans. on Antennas and Propagation,* Vol. AP-32, No. 8, August 1984, pp. 822–829.

[7] Van Dooren, G. A. J., A Deterministic Approach to the Modeling of Electromagnetic Wave Propagation in Urban Environments, University of Eindhoven, Ph. D. Thesis, 1994.

[8] Loew, K., "UHF Field-Strength Measurements for Determination of the Influence of Buildings and Vegetation in Land Mobile Radio Service," IEEE Veh.Tech. Conference, 1983.

[9] Lee, W. C. Y., *Mobile Communications Design Fundamentals,* New York: John Wiley & Sons, 1993.

[10] Kozono, S., and K. Watanabe, "Influence of Environmental Buildings on UHF Land Mobile Radio Propagation," *IEEE Trans. on Communications,* Vol. COM-25, No. 10, Nov. 1977, pp. 1133–1143.

[11] Propagation data and prediction methods for terrestrial land mobile service using the frequency range 30 MHz to 3 GHz, ITU-Report 567-4, Geneva, 1990.

[12] Reudink, D. O., "Properties of Mobile Radio Propagation Above 400 MHz," *IEEE Trans. Veh. Tech,* VT-23, Nov. 1974, pp. 595–611.

[13] Epstein, J., and D. W. Peterson, "An Experimental Study of Wave Propagation at 850 Mc/s," *Proc. IRE,* October 1953.

[14] COST 231, Urban transmission loss models for mobile radio in the 900 and 1,800 MHz bands (revision 2), COST 231 TD(90)119 Rev. 2, The Hague, The Netherlands, Sept. 1991.

[15] Walfish, J., and H. L. Bertoni, "A Theoretical Model of UHF Propagation in Urban Environments," *IEEE Trans. on Antennas and Propagation,* Vol. AP-36, No. 12, December 1988, pp. 1788–1796.

5

Mobile Network System Engineering

5.1 Introduction

Previous chapters have dealt with the mobile propagation channel (shadowing plus multipath) in some detail. Here, propagation calculations are applied to the project of a radio network. Two main studies are carried out for the planning of any mobile radio network:

1. Traffic computations;
2. Coverage computations.

These calculations are directly related to the quality objectives specified for the radio communications network to be planned:

1. The *traffic quality objective* is defined in terms of the probability that a call cannot be served because all traffic channels are busy. This is the case for *lost call systems* (e.g., cellular networks). In the case of *waiting or queuing systems* (for example, trunking networks), quality is defined in terms of the probability of exceeding a given waiting time before the connection is available. The traffic quality objective is usually termed *grade of service* (GoS).
2. The *coverage quality objective* is defined as the percentage of time and locations within the service area in which an adequate received signal level is received.
3. A third *objective* is called *transmission quality*. This objective is specified in terms of the fidelity of the received message. For voice communica-

tions, quality levels are given in terms of the so-called mean opinion score (MOS)—see Table 5.1. For digital systems the quality objective is defined by the BER.

Traffic calculations will be studied in Chapters 8 and 9. This chapter deals with coverage quality issues. Additionally, there are a number of complementary studies that must also be performed when carrying out the planning of any radio system. Specifically, it is of fundamental importance to perform an adequate selection of the transmit and receive frequencies to avoid intermodulation and cochannel and adjacent channel interference (frequency planning).

5.2 Coverage Computations

The first element to be taken into account to perform coverage computations is the evaluation of the *minimum necessary received field strength* in the neighborhood of the antennas at both ends of the mobile radio link (mobile-to-base and base-to-mobile directions). A field strength value is sought that insures that, at the fringe (coverage contour) of the service area, a given received signal level is exceeded for a specified percentage of time and locations. The necessary field strength value represents the median of the distribution of received signal levels at the fringe of the coverage area. This value is usually called *median necessary field strength*, $\tilde{E}_n$.

In this section, noise-limited systems are analyzed. The procedure for the computation of $\tilde{E}_n$ for voice communication systems is now described. A similar procedure can be followed for digital systems. In Section 5.4, interference-limited systems are studied.

Table 5.1
Mean Opinion Score (MOS)

Grade	Interfering Effect	Intelligibility
5	Almost nil	Speech understandable, but with increasing effort as the grade decreases
4	Noticeable	
3	Annoying	
2	Very annoying	
1	So bad that the presence of speech is barely noticeable	

The median necessary field strength is computed by increasing the minimum field strength value, E_m, corresponding to the receiver sensitivity, S (expressed in voltage units) or P_m (expressed in power units) to account for:

- Noise and multipath effects ($\Delta_r E$);
- Time and locations variability ($\Delta_e E$).

Thus, the value of $\bar{E}_n$ is given by

$$\bar{E}_n = E_m + \Delta_r E + \Delta_e E \tag{5.1}$$

This calculation can also be performed using power units (*dBm*) instead of field strengths. The corresponding parameter, $\bar{P}_n$, is:

$$\bar{P}_n = P_m + \Delta_r E + \Delta_e E \tag{5.2}$$

E_m is the field strength value in the vicinity of the receive antenna that produces, at the receiver input, the sensitivity voltage level, S, (Figure 5.1). For the computation of E_m the following data are required:

- Receiver sensitivity: S (dBμV) or P_m (*dBm*);
- Receiver antenna system characteristics.

and two approaches may be followed:

1. Using the *antenna effective length* (l_{ef}) that relates the open circuit voltage v_{oc} at the antenna connector with the value of the field strength impinging on the antenna

$$e \cdot l_{ef} = v_{oc} \tag{5.3}$$

2. Using the *effective area* (A_{ef}) of the receiver antenna.

For these computations the antenna gain and the characteristics of the different elements between the receiver and the antenna must be taken into account. These will include the feeder, which is characterized in terms of a *linear attenuation* parameter α (dB/m).

Here the second approach (i.e., the antenna effective area parameter) is used. In this case, the following relationships hold:

194 Introduction to Mobile Communications Engineering

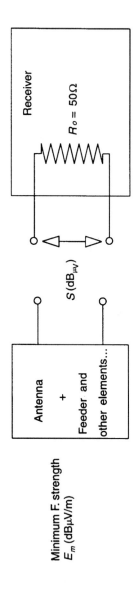

Figure 5.1 Schematic representation of a receiver system.

$$p_r = \Phi \cdot A_{ef} \text{ with } A_{ef} = \frac{\lambda^2}{4\pi} g_i \quad p_r = \frac{v_{Rx}^2}{R_o} \quad \Phi = \frac{e^2}{120\pi} \quad (5.4)$$

where

Φ is the power flux density (W/m^2);
e is the field strength in the vicinity of the antenna (V/m);
R_o is the receiver and antenna impedance (Ω);
v_{Rx} is the voltage at the receiver input (V);
g_i is the isotropic gain of the receive antenna (in linear units).

Now, equating the previous expressions

$$\frac{e^2}{120\pi} \frac{\lambda^2}{4\pi} g_i = \frac{v_{Rx}^2}{R_o} \quad (5.5)$$

and including the losses between the receiver and the antenna L_c (dB) = 10 log (l_c), then

$$\frac{e^2}{120\pi} \frac{\lambda^2}{4\pi} g_i \frac{1}{l_c} = \frac{v_{Rx}^2}{R_o} \quad (5.6)$$

the expression for e is thus

$$e = v_{Rx} \frac{\pi}{\lambda} \sqrt{\frac{480 l_c}{g_i R_o}} \quad (5.7)$$

Note that the values of e and v_{Rx} in the previous equations are rms values.

The receiver's *sensitivity*, s (μV) is defined as the minimum voltage level at the receiver input, v_{Rx}, required to obtain a specified SINAD value (usually a value between 12 and 20 dB). The SINAD parameter is defined as the power ratio: signal + noise + distortion over noise + distortion expressed in decibels.

$$\text{SINAD (dB)} = 10 \log\left(\frac{s + n + d}{n + d}\right) \quad (5.8)$$

Equation (5.7) can be rearranged and expressed in decibels using practical units: S (dBμV), E_m (dBμV/m), f (MHz), L_c (dB) yielding

$$E_m \text{ (dB}\mu\text{V/m)} = S(\text{dB}\mu\text{V}) + 20 \log f(\text{MHz}) - 10 \log R_o (\Omega) \quad (5.9)$$
$$- G_i \text{ (dB)} + L_c \text{ (dB)} - 12.8$$

If hand-held terminals are used, as is usual nowadays, the antenna gain must be corrected as a function of the terminal usage, thus taking into account the influence of the human body and the terminal antenna position. Table 5.2 provides values for a *usage correction parameter, L*. Table 5.3 shows gain values for reference antennas and for the most commonly used antennas in mobile communications.

Table 5.2
Antenna Gain Usage Correction L (dB) for Handheld Terminals

Band	Usage and Antenna Type	Losses (dB)
VHF 150–174 MHz	Manual (vertical)	
	Telescopic $\lambda/4$	1
	Helix	4–5
	Manual (tilted)	
	Telescopic $\lambda/4$	5–6
	Helix	8–9
	Belt	
	Telescopic $\lambda/4$	20–30
	Helix	10–20
UHF 450–470 MHz	Manual (vertical)	
	Telescopic $\lambda/4$	2–4
	Helix	8–10
	Manual (tilted)	
	Telescopic $\lambda/4$	12–15
	Helix	17–20
	Belt	
	Telescopic $\lambda/4$	25–35
	Helix	8–10

Table 5.3
Gains of Reference and Typical Mobile Communications Antennas

Antenna	g_i	G_i (dBi)
Isotropic	1	0
Hertz dipole	1.5	1.75
$\lambda/2$ dipole	1.65	2.15
Short monopole on a ground plane	3	4.8
$\lambda/4$ monopole on a ground plane	3.3	5.2

Example 5.1. For $s = 0.35~\mu\text{V}$ and assuming an ideal dipole antenna with $G_i = 2.15$ dB$_i$, the sensitivity in dBμV is

$$S = 20 \log(0.35) = -9.11 \text{ dB}\mu\text{V}$$

if the sensitivity is now expressed in power units (*dBm*), the threshold power is

$$p_{min} = 10 \log\left(\frac{(0.35 \cdot 10^{-6})^2}{50} 1000\right) = -116 \text{ dBm}$$

An antenna, feeder, and receiver impedance of 50Ω were assumed. Also, a value of $L_c = 0$ dB was assumed (negligible feeder losses). Using (5.9), the minimum field strength is

$$\begin{aligned} E_m \text{ (dB}\mu\text{V/m)} &= S \text{ (dB}\mu\text{V)} + 20 \log f (\text{MHz}) - 10 \log R_o (\Omega) \\ &\quad - G_i \text{ (dB)} + L_c \text{ (dB)} - 12.8 \qquad (5.10) \\ &= -41 + 20 \log f (\text{MHz}) \end{aligned}$$

Example 5.2. Now assume a hand-held terminal with a $G_i = 2.15$ dBi helix antenna operating in the 450-MHz band. Also assume that its sensitivity is $s = 0.25$ μV (−119 dBm) and that the usage losses are $L_c = 18$ dB. In this case the minimum field strength value is

$$\begin{aligned} E_m &= S \text{ (dB}\mu\text{V)} + 20 \log f (\text{MHz}) - 10 \log R_o (\Omega) \\ &\quad - G_i \text{ (dB)} + L_c \text{ (dB)} - 12.8 \\ &= 27 \text{ dB}\mu\text{V/m} \end{aligned}$$

Noise and Multipath Correction ($\Delta_r E$)

This *correction parameter* is used to account for the signal *degradation* due to the fast signal variations caused by multipath (Rayleigh fading) and man-made noise (vehicle ignition noise) in the area where the receiver is located. The minimum field strength, E_m, corresponding to a given receiver sensitivity level, S, must be incremented by several decibels in order to compensate for these degradations.

For the planning of a voice radiocommunications system, several quality levels have been defined to quantify the received message fidelity. ITU-R Report 358 specifies five different quality levels (*MOS*) (Table 5.1). Also this report provides empirical curves for the correction parameter $\Delta_r E$. In these curves, several different receiving conditions are considered: moving or stationary vehicles, mobile-to-base and base-to-mobile directions and traffic density. Here, the curves for MOS 4 are illustrated in Figures 5.2 and 5.3.

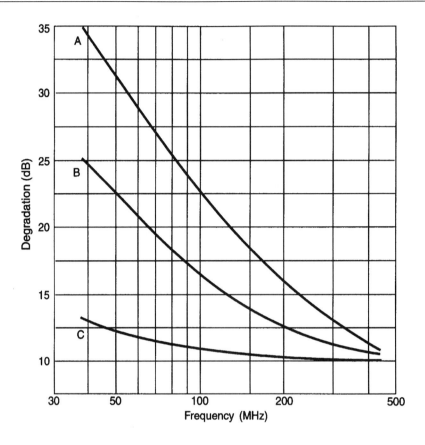

Figure 5.2 Degradation $\Delta_r E$ experienced by a mobile receiver for voice MOS 4 communications. (ITU.)

Figure 5.2 shows the degradation value $\Delta_r E$ for the reception in a mobile for MOS 4. The curves in this figure correspond to the following cases:

- A: Stationary vehicle located in a high noise area;
- B: Moving vehicle within a high noise area;
- C: Moving vehicle within a low noise area.

In Figure 5.3 the degradation value $\Delta_r E$ is provided for base station reception for *MOS 4* voice communications. The curves in this figure correspond to the following cases:

- A: Moving vehicle, traffic density 2 vehicles/s;

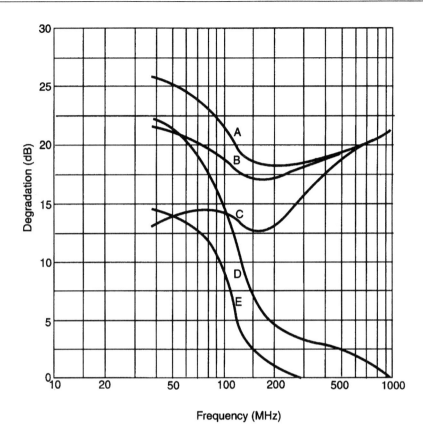

Figure 5.3 Degradation $\Delta_r E$ experienced by a base station receiver for voice MOS 4 communications. (ITU.)

- B: Moving vehicle, traffic density 1 vehicle/s;
- C: Moving vehicle, low noise area;
- D: Stationary vehicle, traffic density 2 vehicles/s;
- E: Stationary vehicle, traffic density 1 vehicle/s.

Example 5.3. From the figures for MOS 4 the following values can be read:

1. For a mobile receiver at 150 MHz in a low noise area (Curve C, Figure 5.2): $\Delta_r E = 10.5$ dB.

2. For base station reception at 450 MHz from a moving vehicle in a high traffic density area (Curve A, Figure 5.3): $\Delta_r E = 19.2$ dB.

Statistical Correction To Account for Time and Locations Variability ($\Delta_e E$)

Received signal time and locations variations, when expressed in logarithmic units, follow a normal distribution (slow variations). Coverage quality is specified for a given percentage of time ($T\%$) and locations ($L\%$). Generally, coverage is evaluated at the fringe (contour or perimeter) of the service area. Overall coverage (area coverage) quality within the coverage contour is then obtained by extrapolation of the fringe coverage quality to the whole area within the coverage contour.

As terrain databases became widely used in the 1980s for radio planning purposes, particularly for cellular networks, coverage calculations are now being performed for small individual surface elements of, for example, $250 \times 250\text{m}^2$ or smaller ($50 \times 50\text{m}^2$ is a typical value nowadays).

Here the classical approach is followed; that is, the procedure to perform fringe coverage studies is presented. The overall *area coverage probability* is obtained from the *fringe coverage probability* as indicated below.

If R is the radius of the coverage area (distance from the base station to the fringe of the coverage area), the parameter F_u (area coverage probability) is defined as the fraction of the total area within a circle of radius R for which the signal level exceeds a given threshold level x_0. If $P(x_0)$ is the probability that the received signal x exceeds x_0 for an infinitesimal (incremental) area dA, then [1]

$$F_u = \frac{1}{\pi R^2} \int_{\text{Area: } \pi R^2} P(x_0) dA \quad \text{with} \quad P(x_0) = \text{prob}(x > x_0) \quad (5.11)$$

If terrain databases were used, the area of interest would be subdivided into smaller study areas for which the coverage probability $P(x_0)$ could be computed. Later, all individual probabilities could be added up and averaged to compute the overall area coverage.

Assuming now that the propagation model used provides the median received power, $\bar{x}$, and that the propagation law with the distance from the transmitter to the receiver is r^{-n}, the median received power can be calculated using an expression of the form

$$P_r = A - B \log(r) \quad \text{with} \quad B = 10n \quad (5.12)$$

as, for example, in the Okumura-Hata model (Chapter 4). The median received power value, $\bar{x}$, will decrease with distance following the law

$$\bar{x} = \alpha - 10n \log\left(\frac{r}{R}\right) \tag{5.13}$$

where α is a constant depending on such factors as the transmitted power, and the antenna heights, and gains. The reader is again reminded that the slow variations of the received signal, expressed in decibels, follow a Gaussian distribution defined by its median value, $\bar{x}$, and its standard deviation (locations variability), σ_L, with a probability density function

$$p(x) = \frac{1}{\sigma_L\sqrt{2\pi}} \exp\left[-\frac{(x-\bar{x})^2}{2\sigma_L^2}\right] \tag{5.14}$$

The probability that x exceeds a given threshold x_o at a distance $r = R$ (fringe) from the base, is

$$P(x_o, R) = \text{prob}(x \geq x_o) = \int_{x_o}^{\infty} p(x)\,dx = \frac{1}{2} - \frac{1}{2}\text{erf}\left(\frac{x_o - \bar{x}}{\sigma_L\sqrt{2}}\right) \tag{5.15}$$

Example 5.4. If the values of $\bar{x}$ and σ_L at a distance R are -100 dBm and 10 dB respectively [1], and if the threshold is set to $x_o = -110$ dBm, the probability of exceeding the threshold level at a distance R will be (Figure 5.4)

$$P(x_o, R) = \frac{1}{2} + \frac{1}{2}\text{erf}\left(\frac{1}{\sqrt{2}}\right) = 0.84 \equiv 84\%$$

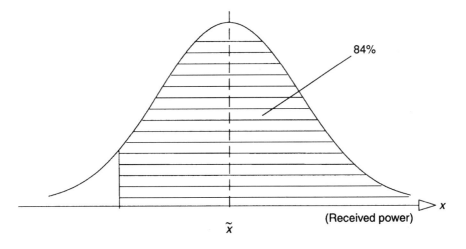

Figure 5.4 Area coverage probability in Example 5.4.

Now, developing further the expression of $P(x_o, r)$, the following expression is reached for any distance, r, from the base station

$$P(x_o, R) = \frac{1}{2} - \frac{1}{2}\text{erf}\left(\frac{x_o - \alpha + 10n \log\left(\frac{r}{R}\right)}{\sigma_L\sqrt{2}}\right) \qquad (5.16)$$

The area coverage probability, F_u, is defined by the integral

$$F_u = \frac{1}{\pi R^2} \int_{\text{Area: } \pi R^2} P(x_o, r)dA \text{ with } dA = r\, dr\, d\varphi \qquad (5.17)$$

inserting (5.16) in (5.17) and solving the integral gives [1]:

$$F_u = \frac{1}{2}\left\{1 + \text{erf}(a) + \exp\left(\frac{2ab + 1}{b^2}\right)\left[1 - \text{erf}\left(\frac{ab + 1}{b}\right)\right]\right\} \qquad (5.18)$$

where $a = (x_o - \alpha)/\sigma_L\sqrt{2}$ and $b = 10n \log(e/\sigma_L\sqrt{2})$.

For the case where $\bar{x} = x_o$ at the fringe $r = R$, i.e., if a 50% probability level is assigned to the coverage contour, the expression for F_u is greatly simplified

$$F_u = \frac{1}{2} + \frac{1}{2}\exp\left(\frac{1}{b^2}\right)\left[1 - \text{erf}\left(\frac{1}{b}\right)\right] \qquad (5.19)$$

In Figure 5.5, curves for F_u (area coverage probability) as a function of the propagation parameters σ_L and n are shown. In this figure, the fraction of the area within a circle of radius R for which the received signal exceeds the threshold for several fringe probability values $P(x_o, R)$ is given. For example, for $\sigma_L/n = 3$ and $P(x_o, R) = 0.5 \equiv 50\%$ a value of $F_u > 0.7 \equiv 70\%$ (*area coverage*) can be read.

To further clarify the concept of *"coverage contour"* associated with a given probability level, Figure 5.6 is used. Figure 5.6(a) illustrates three contours corresponding to three probability levels of *50%, 90%*, and *95%*. These probabilities mean that if it were possible to drive along the corresponding contours, availability percentages of 50%, 90%, and 95% would be observed. Figures 5.6(b)(c) schematically illustrate such test drives for the 50% and 90% probability levels.

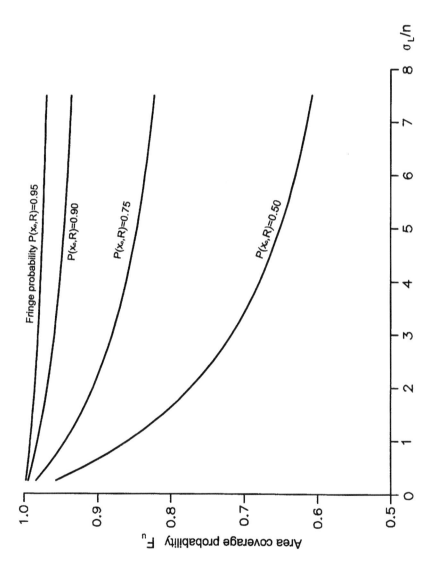

Figure 5.5 Relation between fringe coverage and area coverage probabilities for different propagation laws.

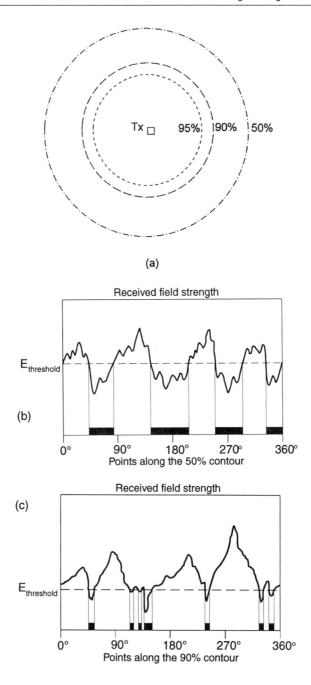

Figure 5.6 (a) Coverage contours for 50%, 90%, and 95% of locations. Received signal levels while driving along the (b) 50% and (c) 90% contours.

After presenting the concepts of fringe or contour coverage and area coverage, now the procedure for the evaluation of the correction parameter $\Delta_e E$ is explained. Two data are required:

- The *time variability*, σ_T;
- The *locations variability*, σ_L.

These data are the standard deviations of two Gaussian distributions representing the time and locations slow variations of the received signal. Table 5.4 provides values for these two parameters quoted from ITU-R Report 567. From Table 5.4, it can be observed that the locations variability is significantly greater than the time variability. Thus, in many instances, only the locations variability is considered in coverage (and interference) calculations.

Table 4.1 provides correction values to move within a Gaussian distribution from the median value (50% probability level) to other selected probability levels, such as 90% and 95%.

For the computation of the correction parameter $\Delta_e E$ the following expression is used to consider the combined effects of time and locations variability

$$\Delta_e E = \sqrt{(k(L\%)\sigma_L)^2 + (k(T\%)\sigma_T)^2} \qquad (5.20)$$

where $L(\%)$ and $T(\%)$ are the locations and time percentages, respectively.

Example 5.5. Assume a MOS 4 mobile communication link in the 450-MHz UHF band at the fringe of the coverage area with availability levels of $L = 90\%$ and $T = 90\%$. The receivers used have a sensitivity of $s = 0.35 \mu V$ ($S = 20 \log (0.35) \approx -9$ dBμV), and the reception conditions are the following:

- Base station reception (Curve A, Figure 5.3): moving vehicle, traffic density 2 vehicles/s;

Table 5.4
Typical Time and Locations Variability Values

Band	σ_L (dB)	σ_T (dB)
VHF	8	3 (over land and sea)
UHF	10	2 (over land)
		9 (over sea)

- Mobile reception (Curve B, Figure 5.2): moving vehicle in a high noise area.

The base station antenna gain is $G = 8$ dB$_i$ and the feeder losses are $L_c = 2.5$ dB. Mobile stations have antenna gains of $G = 2.15$ dB$_i$ and $L_c = 0$ dB (short feeders). The hand-held terminal antenna gain is also $G = 2.15$ dB$_i$ while the usage losses are assumed to be $L = 9$ dB (Table 5.2).

Now, $\bar{E}_n$ is computed for each of the three possible receiver types in the network: base station, mobile stations, and hand-held terminals.

For the base station, the first step is to evaluate the minimum field strength E_m value

$$E_m = S + 20 \log f - 10 \log R_o - G_i + L_c - 12.8$$
$$= -9 + 20 \log(450) - 10 \log(50) - 8 + 2.5 - 12.8 = 8.7 \text{ dB}\mu\text{V/m}$$

From Curve A in Figure 5.3 it can be read that $\Delta_r E = 19$ dB. Next, the value of $\Delta_e E$ is computed for $L = 90\%$ and $T = 90\%$ bearing in mind that $k(90\%) = 1.28$ (Table 4.1), σ_L (UHF) $= 10$ dB (Table 5.4) and σ_T (UHF, over land) $= 2$ dB (Table 5.4), then

$$\Delta_e E = \sqrt{(1.28 \cdot 10)^2 + (1.28 \cdot 2)^2} = 13.1 \text{ dB}$$

Finally, applying 5.1, the median necessary field strength is

$$\bar{E}_n = E_m + \Delta_r E + \Delta_e E = 8.7 + 19 + 13.1 = 40.8 \text{ dB}\mu\text{V/m}$$

For the mobile stations

$$E_m = S + 20 \log f - 10 \log R_o - G_i + L_c - 12.8$$
$$= -9 + 20 \log(450) - 10 \log(50) - 2.15 + 0 - 12.8 = 12 \text{ dB}\mu\text{V/m}$$

From Curve B, in Figure 5.2 it can be read that $\Delta_r E = 11$ dB. The degradation parameter $\Delta_e E$ has the same value as for the base station, that is, 13.1 dB, then

$$\bar{E}_n = E_m + \Delta_r E + \Delta_e E = 12 + 11 + 13.1 = 36.1 \text{ dB}\mu\text{V/m}$$

For hand-held terminals

$$E_m = S + 20 \log f - 10 \log R_o - G_i + L_c - 12.8$$
$$= -9 + 20 \log(450) - 10 \log(50) - (2.15 - 9) + 0 - 12.8$$
$$= 21 \text{ dB}\mu\text{V/m}$$

A usage correction factor $L = 9$ dB has been introduced. The two correction values do not change, that is, $\Delta_r E = 11$ dB and $\Delta_e E = 13.1$ dB, then

$$\tilde{E}_n = E_m + \Delta_r E + \Delta_e E = 21 + 11 + 13.1 = 45.1 \text{ dB}\mu\text{V/m}$$

5.3 Mobile Station Coverage Evaluation

Area coverage radiocommunication systems can be classified into two groups:

1. Noise-limited systems;
2. Interference-limited systems.

For noise-limited systems the required studies will only be those oriented to the identification of the minimum transmission powers to achieve a given range by exceeding, at the receiver, a threshold level accounting for noise effects as well as for the fast and slow signal variations of the received signal. The equipment parameters to be computed are either the equivalent isotropically radiated power EIRP or the equivalent radiated power (ERP), which are defined by the following expressions:

$$EIRP \text{ (dBW or dBm)} = P_t \text{ (dBW or dBm)} + G_i \text{ (dBi)} - L_c \text{ (dB)} \quad (5.21)$$

$$ERP \text{ (dBW or dBm)} = P_t \text{ (dBW or dBm)} + G_d \text{ (dBd)} - L_c \text{ (dB)} \quad (5.22)$$

For interference-limited systems, additional verifications must be carried out, specifically, a computation of the cochannel reuse distance must be performed.

Case (1) represents the situation typically found in PMR systems, while case (2) will be that of, for example, cellular systems.

Evaluation of the ERP or EIRP of a Radio Station

In order to achieve the required coverage range from a given base station with a specified coverage quality at the service area contour, a sufficient transmit power must be used. In this section the base-to-mobile direction (downlink) will be studied. However, it must always be born in mind that mobile communications are bidirectional, and thus, the mobile-to-base link (uplink) must also be analyzed.

In order to achieve a given coverage range, it is necessary to compensate for the basic propagation losses, L_b. The computation of L_b can be carried out by using any of the models described in Chapter 4. The basic propagation losses can be divided into two components, free-space losses and excess losses

$$L_b = L_{\text{free space}} + L_{\text{excess}} \tag{5.23}$$

If the threshold is given by the median necessary field strength value, $\bar{E}_n$, then its corresponding value in linear units, $\bar{e}_n$, is given by

$$\bar{e}_n = \frac{\sqrt{30 \frac{p_t g_t}{l_c}}}{d} \frac{1}{\sqrt{l_{\text{excess}}}} = \frac{\sqrt{30 \text{ eirp}}}{d} \frac{1}{\sqrt{l_{\text{excess}}}} \tag{5.24}$$

with

$$\frac{\sqrt{30 \frac{p_t g_t}{l_c}}}{d} = \frac{\sqrt{30 \text{ eirp}}}{d} = e_o \tag{5.25}$$

where e_o represents the received field strength under free space conditions. Using practical units, the following expression is obtained

$$E \, (\text{dB}\mu\text{V/m}) = 74.77 + \text{EIRP (dBW)} - 20 \log d \, (\text{km}) - L_{\text{excess}} \, (\text{dB}) \tag{5.26}$$

On the other hand, the basic propagation losses can be expressed as

$$L_b = L_{\text{free space}} + L_{\text{excess}} \tag{5.27}$$
$$= 32.4 + 20 \log d \, (\text{km}) + 20 \log f \, (\text{MHz}) + L_{\text{excess}}$$

introducing (5.27) into (5.26) yields

$$E \, (\text{dB}\mu\text{V/m}) = 107.17 + \text{EIRP (dBW)} + 20 \log f \, (\text{MHz}) - L_b \, (\text{dB}) \tag{5.28}$$

or, in terms of the ERP,

$$E \text{ (dB}\mu\text{V/m)} = 109.32 + ERP \text{ (dBW)} + 20 \log f \text{(MHz)} - L_b \text{ (dB)} \quad (5.29)$$

Example 5.6. In this example the coverage range of a base station is computed, assuming a base station with a transmit power of 10W and a 6-dBd an omni-directional antenna. If the effective antenna height is 50m and the transmit frequency is in the 450-MHz band and assuming also a 1.5-dB feeder loss, the value of $\bar{E}_n$ will be that computed earlier in Example 5.5 (36.1 dBμV/m) for a mobile station. For the calculation of L_b the Okumura-Hata model (Chapter 4) is used

$$L_b \text{ (dB)} = 69.55 + 26.16 \log f - 13.82 \log h_t - a(h_m) \quad (5.30)$$
$$+ (44.9 - 6.55 \log(h_t)) \log d$$

and

$$a(h_m) = (1.1 \log f - 0.7) h_m - (1.56 \log f - 0.8) \quad (5.31)$$

for small-medium cities. Introducing the numeric parameters in the formulas

$$a(h_m = 1.5) = 0 \text{ dB and } L_b = 115.5 + 33.8 \log d \text{ dB}$$

if the field strength is expressed as a function of the EIRP

$$E \text{ (dB}\mu\text{V/m)} = 107.17 + EIRP \text{ (dBW)} + 20 \log f \text{(MHz)} - L_b \text{ (dB)} \quad (5.32)$$

and since $EIRP = 10 \text{ dBW} + (6 + 2.15) - 1.5 = 16.65 \text{ dBW}$, replacing this value in the field strength formula, results in

$$36 \text{ dB}\mu\text{V/m} = 107.17 + 16.65 \text{ dBW} + 20 \log 450 - 115.5$$
$$- 33.8 \log d \text{ (km)}$$

Solving for d we get a maximum coverage range of $d = 5.6$ km.

Example 5.7. Now the inverse calculation will be performed. A coverage range of 8 km is required, and the same conditions as in the previous example will be assumed. The parameter to be computed in this case is the required transmitter power, P_{Tx}.

First, the basic propagation losses are computed for a distance of 8 km

$$L_b \text{ (dB)} = 115.5 + 33.8 \log 8$$

Now, using (5.28) an EIRP = 21.86 dBW is obtained. Now, taking into account the definition of the EIRP a transmitter power P_{Tx} = 15.25 dBW ≡ 33W is obtained.

Nowadays, given the widespread availability of computers, coverage studies have changed substantially. The terrain influence can be taken into consideration much more easily with the aid of terrain databases and computerized radio network planning tools. Moreover, it is extremely important to take into account the influence of the type of land usage (urban, suburban, wood, open, etc.) in the vicinity of the mobile terminal (base stations are assumed to be clear from the surrounding clutter). In Figure 5.7 a terrain profile is schematically illustrated together with different land-usage types along the radio path.

The methodology presented in this section has assumed constant terrain irregularity and land-usage conditions. However, this is not usually the case; thus coverage studies should at least be carried out for several sectors around the radio station using radial terrain profiles as schematically illustrated in Figure 5.8(a). For each radial, a maximum range is calculated. The coverage contour will then be that defined by linking the maximum coverage distances for each radial.

If more accurate coverage plots are required, a detailed study can be carried out by subdividing the area of interest into small study areas of 250 × 250 m² or smaller. These study areas can be those defined by the terrain database grid elements as shown in Figure 5.8(b).

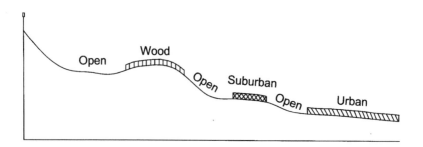

Figure 5.7 Land-usage superposed on a terrain profile.

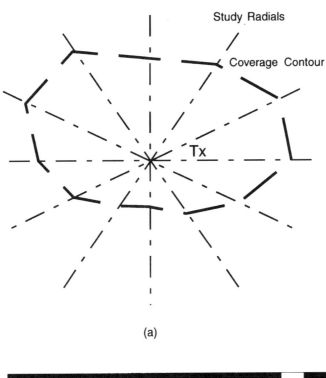

(a)

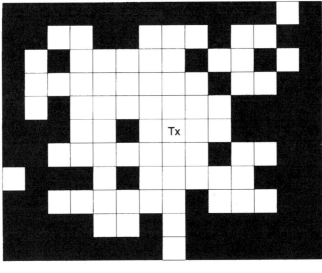

(b)

Figure 5.8 (a) Computation of the coverage contour by means of twelve radials. (b) Coverage study elements in a terrain database.

5.4 Interference-Limited Systems

For those radio systems with a high frequency reuse (cellular, for example), the main coverage limitation is that caused by interference. The basic transmission quality parameter for such systems is the so-called *protection ratio*, R_p. In the same way as in the case of noise-limited systems, the protection ratio shall be exceeded for a large percent of time and locations. As it was observed for the case of noise-limited systems, time variations are significantly smaller than locations variations. Here, only a locations probability objective $L\%$ will be specified.

To ensure that the carrier-to-interference ratio (c/i) exceeds the threshold level R_p for $L\%$ of locations, both the variability of the wanted signal c and the interfering signal i must be taken into account. As in the noise-limited case, Gaussian distributions will be assumed for both signal variations when expressed in decibels

$$C \equiv \text{Normal}(\tilde{C}, \sigma_C) \text{ and } I \equiv \text{Normal}(\tilde{I}, \sigma_I) \tag{5.33}$$

Usually, the assumption is made that both wanted and interfering signals' standard deviations are equal, i.e., $\sigma_C = \sigma_I = \sigma$. The carrier-to-interference ratio expressed in logarithmic units $C - I$ (dB) also follows a Gaussian distribution, that is,

$$C - I \equiv \text{Normal}(\tilde{C} - \tilde{I}, \sigma_{C-I} = \sqrt{\sigma_C^2 + \sigma_I^2} = \sqrt{2}\sigma) \tag{5.34}$$

The median value of $C - I$, $\tilde{C} - \tilde{I}$ must be such that R_p is exceeded for $L\%$ of locations.

The main design parameter for interference-limited systems is the frequency reuse distance, D. If a propagation law with exponent n and only one interfering station are assumed, the carrier-to-interference ratio will be

$$\frac{\tilde{c}}{\tilde{i}} = \tilde{r}_p = \frac{\dfrac{k}{R^n}}{\dfrac{k}{(D-R)^n}} = \frac{(D-R)^n}{R^n} \tag{5.35}$$

Then, the reuse distance is

$$D = R(1 + \tilde{r}_p^{1/n}) \text{ where } \tilde{R}_p = 10 \log(\tilde{r}_p) \tag{5.36}$$

In cellular systems, in which each transmitter is surrounded by at least six cochannel stations, the reuse distance will be

$$D = R(1 + (6\bar{r}_p)^{1/n}) \qquad (5.37)$$

5.5 Intermodulation

The deployment of mobile radio stations operating on the VHF and UHF frequency bands has increased drastically in the past few years. The presence of a number of transmitters on the same site gives rise to intermodulation effects. Two typical cases may be considered with regard to intermodulation interference:

- Existing sites where several one-channel systems for individual or shared use are located. This type of station corresponds to PMR applications. PMR repeaters and base stations tend to concentrate on high rise sites where a good coverage of the service area (a city, for example) can be achieved. These sites are not abundant, and thus, radio stations tend to concentrate on the same location or even share the same tower mast.
- The other case corresponds to multichannel systems such as cellular or trunked (PAMR) systems. These systems use the same transmit and receive antennas for all radio channels. In some instances where there are space limitations, a single antenna may be used for both transmission and reception.

These two typical configurations give rise to intermodulation phenomena, and thus, emissions will be radiated that can interfere with existing communications.

In this section, a revision is made of the intermodulation phenomenon and the different options to overcome it. These techniques allow the efficient use of the available spectrum. Two main approaches may be followed depending on the system type; one of them is to adequately select those frequencies that do not produce intermodulation products "falling" on the channels operating on the site itself. This technique is not efficient since it requires to remove a large number of channels to avoid intermodulation. Later, an example is shown in which 62 channels must be reserved in order to find 10 channels that are mutually compatible on the same site.

The other approach is to increase the isolation between cosited transmitters to reduce the possibility of intermodulation products being generated. Such an approach includes the use of several elements such as cavity filters, ferrite isolators, and duplexers. Additionally, the isolation can be increased by physically separating the interfering transmitters.

Generation of Intermodulation Products

Intermodulation products are generated when two or more signals are mixed on a nonlinear device. When this happens, a number of frequency components are generated incorporating the original modulating signals. Some of the frequencies in this set are far from the used band whereas others may coincide with the operating frequency of one or more receivers, thus producing cochannel interference.

A nonlinear device is characterized by means of its *transfer characteristic function*, which can usually be expressed in a polynomial form

$$y = c_o + c_1 x + c_2 x^2 + \ldots + c_n x^n \tag{5.38}$$

From this expression and considering that several cosited transmissions defined by frequencies f_1 through f_n are present, a number of previously nonexistent frequencies are generated. These frequencies can be expressed in a generic form by the formula

$$f_i = a_1 f_1 + a_2 f_2 + \ldots + a_n f_n \tag{5.39}$$

where the coefficients a_1 are integers.

These new frequencies are called *intermodulation products* (IMs). The *order* of an IM is given by the sum $|a_1| + |a_2| + \ldots + |a_n|$.

Intermodulation can be reduced to moderate levels by lowering the levels of the unwanted signals reaching the nonlinear element. Intermodulation products of any order will only appear when the unwanted signals are present simultaneously.

There are three main sources of intermodulation in radiocommunication systems:

- Intermodulation generated at the transmitter;
- Intermodulation generated at non-linear external elements;
- Intermodulation generated at the RF stages of the receiver.

Intermodulation generated at the transmitter. Intermodulation products will be generated at the power stages of transmitters due to their nonlinear characteristics (class C amplifiers). These products will be radiated by the transmitter antenna.

Transmitter's external elements intermodulation. Typical examples of external elements are supporting guys, trellis type tower elements, metallic floors, acting as rectifying diodes; thus, these cause strong signals in their vicinity mix and are radiated producing interference ("rusty bolt" effect).

Receiver-generated intermodulation. This is produced by two or more high-level signals reaching the RF stages of a receiver.

Intermodulation Products

An example of how intermodulation products are generated and may cause interference is presented here. Assuming that three adjacent channels a, b, and c with the central frequencies $a = f_o - \Delta f$, $b = f_o$, $c = f_o + \Delta f$ are transmitted, the following 3rd-order intermodulation products may be generated:

$$a + b - c = f_o - 2\Delta f \qquad 2a - c = f_o - 3\Delta f$$
$$a + c - b = f_o \qquad 2a - b = f_o - 2\Delta f$$
$$b + c - a = f_o + 2\Delta f \qquad 2b - c = f_o - \Delta f$$
$$2b - a = f_o + \Delta f$$
$$2c - b = f_o - 2\Delta f$$
$$2c - a = f_o + 3\Delta f$$

Figure 5.9(a) presents the original frequencies and the intermodulation products listed above broken down into two groups for clarity: (A) those produced by the combination of two frequencies and (B) those generated by the combination of the three frequencies. It can be clearly observed how some intermodulation products are generated that coincide with the frequency of an existing cosited transmitter thus causing interference problems.

If the frequency separation between the cosited transmissions is selected carefully, an interference-free system can be achieved. In this way, for example, if again three channels a, b, and c are used on the same site but their separations are slightly different to the previous case (for example, $a = f_o - \Delta f$, $b = f_o$, $c = f_o + 2\Delta f$) the following intermodulation products may be generated:

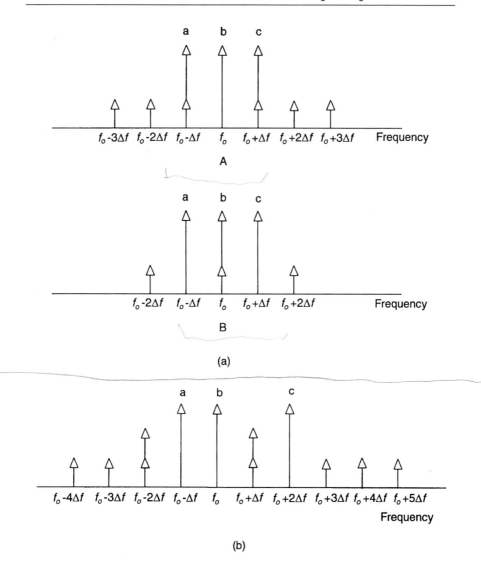

Figure 5.9 Third-order IMs: combinations of two and three frequencies.

$$a + b - c = f_0 - 3\Delta f \qquad 2a - c = f_0 - 4\Delta f$$
$$a + c - b = f_0 + \Delta f \qquad 2a - b = f_0 - 2\Delta f$$
$$b + c - a = f_0 + 3\Delta f \qquad 2b - c = f_0 - 2\Delta f$$
$$2b - a = f_0 + \Delta f$$
$$2c - b = f_0 + 4\Delta f$$
$$2c - a = f_0 + 5\Delta f$$

In this case, none of the produced intermodulation products will "fall" on one of the existing transmitter frequencies. However, there is an important drawback: Six contiguous channels and one intermediate channel are occupied and cannot be used on the same site. Figure 5.9(b) illustrates this situation.

For two transmitters using frequencies f_1 and f_2 with powers P_1 and P_2 (dBW), intermodulation products with frequencies $f_{IM} = nf_1 + mf_2$ may be generated. The power of these intermodulation products P_{IM} will follow the expression

$$P_{IM} = nP_1 + mP_2 + K \qquad (5.40)$$

(where $K(dB)$ is a negative number). From this expression it can be drawn that, if P_1 is incremented by 3 dB, then P_{IM} will increase by $3n$ dB. The contrary will also hold (isolation), that is, if P_1 is reduced by half, the intermodulation product will be $3n$ dB lower. For example, let $n = 2$ and $m = 1$, if P_1 is reduced by half, then the intermodulation product will be 6 dB lower. If P_1 and P_2 are both reduced by half, then the level of the intermodulation product will be 9 dB lower. This explains the importance of reducing the power of the unwanted signals to keep intermodulation products low, using isolation devices (e.g., transmitter combiners).

Second-order products are produced as the sum or difference of two unwanted frequencies:

$$a \pm b$$

This product type is usually of little importance since they are located far from the band of interest. For example [2], two mobile communications at the low-VHF band, one at 75 MHz and the other at 80 MHz will generate second-order intermodulation products at 5 MHz and 155 MHz. The receiver's RF filter selectivity and the transmitter output stages will provide sufficient attenuation. However, problems may arise in the vicinity of high-power broadcast stations. For example [2], a FM/VHF transmitter operating at 92 MHz could beat with a short wave (HF) transmission at 10 MHz giving rise to an 82-MHz product (mobile communications low-VHF band). This interference could spread over several mobile communication channels since broadcast transmissions use a frequency deviation of 75 kHz.

Similarly, second-order intermodulation could be a problem in sites hosting multiple mobile systems such as, for instance, 450 MHz and 900 MHz (2×450) or 900 MHz and 1,800 (2×900).

Third-order products are the most harmful ones. Third-order products may follow two expressions

$$2a - b$$
$$a + b - c$$

The main reason why these products cause problems is that they can easily access the most critical parts of the interfered system. This is due to the fact that these products do not present great frequency differences with respect to the original frequencies and tuned circuits (RF and IF filters) provide little selectivity in these cases.

Fourth-order products follow the expressions:

$$3a \pm b$$
$$2a \pm 2b$$
$$a \pm b \pm c \pm d$$
$$\cdots\cdots$$

These products present similar characteristics to second order products (i.e., they are produced at frequencies far from the operating band); thus they are much more easily eliminated than odd order products.

The frequencies of fifth-order products follow the expressions:

$$3a - 2b$$
$$2a - 3b$$
$$\cdots\cdots$$

They present similar characteristics as third-order products; that is, they are generated close to the original frequencies, and thus, they are difficult to eliminate. However, since they are of a higher order, they are of a lower level and, consequently, less harmful.

Higher-order even products present similar characteristics to those of lower even order except that they are of a much lower level. The effects of higher-order odd products decrease rapidly with increasing order. Seventh or higher order products will hardly pose relevant problems.

Figure 5.10 [3] illustrates the frequency ranges where the different intermodulation product orders may appear for two-frequency systems as a function of the channel spacing, Δf, the transmit-receive frequency separation, D, and the size of the guard band, G. If these parameters are carefully selected, as

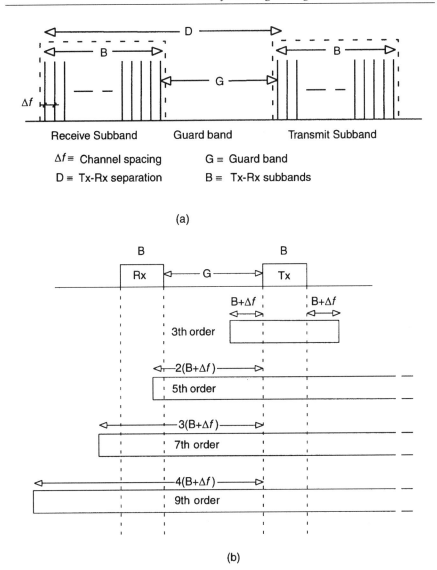

Figure 5.10 (a) Parameters for two-frequency operation. (b) Frequency ranges for the different IM product orders [3].

illustrated in Figure 5.10, it is possible to limit the frequency range for which intermodulation products are generated.

Characterization of Transmitter-Generated Intermodulation

To characterize the effects of intermodulation, a methodology described in ITU-R Report 739 is presented. The production process of IM interference is illustrated in Figure 5.11. Different parameters are used for this purpose; they are related to the generation process of intermodulation products.

The *coupling loss*, A_C (*dB*), is defined as the power ratio between the power radiated by the transmitter producing the unwanted signal giving rise to intermodulation and the power at the output of the transmitter where the intermodulation product is actually generated. A typical value of A_C is on the order of 30 dB for transmitters sharing the same site.

The *conversion loss*, A_I (*dB*), is defined as the power ratio between the unwanted signal from an external source and the intermodulation product itself, both measured at the output of the transmitter where it is generated. If no special precautions are taken, this parameter may be in the range of 5–20 dB for solid-state transmitters for the third-order product ($2f_1 - f_2$). In Figure 5.12 conversion loss values are given for a typical class C transmitter for the 150 MHz band.

From this curve it can be observed how, for example, for a frequency offset of 3 MHz, the level of the intermodulation product will be 24 dB below that of the unwanted transmission (i.e., A_I = 24 dB). Consequently, if the transmitter power is 30W ($\approx$ 45 dBm), and the coupling loss is A_C = 40 dB, then the unwanted signal enters the nonlinear element with a power of 3 mW (5 dBm), this means that the intermodulation product will be transmitted with a power of 5 − 24 = −19 dBm (approximately 12 μW). This intermodulation product may reach a mobile receiver near the site or a receiver in the same site, thus causing interference.

Finally, the propagation losses, A_P (*dB*), must be considered. These losses account for the signal decay with the distance from the transmitter to the interfered receiver.

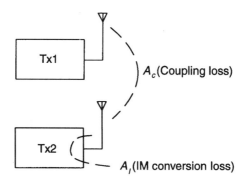

Tx1:1 IM inducing transmitter
Tx2:1 IM generating transmitter

Figure 5.11 Intermodulation interference generation process.

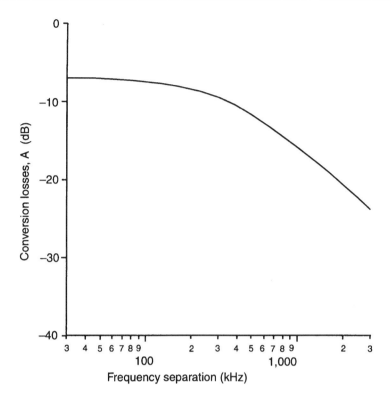

Figure 5.12 Conversion losses for a class C transmitter for the 150-MHz band.

The total losses, A, between a transmitter generating an unwanted transmission giving rise to intermodulation and the interfered receiver operating at the same frequency as the intermodulation product will be

$$A = A_c + A_I + A_p \qquad (5.41)$$

It can be observed that for the previously described methodology (ITU-R Report 739), the parameters used do not account for the effect of the power levels of both the transmitter where the intermodulation product is generated and the power of the unwanted signal. These values are important and influence the value of the conversion loss parameter A_I.

The coupling or isolation between antennas depends on the type of mounting (vertical or horizontal). Figure 5.13 [4] provides isolation curves for separations in the "H" and "E" planes.

Example 5.8. Here, the link budget for an intermodulation product is presented. Let Tx_1 be a transmitter at frequency f_1 where an intermodulation

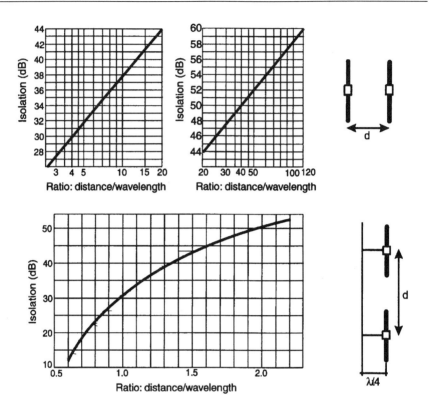

Figure 5.13 (a) Isolation between antennas. Plane "H" separation. (b) Isolation between antennas. Plane "E" separation [4].

product is generated and assume that another transmitter, Tx_2, operating at a frequency f_2 is coupled to the first one producing an intermodulation product. Typical parameters for the link budget of an intermodulation product are the following:

Tx_2 at f_2 transmission power	+44 dBm
Coupling loss A_C	30 dB
Conversion loss A_I	15 dB
Interfered receiver threshold	−116 dBm

The receiver will suffer interference when the propagation losses A_P fulfill the equation

$$44 - A_C - A_I - A_P \geq -116$$

using standard values for A_C (30 dB) and A_I (15 dB), then $A_P \leq 115$ dB.

These losses correspond to a considerable distance for free-space propagation conditions (Figure 5.14, curve A). For mobile propagation conditions using the ITU-R Recommendation 370 propagation model and assuming an effective base station height of 37.5m and a mobile antenna height of 2m (Figure 5.14, curve B), a distance of 5 km is obtained. It can be observed that it is relatively easy to produce interference to receivers at close distances.

It can also be concluded that two-frequency operation produces much better conditions for the reduction of intermodulation effects (Figure 5.10) given that base station receivers will be far from the transmit sub-band.

In the inverse direction (that is, from the mobiles to the base station,) the intermodulation caused by two or more mobile transmitters will be more harmful when the interfering mobiles are near the base station and the wanted signal comes from the edge of the service area. In this case, the base station will be receiving both the intermodulation product and the wanted signal, this one coming from the edge of the service area. The wanted signal will be subject to shadowing and multipath and, in some instances, it will be below the interfering signal (ITU-R Report 739).

Intermodulation Generated at Nonlinear External Elements

Intermodulation may be generated at external nonlinear elements of a radio station such as, for example, metal-metal unions of mast elements, feeders,

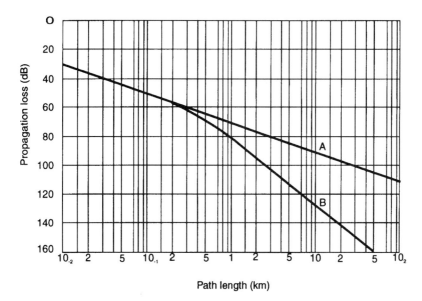

Figure 5.14 Propagation attenuation between two half-wave dipoles at 150 MHz for (a) free-space conditions and (b) ITU-R Rec. 370 for h_1 = 37.5 m and h_2 = 2m (ITU).

and other antennas in the vicinity of transmitters. A similar characterization methodology to that used for transmitter-generated intermodulation could be employed to quantify external element intermodulation. However, this is a difficult task and the parameters A_C and A_I cannot be easily measured. It has been observed, however, that these intermodulation phenomena present lower levels than transmitter-generated intermodulation products, and thus, their influence range is smaller. Of course, this level depends on the transmitters' powers, the distance to the nonlinear elements, and their efficiency as rectifiers and radiators.

The main cause of these intermodulation phenomena is the use of metallic masts for the installation of numerous antennas with small separations between them. It is extremely difficult to ensure perfect metal-metal unions at the numerous contact points in trellis type masts. The best approach to prevent this type of intermodulation is to thoroughly check each metal-metal joint. Corrosion may be reduced by adequate painting of the tower mast. Alternative contacts may also be provided using metal strips to prevent this problem. Other sources of external intermodulation can be supporting guys and other metallic structures nearby.

This type of intermodulation generation mechanism is mentioned here as general information although, given the power levels used in mobile radio systems, external element-generated intermodulation will rarely take place.

Receiver-Generated Intermodulation

Intermodulation products can also be generated at the receiver due to the presence of two or more high-level signals in a nonlinear section in the RF circuits. As occurs with transmitters, the unwanted signals (two or more) can generate an intermodulation product "falling" within the receiver RF passband.

VHF and UHF receivers typically experience overload when the unwanted signals are on the order of 80 dBμV. For these levels, receiver blockage can be observed [2]. However, nonlinear effects can already be present when the interfering signals are on the order of 60 dBμV (1 mV on 50Ω). This is an unwanted signal level that must be avoided using either of the two techniques indicated above (that is, either by selecting intermodulation-free frequencies or by increasing the isolation by using appropriate circuits as explained in Chapter 7).

For the characterization of the intermodulation phenomenon at the receiver, a single measurement is usually performed using two unwanted signals of the same level. Intermodulation is specified as the power ratio between the level of the two interfering signals and that of the intermodulation product, all referred to the input of the receiver.

In real-life situations, the unwanted signals do not present the same level, but still, they may produce significant intermodulation products. In Figure 5.15, three theoreticals taken from ITU-R Report 793 that give the overall third-order intermodulation characteristic of a receiver are illustrated. It is clearly shown how intermodulation can be generated by two unwanted signals with different levels even when one of them is not excessively high.

5.6 Frequency Planning

The different mechanisms producing intermodulation have been reviewed both for multichannel systems and for simple single-channel systems sharing the same site or even the same antenna mast. Now, different approaches for the selection of interference-free channels are described. As a first caution, it is recommended that systems operate using two-frequency channels whenever possible (Figure 5.10 and Chapter 2). Additionally, two main options can be identified:

1. Selection of intermodulation-free channels;
2. Adequate base station engineering (increased isolation).

In general, the first option will be used to make several individual single-channel systems sharing the same site or tower mast *mutually compatible*. This will be the case for a given geographical area where there are not many good sites and PMR radio stations will tend to be concentrated on the same site overlooking a city. The main problem of this option is that its spectral efficiency is very low as illustrated in Table 5.5. In Table 5.5, third-order IM interference-free channel combinations are listed for different numbers of required cosited transmitters. For example, it can be read that, if 10 channels are required to share the same site, a band of 62 channels must be reserved. It is clear that this approach is not efficient at all.

Further information on this approach can be found in [5], where channel spacings to avoid third- and fifth-order intermodulation are provided.

The second approach is based on an attempt to adequately isolate the different elements in multichannel systems. This technique will be much more efficient as regards frequency usage. As it is described in Chapter 7, isolation can be achieved by means of devices such as isolators and cavity filters. These devices will attenuate the unwanted signal reaching nonlinear elements of closely located transmitters.

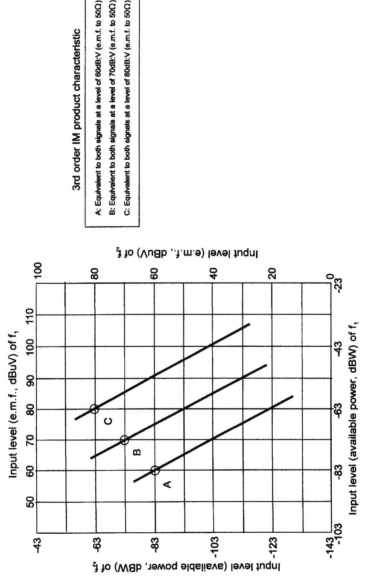

Figure 5.15 Receiver-generated intermodulation characteristics (ITU).

Table 5.5
Intermodulation-Free Channel Combinations [2]

No. of Wanted Channels	Necessary No. of Channels	Third-Order IM-Free Channels
3	4	1,2,4
4	7	1,2,5,7
5	12	1,2,5,10,12
6	18	1,2,5,11,13,18
7	26	1,2,5,11,19,24,26
8	35	1,2,5,10,16,23,33,35
9	46	1,2,5,14,25,31,39,41,46
10	62	1,2,8,12,27,40,48,57,60,62

It must be noted that, in the appropriate selection of frequencies, it is not only necessary to prevent intermodulation effects; an important role is also played by aspects like receiver blocking or desensitization. Additionally, cochannel (and adjacent channel) interference will have to be accounted for when carrying out frequency assignments.

Frequency Assignment Methods for Multichannel Systems

In this section the methodology described in ITU-R Report 901 is presented. The guidelines provided in this report can be used for the planning of multichannel systems (for example, trunked systems). Channel assignment proposals made by different administrations for the reduction of the effects of intermodulation products generated within a system or by nearby systems are summarized here.

One method, used in Canada for the 400-MHz band and in the United States and Canada for the 800-MHz band, splits channels in blocks separated by a fixed number of channels. It is also possible to identify channel sub-blocks to provide flexibility in the assignment of channels to individual systems. The minimum frequency spacings in multichannel systems are 250 kHz and 100 kHz, for the 800-MHz and 400-MHz bands, respectively.

Table 5.6 provides an example of a frequency plan used in the United States in the 800-MHz band for the assignment of 200 channels. These 200 channels can be split into 10 or 20 blocks with a 10-channel spacing between channels belonging to the same block. Additionally, this arrangement subdivides each 20-channel block into five-channel groups.

A similar approach is shown in Tables 5.7 and 5.8, where two plans for the allocation of channels used in Canada for the 800- and 400-MHz bands are presented.

Table 5.6
Frequency Assignment Plan for Trunked Systems in the 800-MHz Band in the United States (ITU)

Block	Channels	Block	Channels
1	1-41-81-121-161	6	6-46-86-126-166
	21-61-101-141-181		26-66-106-146-186
	11-51-91-131-171		16-56-96-136-176
	31-71-111-151-191		36-76-116-156-196
2	2-42-82-122-162	7	7-47-87-127-167
	22-62-102-142-182		27-67-107-147-187
	12-52-92-132-172		17-57-97-137-177
	32-72-112-152-192		37-77-117-157-197
3	3-43-83-123-163	8	8-48-88-128-168
	23-63-103-143-183		28-68-108-148-188
	1313-53-93-133-173		18-58-98-138-178
	33-73-113-153-193		38-78-118-158-198
4	4-44-84-124-164	9	9-49-89-129-169
	24-64-104-144-184		29-69-109-149-189
	14-54-94-134-174		19-59-99-139-179
	34-74-114-154-194		39-79-119-159-199
5	5-45-85-125-165	10	10-50-90-130-170
	25-65-105-145-185		30-70-110-150-190
	15-55-95-135-175		20-60-100-140-180
	35-75-115-155-195		40-80-120-160-200

Table 5.7
Frequency Assignment Plan for Trunked Systems in the 800-MHz Band in Canada (ITU)

Block	System No.	Channels
1	1	1-11-21-31-41
	2	2-12-22-32-42
	3	3-13-23-33-43
	4	4-14-24-34-44
	5	5-15-25-35-45
	6	6-16-26-36-46
	7	7-17-27-37-47
	8	8-18-28-38-48
	9	9-19-29-39-49
	10	10-20-30-40-50

Table 5.8
Frequency Assignment Plan for Trunked Systems in the 400-MHz Band in Canada (ITU)

System No.	Channels
1	1-5-9-13-17
2	2-6-10-14-18
3	3-7-11-15-19
4	4-8-12-16-20

5.7 Cochannel Interference Compatibility

The last stage in the frequency planning of a mobile radio system is to perform an interference study, mainly addressing cochannel interference. This study will provide an indication of what stations in the network are compatible, and thus, if the same frequencies can be used to provide radio coverage to their service areas.

In Figure 5.16 some relevant concepts are illustrated. The *service area* of a radio station is the geographical area where its terminals operate. Inside this area an adequate spectrum management must insure that the RF protection ratio (R_p) is exceeded for high locations and time percentages. In Figure 5.16,

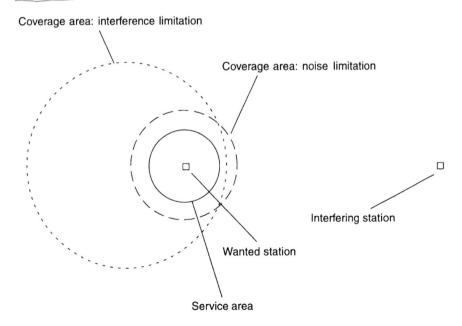

Figure 5.16 Service and coverage areas, for the interference- and noise-limited cases.

the noise-limited and interference-limited coverage areas are also shown for a single or dominant interferer. Within these areas the noise threshold and the protection ratio are respectively exceeded. In the case of a *multisite system,* the *service area* of the global network encompasses the individual service areas of each base station (possibly with some overlapping).

The operator of a mobile radio network will establish groups of frequencies typically with a separation between the frequencies belonging to the same group of a fixed number of channels. Figure 5.17(a) illustrates such a scheme with a spacing of 12 channels between frequencies of the same group. The groups will be reused in other locations so that the protection ratio criteria will be met (Figure 5.17(b)).

ITU-R Report 319 specifies, for PMR and PAMR services using narrowband angle modulation, a protection ratio of 8 dB for 25-kHz channel separations and 12 dB for 12.5-kHz channel separations. For the sake of comparison, it can be mentioned here that in the TACS analog cellular system the protection ratio is 17 dB and in the GSM digital system this ratio is 9 dB.

An exhaustive check for the verification that the protection ratio is exceeded can be made by making use of terrain databases and by defining surface study areas of 250×250 m^2 or smaller (Figure 5.18(a)). For each individual study area a protection ratio test must be made both for the base-to-mobile and mobile-to-base directions. Then, statistics of the number of individual study areas where the protection ratio is exceeded can be compiled and compared with the specified interference quality objective.

However, this test is somewhat complex and time consuming so, in most cases, some simplifications are called for. For example, a simpler study can be made for the downlink only along a limited number of radials (Figure 5.18(b)). A $C - I = R_p$ contour can be drawn by linking the distances for each radial for which $C - I = R_p$. In case this contour includes the whole service area, this will mean that the two stations being tested are compatible from the cochannel interference point of view.

So far, only one dominant interfering station has been considered, but, in some instances, this will not be the case. For multiple interferers a simplified analysis could be made at points located on the boundary of the service area, as shown in Figure 5.18(c).

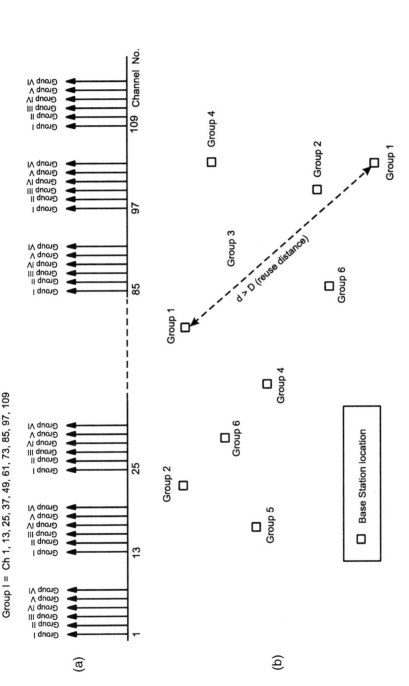

Figure 5.17 (a) Arrangement of available channels in groups. (b) Schematic representation of a radio network layout and frequency group assignments.

232   Introduction to Mobile Communications Engineering

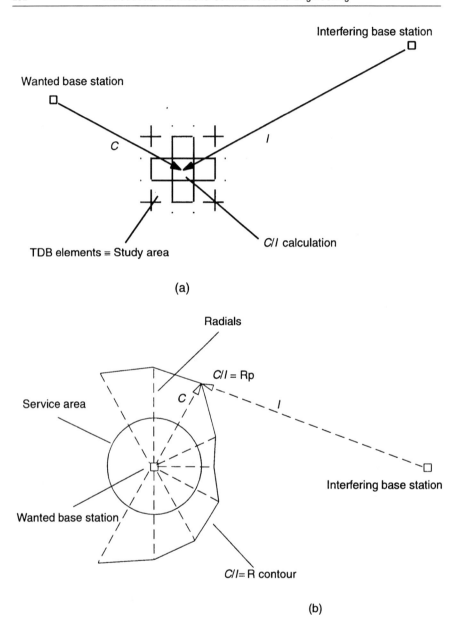

Figure 5.18 (a) Interference study for each terrain database surface element. (b) Interference study by radial lines. (c) Multiple interference study.

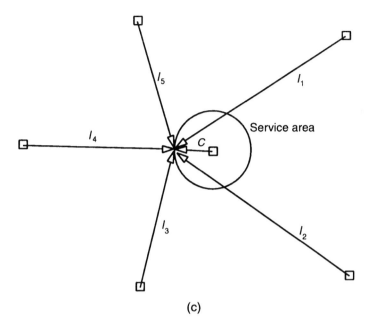

(c)

Figure 5.18 (continued).

References

[1] Jakes, W. C. (editor), *Microwave Mobile Communications,* New York: John Wiley & Sons, 1974.

[2] Pannell, W. M., *Frequency Engineering in Mobile Radio Bands,* Cambridge, Great Britain: Granta Technical Editions & Pye Telecommunications Ltd., 1979.

[3] Mehrotra, A., *Cellular Radio Performance Engineering,* Norwood, MA: Artech House, 1994.

[4] 350-512 MHz Base Station Antennas for Mobile Communications, Edition 12, Rosenhein, Germany: Kathrein-Werke KG.

[5] Edwards, R., J. Durking, and D. H. Green, "Selection of Intermodulation-Free Frequencies for Multiple-Channel Mobile Radio Systems," *Proc IEE,* Vol. 116, No. 8, Aug. 1969, pp. 1311–1318.

Selected Bibliography

ITU-R Recommendation 370, VHF and UHF propagation curves for the frequency range from 30 MHz to 1000 MHz.

ITU-R Report 319, Characteristics of equipment and principles governing the allocation of frequency channels between 25 and 1000 MHz for land mobile services.

ITU-R Report 358, Protection ratios and minimum field strengths required in the mobile services.

ITU-R Report 567, Propagation data and prediction methods for the terrestrial land mobile service using the frequency range 30 MHz to 3 GHz.

ITU-R Report 739, Interference due to intermodulation products in the land mobile service between 25 and 1000 MHz.

ITU-R. Report 901, Methods of frequency assignment for trunking systems in mobile services.

ITU-R Report 1019, Sources of unwanted signals in multiple base station sites in the land mobile service.

6

Propagation in New Scenarios

6.1 Introduction

This chapter deals with propagation and system planning aspects related to new environments in mobile and wireless communications, namely, urban and highway microcells, and indoor picocells. Also propagation models for indoor coverage from outdoor base stations will be analyzed. These scenarios do not complete the number of possible environments where wireless communications may take place. In this respect, other paths such as those in mines or tunnels, land mobile satellites, or millimetric wideband communications are also important. However, they will not be dealt with here.

6.2 Microcell Scenarios

In most developed countries during the second half of the 1980s and through the 1990s, there was a tremendous growth in the number of users of analog, first-generation cellular systems. These systems, in order to increase their capacity to handle new subscribers, steadily reduced their urban cell sizes down to roughly a 1-km radius. These small cells may still be considered standard, conventional cells with base station antennas above rooftops. Using these conventional cells, which are also called macrocells, analog cellular systems were already close to their traffic-handling capacity limits.

One possible solution to this saturation problem, as explained in Chapters 10 and 11, was to resort to new digital, second-generation technologies (e.g., GSM, D-AMPS, and IS-95). Another alternative, valid both for analog and digital systems, is the deployment of microcells serving significantly smaller areas. Standard hand-over procedures and all other cellular network mechanisms would still be employed. The possibility of frequency reuse at closer distances permits an increase in network capacity. Radiation confinement to achieve the smaller coverage ranges of microcells is obtained thanks to the shielding effect of buildings. For this purpose, base station antennas must be placed well below building rooftops, and transmitted powers should be smaller than those of standard macrocells. However, some energy propagates behind corners at the beginning of perpendicular streets (due to diffraction and reflection), thus allowing the possibility of implementing hand-over procedures to other neighboring microcells or macrocells. In Figure 6.1, a cellular network layout in which both conventional macrocells and microcells co-exist is shown.

It is expected that mature configurations of existing or future cellular networks will consist of several different types of cells ranging from standard macrocells with radii of up to 35 km to very small indoor picocells with radii of 15–20m. Wide-area coverage will be achieved using macrocells while large traffic demand locations (hot spots) will be served with highway or street microcells (with radii of 50–200m). Highway microcells will address the coverage of roads and motorways in rural or urban areas. These cells may be from 1 to 3 km long and show cigar-like shapes. Typically, the frequency reuse pattern (Chapter 9) will be of two cells (two-cell clusters) as shown in Figure 6.2.

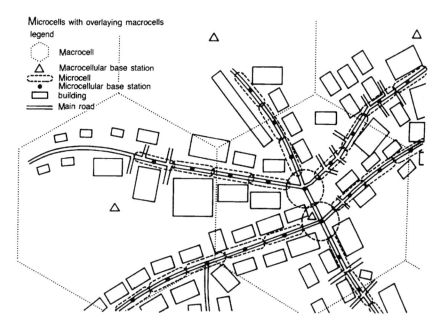

Figure 6.1 Standard macrocells and urban microcells. (*Source:* [1]. Reprinted with the permission of *British Telecom Tech. Journal.*)

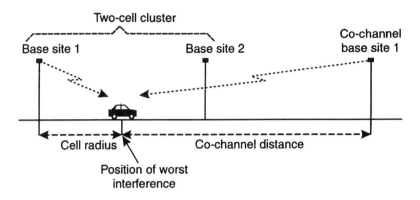

Figure 6.2 Two-cell frequency reuse pattern in a highway microcell system [2].

Avenues and major roads within urban areas with high traffic densities will be served by street microcells. In general, microcells have the advantage of providing better indoor coverage to surrounding buildings. The service demand within buildings may be handled, if necessary, by means of indoor microcells or picocells. Picocells are also suited to set up wireless private branch exchanges (WPABX) or radio LANs (see Chapter 11). Conventional macrocells may be used both to serve rural, low-traffic density areas as well as to provide coverage continuity in urban areas (umbrella cells). Typically, in cellular networks, for hand-over applicability reasons, conventional cells in urban areas tend to serve terminals in driving automobiles with moderate or high speeds while microcells tend to address the pedestrian or quasi-stationary user.

In order to set up a microcell it is necessary to have a good knowledge of its propagation conditions. These conditions differ in a substantial way from those in macrocells. Standard models such as those of Hata, Okumura, or COST 231 (Chapter 4) are not applicable to microcell propagation. In this chapter, experimental results as well as empirical and deterministic propagation models are presented. However, for the computation of such aspects as link budgets, fading margins, and availability levels, similar methodologies to those employed in conventional macrocell scenarios are applicable (i.e., the coverage probability with locations is assessed by comparing the received signal and its variability with a noise threshold or minimum required received power). Similarly, carrier-to-interference computations must be carried out and comparisons made with a given protection ratio. To statistically quantify system availability from the point of view of coverage and interference, an adequate knowledge of the signal variability statistics is also required.

The standard procedure for coverage evaluation is to calculate the link budget, especially at the fringe of the cell and compare it with a threshold. Also, a link budget for the carrier-to-interference ratio is required to verify if the protection ratio is exceeded or not and with which probability. A typical link budget is illustrated in Figure 6.3 [2] where the signal variability and the path loss effects are illustrated. Due to the confinement effect of buildings (low time dispersion), it is expected, in most cases, that the channel will present narrowband conditions.

Figure 6.3 illustrates a microcell link budget. The first parameter of interest for a microcell link budget is the intercept loss point. This is simply a normalization factor shifting up/down the path loss characteristic. For microcells and picocells the intercept point is defined at 1m from the base station. Intercept losses depend on a number of factors such as the frequency of operation and the base station antenna height and pattern. The second element is the propagation path loss characteristic or path loss increment law with distance. Together, the intercept point and the propagation characteristic provide the means to compute the median path loss for any arbitrary point at a distance d from the base station.

Also in Figure 6.3 it can be observed that two margins must be set up in the link budget to provide a statistical protection against signal variability effects. The first margin is set up to account for *shadowing effects*. This margin may also be viewed as a protection to account for the *prediction error* in the propagation model used. The better the propagation model, the smaller the required margin. Typically a *log-normal distribution* is considered for these

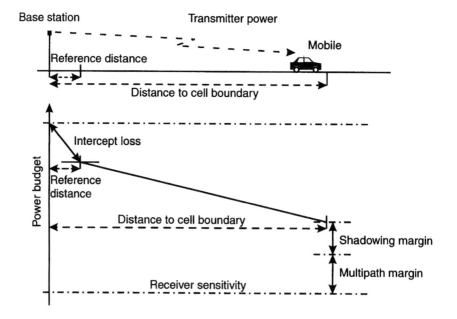

Figure 6.3 Link budget in a microcell system [2].

variations. The final margin takes into account the *fast multipath fading* effect. However, as it is shown in Section 6.2.1, the multipath characteristics are not as "hard" as in the case of conventional cells since LOS conditions will be present most of the time and fading will be *Ricean* instead of *Rayleigh*.

6.2.1 Microcell Propagation Modeling

Microcell propagation may be described with deterministic as well as empirical models [1, 2]. Deterministic models provide further insight into the propagation phenomena involved. Two different propagation environments or two different types of microcells may be defined:

- Urban microcells;
- Highway microcells.

For highway microcells the *two-ray model*—the direct plus a ground reflected ray—may be used. The *four-ray model* is best suited for urban area scenarios. In this model two additional rays are considered which are reflected off buildings located on both sides of the street. Figure 6.4(a) shows the geometry of the four-ray model. It is clear from Figure 6.4 that the buildings at both sides of the street form a canyon-like scenario producing a guiding effect on the transmitted waves.

The received signal may be calculated as the magnitude of the coherent sum of the rays for equally spaced points along the mobile route. The total field strength will follow the expression

$$|e_{Total}| = \left| e_{direct} e^{-j\frac{2\pi}{\lambda}d_{direct}} + R_{ground} e_{ground} e^{-j\frac{2\pi}{\lambda}d_{ground}} \right. \tag{6.1}$$

$$\left. + R_{wall-1} e_{wall-1} e^{-j\frac{2\pi}{\lambda}d_{wall-1}} + R_{wall-2} e_{wall-2} e^{-j\frac{2\pi}{\lambda}d_{wall-2}} \right|$$

Figure 6.4(b) shows simulation results at 900 MHz using the two- and four-ray models. The magnitudes of the reflection coefficients of building walls, R_{wall-i}, and the ground, R_{ground}, were considered to be one. The street width was 20m and the transmitter and receiver antenna heights were 5m and 1.5m, respectively. Also in Figure 6.4, the received signal under free-space ($n = 2$) conditions is shown for comparison.

Between 10m and approximately 300m both models show marked signal variations that approximately follow the free-space losses. Beyond the point where free-space and $n = 4$ losses are equal, the two- and four-ray models

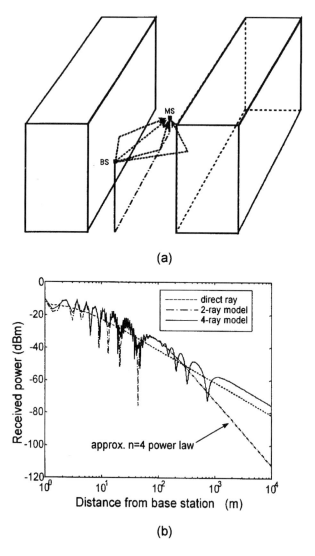

Figure 6.4 (a) Four-ray model geometry. (b) Comparison between the results of the two- and four-ray models [2].

diverge. The received signal profile generated with the two-ray model closely approximates a $n = 4$ power law while the profile generated with the four-ray model continues to follow the free-space curve but with a great variability.

The *breakpoint* or the point where free space losses equal those of the $n = 4$ power law is located at a distance from the base and defined by the expression (see Chapter 4's discussion of the plane Earth model):

$$d_{\text{Break Point}} = \frac{4\pi h_{\text{Base}} h_{\text{Mobile}}}{\lambda} \quad (6.2)$$

which, in the case of the example in Figure 6.4(b), is equal to

$$d_{\text{Break Point}} = \frac{4\pi h_{\text{Base}} h_{\text{Mobile}}}{\lambda} = \frac{4\pi\, 5\, 1.5}{0.33} = 283\text{m} \quad (6.3)$$

Figure 6.5 [2] shows both measured and simulated (two-ray model) received signal profiles for a highway microcell at 900 MHz. Very fast sampling was used in the measurements (every 33 mm). Later, samples were averaged in blocks of 400 (local mean). The same process was applied both to the measured and the simulated series. It can be observed that the model follows very closely the measurements. Also in Figure 6.5, a breakpoint can be observed in which the received signal decay slope changes.

A regression study was performed to verify what power law, d^{-n}, was followed by the path losses with distance. In Figure 6.6 [2] two slopes were fitted to the measured data in a different route, one defining a $d^{-0.51}$ law and the other a $d^{-3.67}$. The intersection of the lines with these slopes is the break point distance.

In Figure 6.7 [2] measurements and modeling results are shown using the four-ray model for an urban microcell route. It can be observed how the received signal follows the trend defined by the model, although the deep null

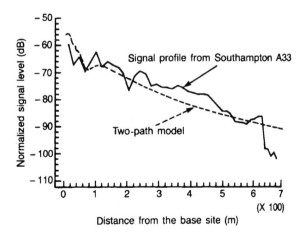

Figure 6.5 Measurements and two-ray model results in a highway microcell. (*Source:* [2]. Reprinted with the permission of *British Telecom Tech. Journal.*)

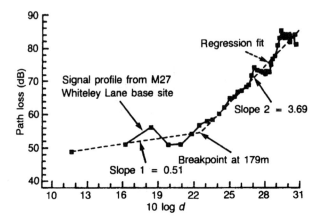

Figure 6.6 Variation of the path loss slope with distance. (*Source:* [2]. Reprinted with the permission of *British Telecom Tech. Journal.*)

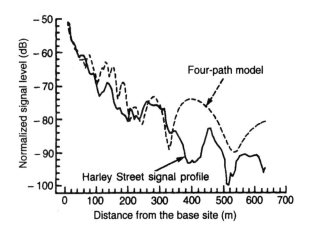

Figure 6.7 Measurements and four-ray model results for an urban microcell. (*Source:* [2]. Reprinted with the permission of *British Telecom Tech. Journal.*)

positions cannot be determined accurately. For the results shown in Figure 6.7, similar small area averaging procedures were used.

The received signal from conventional macrocells undergoes fast and deep fades due to multipath. The worst case appears when there is no direct LOS between the base and the mobile. In microcells, the LOS condition will be the most frequent one. It is thus expected that the received signal variability will be less marked than in conventional cells. The received signal's fast variations follow a Rice distribution that can be characterized by the c/m (carrier-to-multipath) ratio or k-factor:

$$k = \frac{c}{m} = \frac{a^2}{2\sigma^2} \qquad (6.4)$$

where a is the direct signal's magnitude and $2\sigma^2$ is the average multipath power. Ricean probability density functions are shown in Figure 6.8 for different values of k. In case $k = 0$, the Rice distribution becomes a Rayleigh distribution. Values of k of up to 32 (15 dB) have been observed experimentally. As it can be seen in Figure 6.8, the larger k is, the smaller the signal variability will be.

Figure 6.9 [2] shows a measured received signal levels for a highway microcell with a base station located on an overpass and a conventional macrocell located in the same site. Both the received power and the k-factor are plotted. Figure 6.9 clearly shows how the received signal profile for the microcell steadily decreases with distance while the macrocell signal is more constant over the stretch of the route measured. It can also be observed how the k parameter values for the macrocell are close to 0 (*Rayleigh fading*) while, for the microcell, k values will be high (*Rice fading*). This means that signal fades will be smaller.

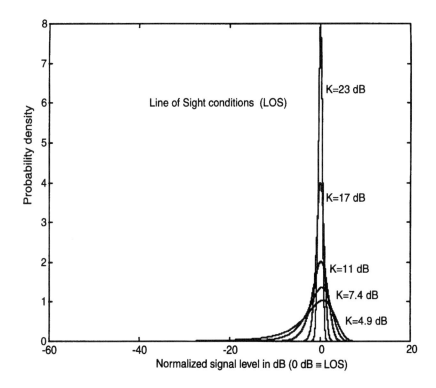

Figure 6.8 Rice distributions for different values of the k-factor [2].

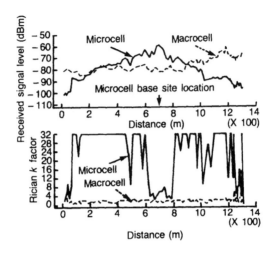

Figure 6.9 Measured signal level and k-factor in a microcell and a macrocell. (*Source:* [2]. Reprinted with the permission of *British Telecom Tech. Journal.*)

Some distance away (100–200m) from the microcell base (cell fringe), k falls sharply.

An urban microcell case is illustrated in Figure 6.10 [2] where occasional locations with low k values are observed. This is basically due to the existence of multipath off building faces which is much more intense than that generated by roadside trees in highway microcells.

6.2.2 Empirical Propagation Models for Urban Microcells

In practice, instead of using the two- or four-ray models for predictions, empirically derived propagation models are normally used. The model parame-

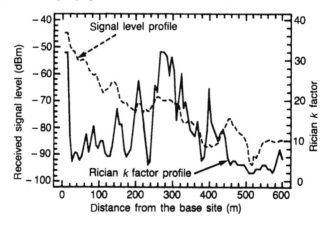

Figure 6.10 Measured signal level and k-factor in an urban microcell. (*Source:* [2]. Reprinted with the permission of *British Telecom Tech. Journal.*)

ters presented in this section have been derived from measurements carried out in the 1,700-MHz band in Stockholm where streets presented a rectilinear pattern with buildings of five to seven stories with parallel streets and occasional open areas. The terrain variations in the measurement areas were very small. The transmission antennas were placed below rooftops at 5m, and the receiving antenna was placed on a van at 2.2m. These measurements correspond to work carried out in the frame of the European Union Project RACE Phase I - R 1043 [3].

A two-slope model for path losses was obtained from the analysis of measured data. In the model, for short distances and LOS conditions (Figure 6.11), path losses can be described using the expression

$$l(x) = Mx^n \qquad (6.5)$$

For free-space conditions, $n = 2$ and $M = 10^{3.7}$ at $f = 1,700$ MHz. The variable x is the base-to-mobile distance. The losses between isotropic antennas expressed in decibels are given by

$$L_b(x) = 37 + 10n \log x(m) \text{ (dB)} \qquad (6.6)$$

To introduce two slopes in the model, a second term must be added for distances beyond the *breakpoint* x_L. Path losses may thus be expressed in the following way:

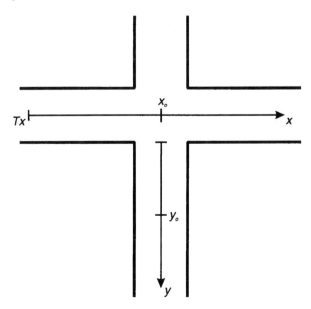

Figure 6.11 Geometry of LOS and non-LOS (NLOS) links in urban microcells [3].

$$l_1(x) \alpha x^{n_1} \qquad \text{for } x < x_L \qquad (6.7)$$

$$l_2(x) \alpha x^{n_1}\left(\frac{x}{x_L}\right)^{n_2-n_1} \qquad \text{for } x > x_L$$

where α means proportional to and the M factor has been left out for simplicity. In order to introduce a smooth transition between the two slopes in the model, an appropriate combination must be used to go from l_1 to l_2:

$$l(x) = [l_1(x)^p + l_2(x)^p]^{1/p} \qquad (6.8)$$

The p parameter is a *shape factor* that determines how abrupt the transition between both slopes is. A typical value may be $p = 4$. Note that at the breakpoint x_L, both losses l_1 and l_2 are equal.

After turning a corner into a side street, the mobile goes into the NLOS condition since the direct ray will no longer be available. Path losses in this case will be modeled in much the same way as in the LOS case assuming the base is placed in a hypothetical position facing the NLOS street at a distance y_{E_1} from the corner. Now, the model will have the following expression, assuming e_1 and e_2 are the path losses in linear units

$$e_1(y) = \left[\frac{y}{y_{E_1}} + 1\right]^{n_{E_1}} \qquad \text{for } y < y_{E_2} \qquad (6.9)$$

where y is the distance taken perpendicularly to direction x (Figure 6.11). After the break point at y_{E_2}, another approximation to the path losses shall be used as indicated by the expression

$$e_1(y) = \left[\frac{y + y_{E_1}}{y_{E_1} + y_{E_2}}\right]^{n_{E_2}} \left[\frac{y_{E_2}}{y_{E_1}} + 1\right]^{n_{E_1}} \qquad \text{for } y > y_{E_2} \qquad (6.10)$$

In the same way as for the LOS case, both equations are chosen such that they present the same value at the breakpoint y_{E_2}, and they are combined using the expression

$$e(y) = [e_1(y)p + e_2(y)p]^{1/p} \qquad (6.11)$$

In this way, a set of simple models are available to compute path losses by selecting different values of y and n for a variety of environments. The total path loss between a base station and a mobile located at y_o (Figure 6.11) will then be

$$l_{\text{Total}}(x_o, y_o) = Ml(x_o)e(y_o) \quad (6.12)$$

where M (intercept loss) represents the total loss at a reference distance $x = 1\text{m}$ (37 dB under free-space conditions at 1,700 MHz and assuming isotropic antennas).

In Table 6.1, experimentally derived parameters for LOS paths are given. For NLOS paths, the situation is extremely variable depending on the configuration of the urban area under study. In Table 6.2, empirically derived parameters are given for different urban area configurations.

Table 6.1
Empirically Derived Model Parameters for LOS Paths [3]

	n_1	n_2	x_L
Mean	2.24	5.58	254.8
Variance	0.24	0.53	72.5

Table 6.2
Empirically Derived Model Parameters for NLOS Paths [3]

Width of Base Station Street (m)	Width of Mobile Street (m)	Distance Base Station–Street Corner (m)	n_{E_1}	n_{E_2}	y_{E_1}	y_{E_2}
18.5	13.6	81	3.80	5.23	14.30	163.0
18.5	18.1	205	2.34	2.84	2.60	65.3
18.5	18.0	291	3.31	6.31	11.60	120.6
18.5	14.0	359	0.10	2.57	0.01	3.1
18.5	18.0	425	0.10	2.78	0.10	5.8
15.8	18.5	32	0.16	2.54	4.44	1.3
13.6	18.5	44	2.60	4.15	17.50	181.0
18.1	18.5	40	1.57	4.60	1.25	130.0
18.5	17.9	535	2.10	2.10	1.69	39.3
17.9	18.5	48	3.17	3.28	8.80	448.1
17.9	18.1	92	3.65	4.60	14.20	121.0

6.3 Indoor Propagation

Good propagation prediction methods [3, 4] for paths where the transmitter and the receiver are both located indoors are not only necessary for cellular systems but also for other wireless applications like cordless systems, radio LANs, or wireless-PABXs. Indoor propagation conditions differ considerably from those found outdoors. For one thing, the distances between the transmitter and the receiver are much smaller due to the high attenuations introduced by walls, floors, and furniture. Furthermore, propagation distances are short since small powers are used. Additionally, the short propagation ranges will help to reduce the time dispersion in the channel. Also, the time variability will be smaller.

Typically, available models mainly concentrate on the 900 and 1,700–1,800 MHz bands. Although models and measurements are available for other bands as for example, 5 GHz for ETSI's HIPERLAN applications. These models can be classified into four categories: narrowband empirical models, wideband empirical models, time variability models, and deterministic models. Empirical models provide simple mathematical expressions to compute path losses for a given distance and environment configuration. These models have been derived by means of regression techniques. This text will deal with these models only. As for wideband models, these provide the user with tables in which, for different building and hall configurations, PDPs and delay spread values are given. Deterministic or site-specific models, on the other hand, are based on ray-tracing or ray-launching techniques. These types of models were introduced in Chapter 4. These models are capable of providing not only path loss but also signal variability and time dispersion predictions.

6.3.1 Indoor Picocell Propagation Models

In much the same way as for urban microcells, for indoor scenarios different path types are acknowledged by empirical propagation models:

- LOS: In these paths there is a total or partial view from the base to the mobile (e.g., open rooms or halls with or without objects blocking the LOS).
- Obscured LOS (OLOS): In this case, thin walls or objects block the signal path.
- NLOS: In this case, there are thick walls blocking the signal path.

It is assumed that the transmitter and the receiver are on the same floor at a distance d in the range of 20–100m. The results presented are valid for the 1,700-MHz band [3]. These losses will vary with distance. Two terms may be considered in the model

$$L \text{ (dB)} = M(d) + X(l, t) \quad (6.13)$$

where $M(d)$ represents the median path loss value at a given distance d and $X(l, t)$ is a zero mean random variable that describes the uncertain behavior of the channel (i.e., time and locations variations). The distance-dependent losses may be expressed in the following way

$$M(d) = 10n \log(d) + K \text{ (dB)} \quad (6.14)$$

where d is the distance, in meters, from the reference point to the receiver. The value n describes the loss increment with distance, and K is a constant. The value of n will depend on the shape of the propagation environment, including room dimensions, wall materials, sizes, and number of objects such as furniture and cabinets, blocking the transmit-receive path. Several models have been proposed to predict the median path loss value:

1. Power law:

$$M_1(d) = 10n_1 \log(d) + K_1 \text{ (dB)} \quad (6.15)$$

2. Free-space losses + linear attenuation:

$$M_2(d) = 20 \log(d) + \alpha d + K_2 \text{ (dB)} \quad (6.16)$$

3. Power law + linear attenuation:

$$M_3(d) = 10n_3 \log(d) + \alpha d + K_3 \text{ (dB)} \quad (6.17)$$

Model 2 assumes free-space losses plus an additional loss factor that increases linearly (dB/m) with distance. This factor depends on the type of environment considered. Model 3 assumes a fixed 0-dB loss value at the reference point (i.e., $K_3 = 0$), and, then, only two parameters must be quantified from experimental data, namely, the propagation law, n, and the linear attenuation coefficient, α. Table 6.3 presents parameters for the three models and for different cell types.

A more sophisticated model is the so-called multiwall model [4], which predicts path losses as the sum of the free-space losses and the losses due to

Table 6.3
Parameters for the Three Models and for Different Types of Cells [3] (at 1,700 MHz)

MODEL Cell Type	Power Law		Free Space + Linear Attenuation		Power Law + Linear Attenuation	
	n_1	K_1	α	K_2	n_3	α
LOS	3.17	−8.07	0.23	−1.05	2.61	0.03
OLOS	4.36	−23.2	0.34	−0.57	2.26	0.24
NLOS	5.84	−34.0	0.63	−1.52	2.77	0.34

all walls and floors traversed in the base-to-mobile path. It was observed experimentally that floor-traversing losses were not a linear function of the number of floors. This feature is introduced in the model by means of an empirically derived factor b. The expression for this model is

$$L = L_{fs} + L_C + \sum_{i=1}^{I} k_{wi} L_{wi} + k_f^{\frac{k_f + 2}{k_f + 1} - b} L_f \qquad (6.18)$$

where

L_{fs}: Free-space losses;
L_C: Constant;
k_{wi}: Number of traversed type-i walls;
k_f: Number of traversed floors;
L_{wi}: Wall type-i losses;
L_f: Losses between adjacent floors;
b: Empirical parameter;
I: Number of different wall types traversed.

The constant term L_C will be close to zero. The third term represents the total losses due to paths through walls between the base and the mobile. For practical reasons the number of wall types considered must be small. In Table 6.4, a possible wall classification is given. It must be borne in mind that the loss factors in the table are not physical penetration loss values; rather they are coefficients obtained from multiple-regression studies. These loss factors implicitly include furniture effects and corridor and hallway guiding effects.

Within the framework of Euro-COST 231, the parameters listed in Table 6.5 for the different models were obtained from experimental data. The values in Table 6.5 correspond to the 1,800-MHz band.

Table 6.4
Wall Type Classification [4]

Type of Wall	Description
Light wall (L_{W1})	A wall that is not bearing load (e.g., plasterboard, particle board, or thin (< 10 cm), light concrete wall
Heavy wall (L_{W2})	A load-bearing wall or other thick (> 10 cm) wall made of concrete or brick, for example

Table 6.5
Indoor Propagation Model Parameters [4]

Environment	One-Slope model (1SM) K_1 (dB)	n	Multiwall Model (MWM) L_{W1} (dB)	L_{W2} (dB)	$L_f(I)$ (dB)	b	Linear Model (LAM)
Dense							
One floor	33.3	4.0	3.4	6.9	18.3	0.46	0.62
Two floors	21.9	5.2	—	—	—	—	—
Multi floor	44.9	5.4	—	—	—	—	2.8
Open	42.7	1.9	3.4	6.9	18.3	0.46	0.22
Large	37.5	2.0	3.4	6.9	18.3	0.46	—
Corridor	39.2	1.4	3.4	6.9	18.3	0.46	—

Other features observed in the received signal are its variability and depolarization:

- *Variability:* The slow signal variations within a building may be described using a log-normal distribution with a standard deviation in the range 2.7–5.3 dB, while the fast variations, in most cases, can be described by means of a Rayleigh distribution. However, Rice-type variations have also been observed.
- *Depolarization:* Typically, the terminals used for indoor applications are hand-held. Generally, their antennas will not be vertically oriented. This means that polarization losses will affect the received signal. Moreover, the transmitted wave as it propagates through walls and other obstacles will undergo polarization leaks, meaning that part of the energy in the vertically polarized wave will reach the receiver in the horizontal component. The following loss parameter may be used to characterize this phenomenon

$$X_{\text{POL}} = P_h - P_v \qquad (6.19)$$

where

P_h is the mean received power for a horizontally oriented antenna.
P_v is the mean received power for a vertically oriented antenna.

The X_{POL} parameter ranges from -2 to -14 dB in indoor environments. Signal depolarization will normally be correlated with the traversed material losses (i.e., the larger the attenuation, the greater the received signal on the cross-polar component will be). This effect will help reduce the influence of hand-held terminal antenna orientation.

6.4 Indoor Coverage Using Outdoor Base Stations— Penetration Losses

For an adequate development of cellular (and other wireless) networks, it is of great interest to provide good coverage levels within buildings using base stations located outdoors. A great effort in the past few years has been carried out to quantify penetration losses into buildings. Two possible situations may be considered:

- The base station defines a macrocell (i.e., its antenna is well above neighboring building rooftops).
- The base station creates an urban microcell.

Current models make use of the geometrical parameters in Table 6.6 whose definitions are clarified with the aid of Figure 6.12.

Macrocell models: For the computation of penetration losses in macrocells, the following expression shall be used

$$L_{\text{Total}} \text{ (dB)} = L(d) + L_{we}(v_i) + n_w L_{wi} - n_f G_h \qquad (6.20)$$

where L_{we} and L_{wi} can be read from Table 6.7 and $G_h = 2$ dB/floor. If there is no direct LOS between the base and the building, the received power at each building face will be considered to be the same. In this case, internal walls must be numbered from the external wall they are closest to (and not with respect to the wall closest to the base). The model shall be used with care when many internal walls separate the mobile and the base. If there exists an

Table 6.6
Indoor Coverage From Outdoor Bases—Model Parameters [3]

L_{tot}	Total path loss from base to mobile
$L(d)$	Path loss up to building; (d) indicates distance dependence
L_{we}	External wall attenuation (Table 6.7)
L_{wi}	Internal wall attenuation (Table 6.7)
n_w	Number of internal walls separating transmitter and receiver
n_f	Floor number; ground floor is zero
G_h	Height gain per floor; height gain defined as the increase in received power when lifting the receiver to a higher floor
v_i	Angle of incidence (Figure 6.12(b))
v_h	Deviation from horizontal plane (Figure 6.12(c))

open area or a corridor parallel to the direct propagation path, the model will overestimate the losses.

Models for microcells with direct LOS between the building and the base station. For the computation of the propagation losses, in this case, the following expression will be used assuming a small horizontal deviation v_h ($v_h < 10$ degrees)

$$L_{\text{Total}} \text{ (dB)} = L(d) + L_{we}(v_i) + n_w L_{wi} \qquad (6.21)$$

where L_{we} and L_{wi} may be read from Table 6.7. If the transmitter were placed very close to a building or the receiver were high up in the building, then v_h would be large, and, in this case, the transmitter vertical radiation pattern should be taken into account together with the actual geometrical distance for the computation of the external losses. The incidence angle increment will increase the penetration losses with respect to those given in Table 6.7, and it is necessary to account for them. The losses in this case are given by the expression

$$L_{\text{Total}} \text{ (dB)} = L(d_{\text{geometric}}, \text{vertical radiation pattern}) + L_{we}(v_i) + n_w L_{wi} \qquad (6.22)$$

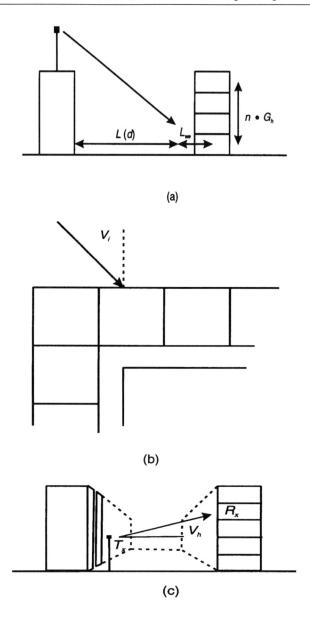

Figure 6.12 (a) Model geometry. (b) Top view, incidence angle. (c) Side view, horizontal plane deviation [3].

Table 6.7
Wall Attenuation for Different Materials [3]

Wall Type	L_{Wall} (dB) Min.	L_{Wall} (dB) Typical	L_{Wall} (dB) Max.
Thick (25-cm) concrete with large windows	4	4	5
Thick (25-cm) concrete with large windows and large incidence angle	9	11	12
Thick concrete, no windows	10	13	18
Double (2 × 20 cm) concrete; indoor	14	17	20
Thin (10 cm) concrete; indoor	3	6	7
Brick wall, small windows	3	4	5
Steel wall (1 cm) with large reinforced windows	9	10	11
Glass wall	1	2	3
Reinforced glass	7	8	9
Concrete (20 cm) large windows		5.4	
Concrete, 30 cm		9.4	
Bricks, 63 cm		4.0	
Bricks, 70 cm		4.5	
Bricks, 70 cm		9.1	
Porous Concrete		6.6	

Note: Unless otherwise noted, values are for 0-degree incidence. In addition, no mathematical formula is provided for the angle of incidence dependency since not enough experimental data were still available.

References

[1] McFarlane, D. A., and S. T. S. Chia, "Microcellular Mobile Radio Systems," *British Telecom Technology Journal*, Vol. 8, No. 1, January 1990, pp. 79–83.

[2] Green, E., "Radio Link Design for Microcellular Systems," *British Telecom Technology Journal*, Vol. 8, No. 1, January 1990, pp. 85–96.

[3] Paulsen, S. E., et al., *Radiowave propagation model document*, Race Project R1043. Ref. RMTP/CC/R1576, London, October 12, 1991.

[4] Evolution of land mobile radio (including personal) communications, Euro-COST 231 Project, Final Report Draft, 1996.

7

Base Station Engineering

7.1 Introduction: The Antenna System

This chapter describes the necessary elements for the connection of several transmitters and receivers to a common antenna. Trunked and cellular systems transmit several control and traffic channels that must be coupled to the same antenna system. To achieve this, specific RF devices, such as transmitter combiners, duplexers, and receiver multicouplers, are used. In Figure 7.1, two typical arrangements for a multichannel system are represented; Figure 7.1(a) shows an arrangement using two antennas, one for the transmitters and the other for the receivers, and Figure 7.1(b) shows an arrangement using a single antenna for both transmitters and receivers, coupled to the antenna by means of a duplexer. The choice between these alternatives is usually made considering economic factors, system gain, and space restrictions on masts or in equipment racks or shelters.

7.2 Transmitter Combiners

Combiners enable several transmitters with relatively close frequencies to be connected to a common antenna to produce the required isolation ensuring that their mutual coupling is very small. This means that the coupling loss A_c (see Chapter 5) for each pair of transmitters must be as high as possible. A combiner is a unidirectional device that selectively transfers energy from each transmitter to a common antenna, thus minimizing the injection of energy into other transmitters.

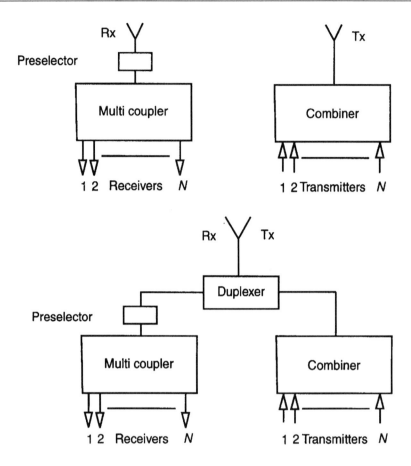

Figure 7.1 Connection options of a multichannel system to common antenna racks: (a) using two antennas and (b) using one antenna.

A typical combiner configuration uses ferrite isolators and hybrid unions. The former are sometimes called RF diodes, because they have a low insertion loss (typically 0.5 dB) in one direction and tens of decibels (on the order of 30 dB) in the other direction. Hybrid unions enable the connection of two transmitters to a common output port with a high isolation (25–40 dB) between them; however, they require a terminating load in which, theoretically, half of each transmitter's power is absorbed. As a consequence, a hybrid union has a minimum insertion loss of 3 dB, which amounts to approximately 3.5 dB in practice. These losses restrict the number of transmitters that can be combined using isolators and hybrids.

In Figure 7.2, a four-transmitter combiner is represented. It has four ferrite isolators and three hybrids. Assuming that these are correctly impedance

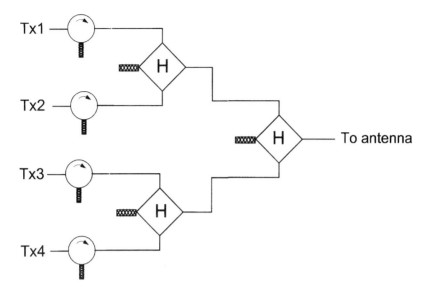

Figure 7.2 Four-Tx combiner with isolators and hybrids.

matched to the antenna, the total isolation between the transmitters connected to the same hybrid will be 70 dB (30 dB provided by the isolators and 40 dB by the hybrid). The insertion loss for each transmitter will be 7.5 dB (3.5 + 3.5 + 0.5) since the signal goes through two hybrids and one isolator. This means that, in the combiner, 82% of the power is lost; only the remaining 18% of the power is transferred to the antenna.

$$7.5 \text{ dB} = 10 \log\left(\frac{p_o}{p_i}\right) \quad \text{or} \quad \frac{p_o}{p_i} \approx 0.18 \qquad (7.1)$$

The use of circulator plus hybrid combiners has the following advantages:

- Flexibility: Transmitters with small frequency separations, including adjacent channels, can be used;
- Low size.

Their disadvantages are the following:

- High insertion losses, with the subsequent reduction of the power delivered to the antenna;
- Unidirectionality, allowing them to be used only for connecting transmitters to the antenna;
- High cost.

Using this type of combining system, four transmitters with correlative channels (that is, for example, channels 1, 2, 3, 4) can be connected to the same antenna. From the table of intermodulation-free channels (Table 5.5) it can be observed how, without this combiner, the channels to be used would be 1, 2, 5, 7. A saving of three channels, 5, 6, 7, is achieved. These channels could be assigned to other users.

Figure 7.3 represents the typical characteristics of an isolator: isolation, and insertion loss.

An alternative configuration for the combination of several transmitters with a low insertion loss is based on the use of resonant cavity filters with a high Q-factor, either in bandpass mode (cavities in series) or bandpass rejection mode (cavities in parallel).

Passband Cavities

The main role of these cavities is to provide impedance matching in such a way that the RF signal coming from each transmitter goes through its corresponding cavity suffering small losses, while the signals from other transmitters suffer high losses. These features are dependent on the cavity selectivity. The value of this parameter determines the minimum frequency separation between the transmitters connected to the same antenna.

Figure 7.4 shows an example of the use of bandpass cavities for the coupling of two transmitters and two receivers to the same antenna. This example has been selected to highlight two important features of cavity combiners:

- The transmitters must work at fairly different frequencies (adjacent channels cannot be used), as in the isolator plus hybrid configuration.
- The device is bidirectional; that is, transmitters and receivers can be attached to it.
- The insertion loss of the cavities is a function of the loaded and unloaded Q values (Q_L, Q_U) according to the expression:

$$L \text{ (dB)} = 20 \log\left(1 + \frac{Q_L}{Q_U}\right) \qquad (7.2)$$

The cavity selectivity increases with Q_L as illustrated in Figure 7.5. Nevertheless, the insertion loss also increases, as can be observed from (7.2). From Figure 7.5, it can be observed how, in general, a single cavity cannot provide the required isolation. For example, for $Q_L = 1,000$ and an offset of 3% from the tuning frequency (13.8 MHz at the 460-MHz band), the isolation is 35 dB, which is half of the isolation achieved with a hybrid combiner.

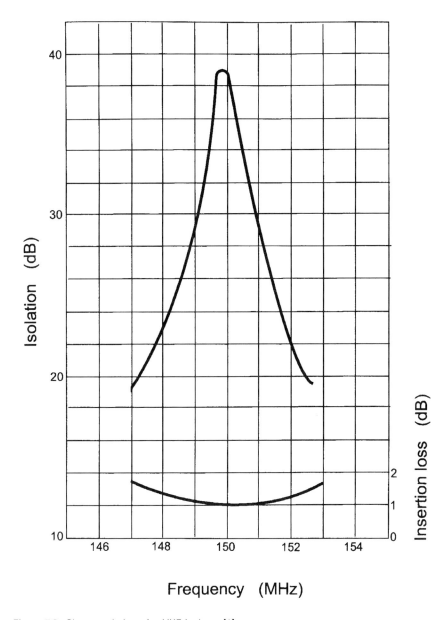

Figure 7.3 Characteristics of a VHF isolator [1].

The isolation can be increased by connecting two or more cavities in series. The total insertion loss is then the sum of the insertion losses of each cavity. For example, with three cavities a total loss of 1.5 dB (0.5 dB per cavity) is obtained. Thus, the use of cavity filters permits a reasonable tradeoff

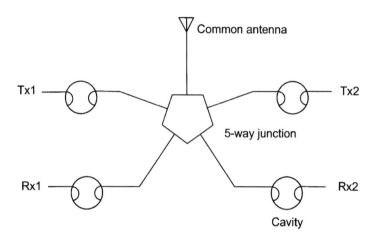

Figure 7.4 Arrangement of four bandpass cavities for the connection of two transceivers.

between selectivity, isolation, insertion loss, and cost. However, this option requires much more space for its installation at a radio station.

In the case of mixed use of transmitters and receivers, the cavities have the advantage of reducing the amount of transmitter noise reaching the receivers. In this case, receiver desensitization (blocking) due to the operation of nearby transmitters is alleviated.

Band-Reject Cavities

These cavities are connected in parallel with the transmitter-antenna path. They present a low impedance at those frequencies that are to be attenuated and a high impedance at the working frequencies (anti-resonance). Their selectivity is higher than that of passband cavities. This means that they can be used by transmitters with lower frequency separations. The insertion loss in this case is given by

$$L \text{ (dB)} = 20 \log\left(1 + \frac{Q_U}{Q_L}\right) \tag{7.3}$$

Figure 7.6 shows an example of the connection of three transmitters to a common antenna through band-reject cavities. Each Tx-antenna branch has in parallel cavities that are "traps" to the frequencies to be eliminated.

As in the case of passband cavities, a single band-reject cavity does not provide a sufficient isolation, and at least two cavities in series must be used for each frequency trap. The total number of cavities in the case of Figure 7.6 would be 10, two more than in the passband case. If the number of transmitters

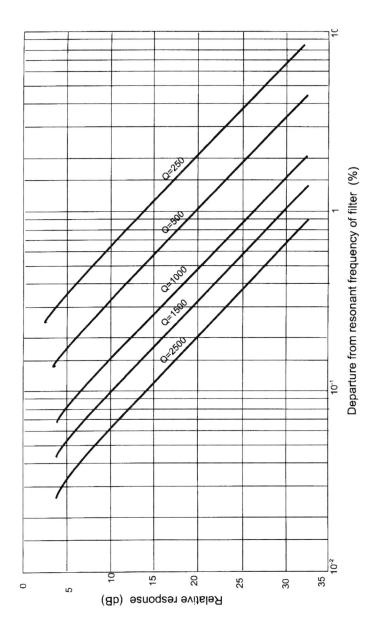

Figure 7.5 Responses of passband cavities (ITU-R).

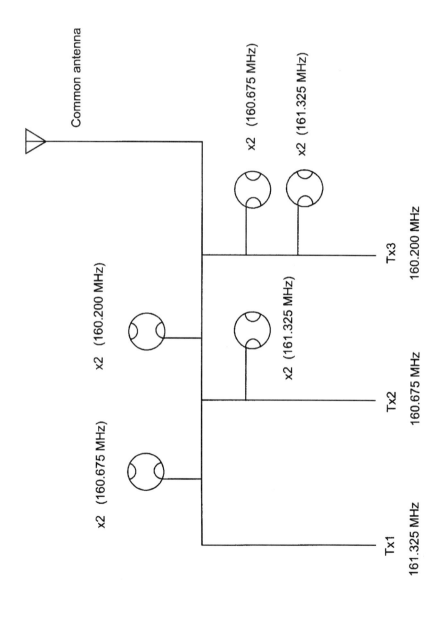

Figure 7.6 Band-reject cavity combiner for three Tx.

were four, then it would be necessary to add four more double cavities. This increment sets a limit to the number of transmitters that can be coupled to a common antenna with the band-reject configuration.

Cavity-Isolator Combiners

This is a mixed approach to transmitter combining. The use of isolators enables the coupling of transmitters with smaller frequency separations, while cavities introduce low insertion losses. This type of combiner is unidirectional; that is, it is only valid for the connection of transmitters to a common antenna. Figure 7.7 shows a combiner of this type for four transmitters. The cavities match the impedance of ports 1-2-3-4 of the five-way junction. The connections of the cavities to the junction are made by means of coaxial cables with a length of $\lambda/4$.

The insertion loss is the attenuation between the output port of each transmitter and the output port of the five-way junction when the frequency offset between transmitters is the same. Figure 7.8 represents this insertion loss as a function of the frequency separation for two four-transmitter combiners, one for VHF (150 MHz) and the other for UHF (450 MHz). A straight line

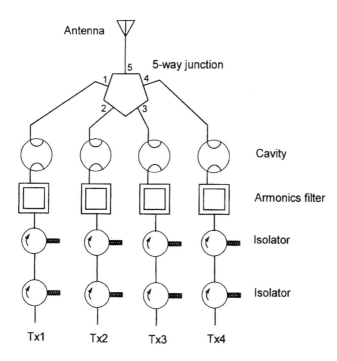

Figure 7.7 Cavity-isolator combiner for four Tx.

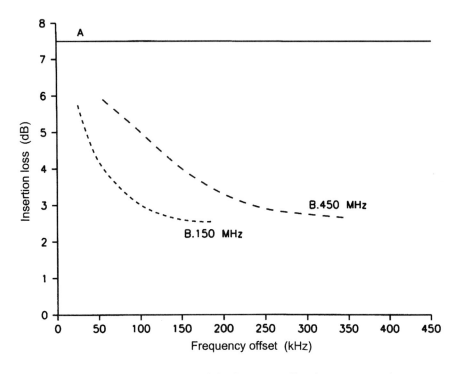

Figure 7.8 Insertion loss as a function of the frequency offset between transmitters.

(A) representing the 7.5-dB losses for a hybrid combiner has also been drawn for comparison purposes. If the frequency offset between the different transmitters is increased, the loss in all the circuits is reduced.

Now, two examples where the insertion losses and the equivalent radiated powers are computed for this type of combiner are presented. This is done to compare its performance with that of a hybrid combiner in the VHF and UHF bands. A channel separation of 50 kHz has been assumed and two to four transmitters have been considered. It is assumed that the transmitters deliver a power of 100W (20 dBW) and that antennas with typical gains in these frequency bands are used. The losses in the antenna feeders have been neglected (Table 7.1).

In addition to their greater flexibility to connect terminals with small frequency separations and with a reduced insertion loss, cavity combiners also attenuate the spurious emissions and noise produced by the transmitters connected to them. This contributes to maintaining a "clean" electromagnetic spectrum in the vicinity of a radio station, so that the nuisance over receiving systems located in the neighborhood is greatly reduced.

Table 7.1
Performance of Different Combiners for 2, 3, and 4 Tx

High VHF band: 150 MHz Antenna gain: 7 dBd Transmitter power: 100W 20-dBW Carrier spacing: 50 kHz				
Combiner type		2 Tx	3 Tx	4 Tx
Hybrid union	Insertion loss (dB)	4	7.5	7.5
	e.r.p. (W) (dBW)	199.5 (23)	89.1 (19.5)	89.1 (19.5)
Cavity	Insertion loss (dB)	4.3 (*)		
	e.r.p. (W) (dBW)	186.2 (22.7) (*)		
(*) Independent of the number of transmitters				
UHF band: 450 MHz Antenna gain: 9 dBd Transmitter power: 100W 20-dBW Carrier spacing: 50 kHz				
Combiner type		2 Tx	3 Tx	4 Tx
Hybrid union	Insertion loss (dB)	4	7.5	7.5
	e.r.p. (W) (dBW)	316.2 (25)	141.2 (21.5)	141.2 (21.5)
Cavity	Insertion loss (dB)	6 (*)		
	e.r.p. (W) (dBW)	199.5 (23) (*)		
(*) Independent of the number of transmitters				

7.3 Antenna Multicouplers

Antenna multicouplers enable the connection of several receivers to a common antenna. They are composed of three basic elements:

- Passband filter;
- Preamplifier;
- Power splitter.

Figure 7.9 shows a multicoupler block diagram. The passband filter preselects the frequencies to be received, rejecting those that are out of the desired band. The preamplifier compensates for the losses of the antenna feeder and the other multicoupler elements. The power splitter delivers the same RF power to each receiver. Figure 7.9 also represents the frequency response of a typical multicoupler. Table 7.2 summarizes some characteristics of a VHF multicoupler.

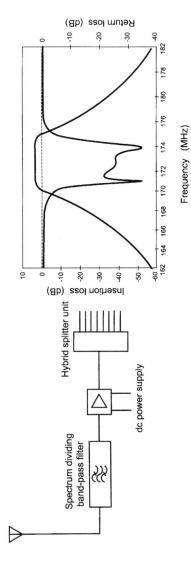

Figure 7.9 Block diagram of a multicoupler and typical multicoupler response [1].

Table 7.2
Typical Multicoupler Characteristics [1]

Frequency range	169.0–173.0 MHz
Bandwidth	3.2 MHz
Impedance	50Ω
V.S.W.R.	Better than 1.2:1
Gain	1–12 dB (preselectable)
Noise figure	4.5 dB
Two-tone intermodulation	−60 dB for −20 dBm inputs
Maximum input levels	+1 dBm
Temperature range	−20 degrees to +50 degrees Celsius
Connectors	N-female
Power	110V, 200–260V one phase, 40–60 Hz and 24 Vdc
Rx-to-Rx isolation	> 20 dB
MTBF	> 150,000 hours
Dimensions	483 × 250 × 88 mm
Weight	6.5 kg

7.4 Duplexers

Duplexers are devices that enable the use of a common antenna for simultaneous transmission and reception (Figure 7.10). Sometimes, due to space restrictions in masts or equipment racks, a single antenna must be used in a transceiver system. A duplexer is basically composed of two bandpass filters corresponding to the transmission and reception sub-bands. Figure 7.10 represents a duplexer showing the selectivity characteristics of the band-dividing filters for a VHF radio system [1]. When a duplexer is installed, the receiver is isolated from the transmitting frequencies, and the transmitter noise is attenuated at the receiving frequencies. Figure 7.10 also illustrates the frequency response of a bandpass duplexer. It is also possible to use band-reject duplexers. As an example, Table 7.3 shows the main features of bandpass and band-reject high-VHF band duplexers.

7.5 Feeders

Transceivers are connected to the antenna by means of feeders. Feeders are low-loss coaxial cables with a 50Ω impedance. These cables must withstand humidity, so they are sometimes pressurized. The cables must have some bending capability (specified as a minimum bending radius) for their laying from the equipment's shelter via the mast to the antenna system. Two kinds of cables can be found in mobile communications installations:

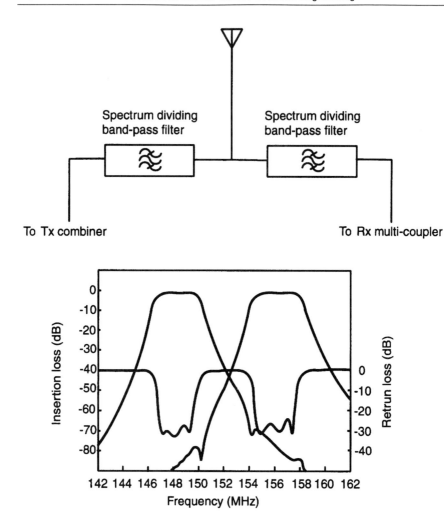

Figure 7.10 Duplexer block diagram and characteristics of a bandpass duplexer [1].

- Semirigid coaxial cables (aluminum or copper);
- Flexible coaxial cables.

As the distance from the transceiver housing to the antenna is large, flexible mesh (braided) coaxial cables cannot be used due to their high attenuation and limited power rating. These cables are mainly used in mobiles. Semirigid cables have greater size and smaller attenuation (Table 7.4) and are used in radio stations (repeaters, base stations). These cables have a corrugated outer conductor and use air as dielectric with a polyethylene helix to maintain the separation

Table 7.3
VHF Duplexer Characteristics [1]

Parameter	Passband	Band-reject
Frequency range	Tx 165 MHz–168.2 MHz Rx 169.9 MHz–173 MHz	140–174 MHz
Bandwidth	3.2 MHz	Tx 1.5 MHz Rx 1.5 MHz
Insertion loss	Tx 1.0 dB Rx 1.7 dB	1.3 dB
Impedance	50Ω input and output	50Ω input and output
V.S.W.R.	better than 1.2:1	better than 1.2:1
Tx/Rx/Tx isolation	>40 dB at both ends of the band >70 dB for a 4.5 MHz separation	>85 dB
Temperature range	−20° to +55°	−20° to +55°
Connectors	N-type	N-type
Maximum power	400W (CW)	50W (CW)
Dimensions	400 × 447 × 176 mm	220 × 200 × 25 mm
Weight	16 kg	1.2 kg

Table 7.4
Characteristics of Feeders

Type	Diameter	Attenuation (dB/100m)			Mean Pwr (kW) @ 40° Ambient		
		30 MHz	400 MHz	1000 MHz	30 MHz	400 MHz	1000 MHz
Semirigid copper foam	1/2"	1.21	4.59	7.54	5.0	1.3	0.80
	7/8"	0.65	2.62	4.49	10.5	2.8	1.7
	1-1/4"	0.49	2.0	3.44	18	4.0	2.4
	1-5/8"	0.39	1.57	2.85	23	5.2	3.0
Semirigid copper air	1/2"	1.49	5.64	9.15	3.7	0.98	0.61
	7/8"	0.66	2.56	4.23	13	3.3	2.1
	1-5/8"	0.36	1.36	2.29	29	7.8	4.9
	3"	0.24	0.89	1.43	70	17	9.15

between the inner and the outer conductors. Alternatively, the dielectric can be of solid foam. Foam dielectric cables are used for moderate lengths, and air dielectric cables are used for longer connections. The linking of the main cable to the equipment or the antenna is made by means of short pieces of flexible cable on account of their greater bending capability.

Figure 7.11 shows a typical configuration of a radio base station. The different components of the antenna system and accessories can be seen in Figure 7.11.

7.6 Complete Antenna Connection System

Finally, to end this chapter, in Figure 7.12 a complete antenna connection system using a bandpass duplexer, a cavity-isolator combiner and a multicoupler is illustrated.

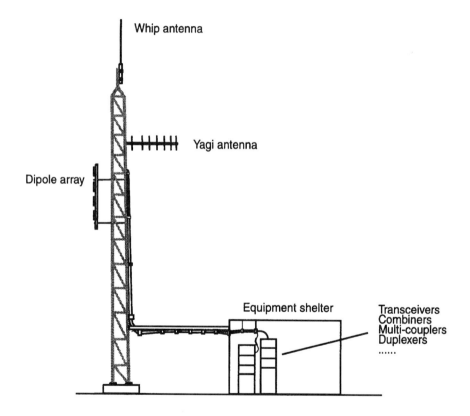

Figure 7.11 Base station components [1].

Base Station Engineering 273

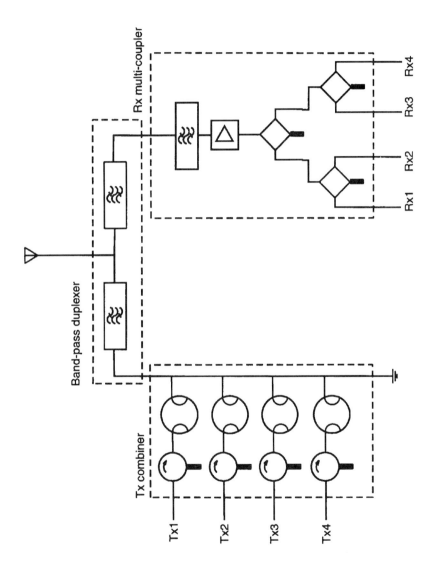

Figure 7.12 Illustration of a Tx-Rx-antenna connection system with a duplexer [1].

References

[1] Catalog, Aerial Facilities Limited, Latimer Park, Chesham, UK: Aerial House, 1995.

Selected Bibliography

ITU-R Report 1019, Sources of unwanted signals in multiple base station sites in the land mobile service.

III

Description of Different Mobile Standards

8

Trunked Systems

8.1 Introduction

Teletraffic theory [1] is used to evaluate the number of lines required to link two switching offices with a specified quality or Grade of Service (GoS) expressed in terms of the blocking probability p. The total offered traffic A, the number of lines N, and the blocking probability p are related through the well-known Erlang-B formula

$$p = B(A, N) \text{ or its inverse } A = B^{-1}(p, N) \qquad (8.1)$$

where B^{-1} denotes the inverse Erlang-B formula. The Erlang-B expression is

$$p = B(A, N) = \frac{\frac{A^N}{N!}}{\sum_{k=0}^{N} \frac{A^k}{k!}} \qquad (8.2)$$

This formula is also used for radio systems. To illustrate the concept and the gain obtained by a trunked radio system with respect to a set of individual systems, a comparison is made between a system in which all communications are served by a single N-channel radio system and the case in which there exist N single-channel systems, each serving its own traffic (A/N). In both cases the same blocking probability p is assumed. A can be calculated using the expression

$$A^1 = NB^{-1}(p, 1) \tag{8.3}$$

Comparing both results, it is clear that A is greater than A^1 for $N > 1$. This is due to the nonlinearity of the Erlang formula.

It can be concluded that it is far more efficient to offer the whole traffic demand to a set of channels (trunked system) than to split the traffic and offer it to individual single-channel systems (Figure 8.1). The traffic-handling capability increment (trunking gain) is also apparent when the traffic per channel is studied. Figure 8.2 represents these magnitudes (i.e., the total offered traffic and the traffic per channel for a blocking probability $p = 0.1$ when N varies from one to 10 channels).

These are the theoretical fundamentals of trunked systems. In these systems there is not a fixed assignment of channels to users but, rather, the assignment is dynamic. A radio channel is assigned to a user only when it is ready to communicate. This scheme minimizes the idle time of the available radio channels, since users only seize the communication channel during their call holding time [2]. When the communication ends, the channel is released and returned to a common pool so it can be assigned to other users. Figure 8.3 illustrates this concept. Trunked systems are fitted with a demand assignment multiple access (DAMA) protocol that makes it possible to implement the trunking concept.

So far, this book has dealt only with lost call trunked systems. In these systems, when all channels are busy, a new incoming communication request will be dropped and a new attempt to set up the communication, sometime later, is required. In PMR/PAMR applications, it is more common to set up waiting or queuing systems in which, if a call attempt is made when all channels are busy, the request is queued and it is served later, when a channel is released.

PMR/PAMR trunked systems are of the waiting type. The Erlang-C formula (see Section 8.4) is used instead of the Erlang-B formula to model their behavior. However, the conclusions presented before lost call systems are still valid as regards the efficiency obtained. Other radio networks (for example, cellular networks) are lost call trunked systems, and, for its dimensioning, the Erlang-B model is generally used to approximate its behavior (see Chapter 9).

The practical implementation of the trunking concept requires a sufficiently fast and intelligent call management system, implemented by means of some standardized protocol. In the United Kingdom, the MPT-1327 signaling protocol, which has become a "de facto" standard throughout Europe, is used for trunked systems. Standardization is necessary so that infrastructure equipment and terminals made by different manufacturers can operate together in the same radio network. Later in Section 8.3 the MPT-1327 signaling standard will be described in some detail.

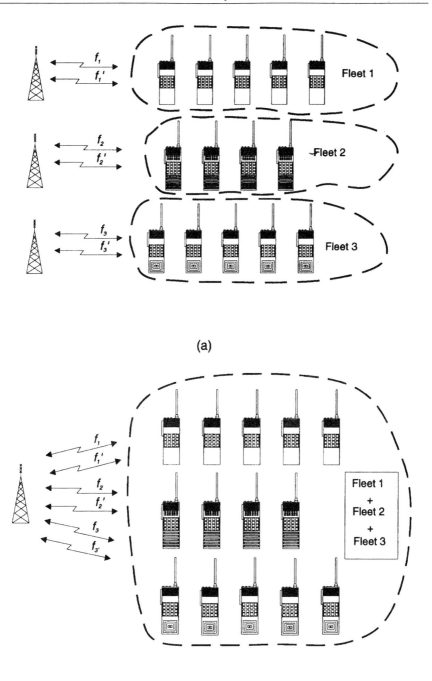

Figure 8.1 (a) Individual single-channel radio systems. (b) Trunked radio system.

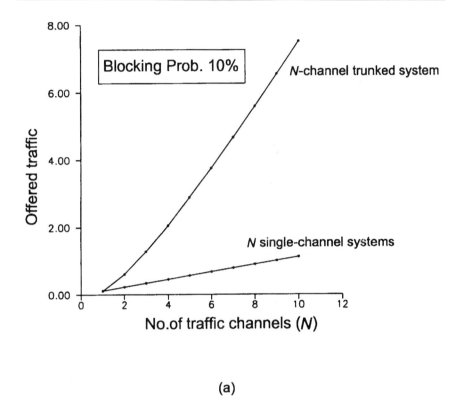

Figure 8.2 (a) Total offered traffic in an *N*-channel system and *N* single-channel systems. (b) Total offered traffic per channel in individual and trunked systems.

Given their efficient traffic handling features, trunked systems contribute to make a much better use of the radio spectrum. Two options are possible when setting up a trunked system: self provision or PAMR systems. Self-provision is justified when the traffic generated or the number of terminals in a single organization is large. Otherwise, a subscription to a PAMR operator is the best choice. In several countries throughout Europe a new band, VHF-Band III (formerly used for television broadcast), has been assigned for PAMR applications.

8.2 Elements in a Trunked Radio Network

Figure 8.4 illustrates a typical trunked radio network structure. Figure 8.4(a) shows its main elements and interfaces.

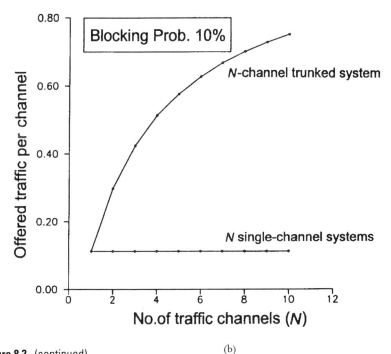

Figure 8.2 (continued). (b)

Subscribers: A, B, C, D, E

Figure 8.3 Demand assignment of radio channels [2].

All radio communications must be performed through a number of base stations operating in half-duplex mode. Communications between subscribers served by different base stations are routed through a network control center (NCC). NCCs are fitted with interfaces to dispatch consoles and public and/or private telephone networks. Figure 8.4(b) illustrates the evolution of a single NCC trunked network to a multi-NCC network with extended coverage and a large number of users (regional or nationwide systems). In Figure 8.4(b) a hierarchical and mesh-interconnected network is shown in which each NCC controls several base stations.

8.2.1 Network Control Center (NCC)

The main tasks carried out by the NCC are the following:

- Call routing between base station sites to allow communications between mobiles being served by different bases. This requires a mesh of links to transport the communications between the NCC and the different bases;
- Management of charging data;
- Updating users' location tables;
- Connection between dispatching positions and their corresponding mobiles;
- Remote supervision of base stations.

Figure 8.5 presents an example of a NCC consisting of three computers. Two of them, known as call control computers, would be configured in active and hot-standby mode for 1+1 redundancy. The third computer is devoted to operation and maintenance (O&M) functions and supplies the user interface for operations and maintenance.

Baseband signals to/from interface units are adequately routed at the switch matrix. This matrix allows the simultaneous handling of a number of individual and group calls. In this way, asymmetric calls (broadcast) can be carried out, and any number of terminals can be addressed in a group call.

8.2.2 Base Stations

A base station (Figure 8.6) consists of several units such as the following:

- Transceivers (one per channel);
- Line interface units (one per channel);

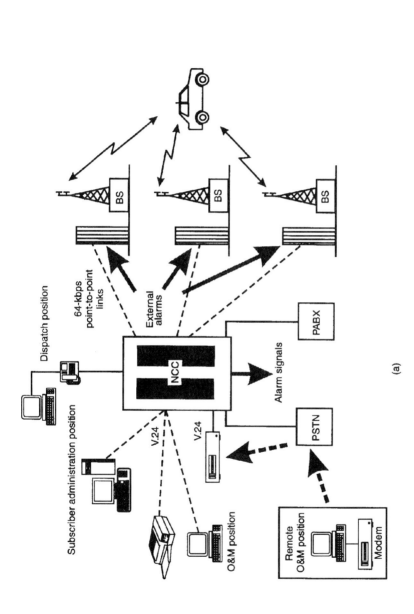

Figure 8.4 (a) Basic elements in a trunked radio network. (b) Evolution of a trunked radio network for greater traffic and extended service area.

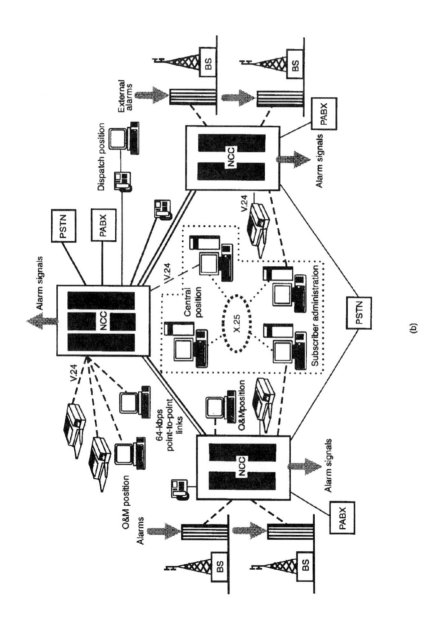

Figure 8.4 (continued).

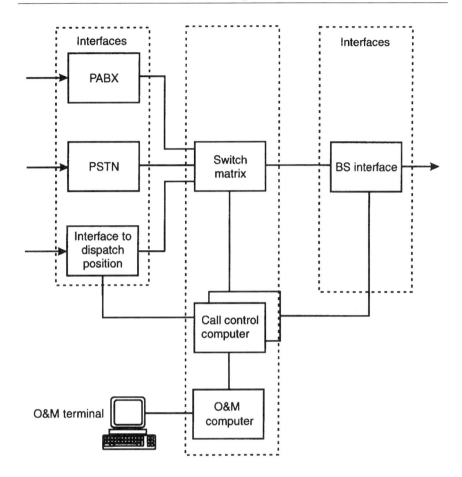

Figure 8.5 Simplified block diagram of a NCC.

- Power supplies (one per channel);
- Combiner(s);
- RF monitoring unit;
- Filter unit;
- Receiver multicoupler.

In the receive chain, a preselector filter is placed before the receiver multicoupler unit to limit the RF frequency band, thus attenuating out-of-band signals. In the case of using a single antenna to transmit and receive, a duplexer is required. The receiver multicoupler amplifies, splits, and routes the received signals to their corresponding receivers. The RF monitoring unit is

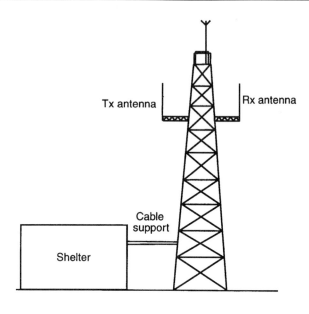

(a)

Figure 8.6 (a) Schematic diagram of a base station. (b) Typical configuration of a four-channel base station.

used to check the transmitted and reflected powers as well as the voltage standing wave ratio (VSWR). A service telephone is used to connect the base station to the NCC or to any of its radio channels for monitoring purposes.

A standard configuration will use two independent antennas in each site, one for the transmitters and the other for the receivers. Figure 8.6 presents a standard base station layout. The combination of different transmitters may be carried out by means of 3-dB hybrids or cavity filters. Using cavity filters, typically up to 24 channels can be coupled to a single antenna. Rarely more than eight channels can be combined by means of hybrid circuits given their high losses (Chapter 7). Normally, each transceiver is equipped with its own

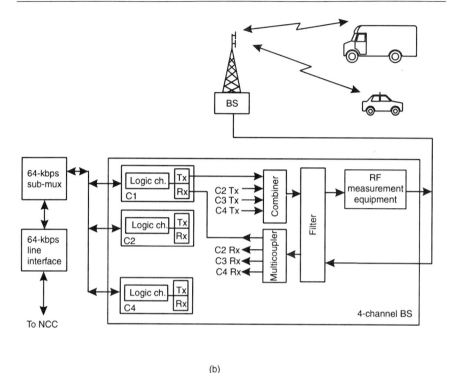

(b)

Figure 8.6 (continued).

power supply unit. Figure 8.6(b) shows a possible configuration of a four-channel base station.

Base stations incorporate remote control functions and also include features that allow advanced frequency management functions controlled from the NCC, such as time-shared control channels and dynamic channel assignment by automatic frequency selection.

In case of failure, the control channel is replaced by any of the other transceivers in the base station. In the case of total link failure between a base station and the NCC, communications may still be set up between mobiles located within the service area of the base station.

Additional functions carried out by base stations are the following:

- Signaling channel management according to the MPT 1327 standard;
- Routing of local traffic between mobiles served by the same base station;
- Storage of mobile registration data;

- Recording of system parameters relative to the base station (call durations, waiting times, etc.);
- Storage of local call tariff charging data;
- Storage of traffic statistical data.

Each traffic channel requires a dedicated transceiver. An additional transceiver is required to set up a dedicated, continually transmitting control channel. In order to prevent untimely aging of the control channel transceiver, the NCC may periodically assign the control channel functions to other transceivers. In this way, all base station transceivers will age in a similar way.

8.2.3 Antenna System

Normally, omni-directional antennas are used. A typical antenna gain value is 6 dBd (relative to the $\lambda/2$ dipole). If necessary, directive antennas may be used to improve coverage ranges and/or reduce interference.

8.2.4 Support Systems

Several support systems are needed for an adequate operation of trunked radio networks. Among these systems the following can be highlighted:

- Configuration assistance;
- Monitoring and maintenance;
- Subscriber administration;
- Subscriber charging.

Configuration assistance. Configuration operations are carried out from the NCC. The configuration data are automatically stored on the local hard disk. Equipment configuration data includes the following:

- Definition of new physical units and their parameters;
- Accessing the status of physical units;
- Modification of physical unit parameters;
- Setting any unit out of service.

Configuration setup affects not only NCC equipment but also those base stations connected to it. Configuration modifications may be carried out without interrupting the operation of other units.

Monitoring and maintenance. Each NCC is equipped with an O&M processor. The following O&M tasks may be carried out:

- Equipment configuration;
- System configuration;
- Subscriber management;
- Failure management;
- Security management;
- Accounting management.

The following network elements are monitored:

- NCCs;
- Base stations;
- Subscriber units (dispatch units and radio terminals);
- Ancillary equipment (subscriber administration center);
- Links between network elements;
- Links to external elements (PABX, PSTN, data networks);
- External alarms.

Subscriber administration. Subscriber administration tasks include the following:

- Definition of mobile and fixed subscribers including allowed call facilities;
- Definition of fleets and related data (such as interfaces with dispatch positions);
- Location update of mobile terminals;
- All subscriber associated parameters, including call facilities and numbering.

Subscriber-related information includes the following:

- Subscribers' individual identity;
- Identification of the fleet to which the subscriber belongs;
- Security number according to the standard used;

- Associated priority level;
- Subscription parameters for PABX access;
- Call barring/restriction facilities.

Subscriber charging. The information from all NCCs may be sent to a central charging center for its processing. Such center may be a PC connected to an NCC. Its main functions include the following:

- Gathering information on all calls served and, optionally, all call attempts;
- Providing the means of data analysis and inspection;
- Generating reports and statistical analyses of traffic to evaluate the GoS;
- Providing the storage capacity needed so that no loss of data can happen;
- Keeping a record of information corresponding to every subscriber such as charge data, generation of statistics and data recovery operations.

Network statistics may include monitoring of the following magnitudes:

- The busy hour (BH);
- The GoS;
- The mean call duration and the distribution of call durations;
- Local calls and calls between base stations.

This allows the realization of a systematic network performance assessment from all points of view, thus allowing to keep network quality parameters and subscriber satisfaction.

8.2.5 Interconnection Network

The interconnection network links the base stations to the NCC. An alternative is to use radio links instead of leased lines from a carrier service provider. Typically, links are digital with a 64-kbps rate. Each 64-kbps channel may be used for three voice links and two 1,200-bps data links by means of a submultiplexer. In this way, with a single line of this type, three voice channels may be connected to the NCC. Also, it is possible to use the connection to send all control, traffic, alarm, and supervision information to the NCC. Figure 8.7 illustrates how a base station may be connected to an NCC.

Trunked Systems 291

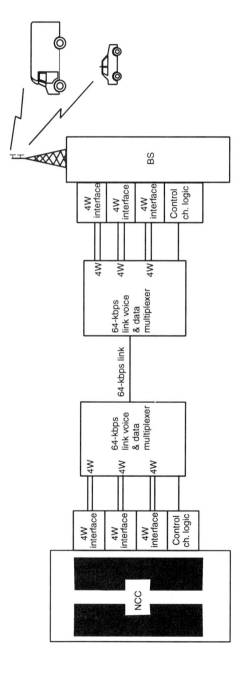

Figure 8.7 NCC-to-base station connection.

8.2.6 Frequency Plan

VHF-Band III trunking radio PAMR systems use two-frequency channels with their nominal carrier frequencies following the expression

$$f_n = 223.000 + n \cdot 0.0125 \text{ MHz and} \quad (8.4)$$
$$f_n' = f_n + 6 \text{ MHz with } n = 1, 2, 3, \ldots, 119$$

A total of 119 half-duplex channels are available in this band, which may be shared between two or more operators. (Additional channels may be made available in each particular country.) A channel spacing of 12.5 kHz has been selected with a transmit-receive separation of 6 MHz.

VHF-Band III presents good propagation characteristics as regards coverage range in comparison to the 450-MHz band. This allows a broad coverage in irregular terrain and urban areas. On the other hand, channel reuse can be accomplished at closer distances when using the 450-MHz band. The 230-MHz band (Band III) presents intermediate characteristics between those found in typical PMR bands: VHF-high (150 MHz) and UHF-low (450 MHz). In order to set up a frequency reuse scheme, the available channels are split into sets. If cavity combiners are used, a minimum channel separation of 150 kHz (twelve 12.5-kHz channels) is required, as is shown in Figure 8.8.

8.3 MPT13xx Standards

8.3.1 Introduction

In this section, a brief description of some of the basic features of the set of MPT-13xx document/specifications [3] related to trunked radio systems is presented. MPT standards originated at the United Kingdom's Department of Trade and Industry (Radiocommunications Agency), and they have been accepted in several European countries as the specification for analog trunked systems, so they can be considered a European "de facto" standard for these systems. The advantage of using standardized systems is that different equipment and terminal manufacturers can develop their products for this type of network, and thus, prices can be reduced.

This book describes the MPT signaling protocol in more detail than other signaling protocols such as AMPS, GSM, IS-95, or DECT. This is because the authors believe that going through a moderate complexity signaling system in some detail will allow readers to become familiar with basic air-interface signaling concepts. This will allow a more fruitful reading of the

Trunked Systems

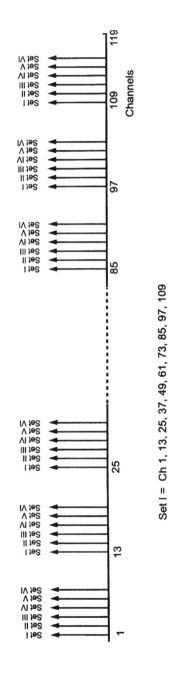

Figure 8.8 Frequency plan. Frequency sets.

following chapters, which discuss complex cellular, cordless, and other wireless systems.

In this section, the standard MPT-1327 (Revised 1991) [4] entitled "A signaling standard for trunked private land mobile radio systems" is presented. This document describes the signaling protocols used to set up calls on trunked radio networks. Other associated documents are listed as follows:

- MPT-1343 (Revised 1991), "System interface specification for radio equipment to be used with commercial trunked networks operating in Band III, sub-bands 1&2";
- MPT-1317 (Revised 1996), "Code of practice for transmission of digital information over land mobile radio systems";
- MPT-1318 (Revised 1994), "Engineering memorandum on trunked systems in the land mobile radio service";
- MPT-1323 (Revised 1996), "Angle-modulated equipment for use at fixed and mobile stations in private mobile radio service operation in the frequency band 174-225 MHz";
- MPT-1331 (Reprinted 1996), "Code of practice for radio site engineering";
- MPT-1347 (Revised 1991), "Radio interface specification for commercial trunked networks operation in Band III, sub-bands 1&2";
- MPT-1352 (Reprinted 1993), "Test schedule for the approval of radio units to be used with commercial trunked networks operating in Band III, sub-bands 1&2";
- MAP-27 (1992) [5], "Mobile access protocol for MPT-1327 equipment."

This section will mainly deal with the signaling protocol defined in MPT-1327. A basic aspect that distinguishes advanced radiocommunication networks from other, more primitive ones (such as those described in Chapter 2) is their capability of handling and interchanging a great variety of signaling messages between intelligent terminals and a control center, the so-called trunking system controller (TSC).

The evolution of mobile radio systems is analyzed in Chapter 2. There it is shown that open channel systems (i.e., those not using any signaling at all) were replaced by systems using primitive signaling schemes like subaudio tones (CTCSS) or five tones. All those systems use the same channel for communication and signaling purposes: traffic channel-associated signaling.

An important step forward has been the introduction of digital signaling, which allowed a wider range of possible messages to be interchanged between the radio terminals and the system. Another important leap forward was the use of a separate control (or signaling) channel common to all communications and traffic channels.

In systems with a common control channel, the first process in any communication is the setup process. A communication request is sent by the mobile terminal to the network through the control channel. Also via this channel, the TSC will assign one free traffic channel to the requesting terminal.

Trunked systems, and specifically MPT-1327 systems, not only allow the setting up of selective and group calls, but also give access to other services and facilities. These include sending status and short data messages, priority calls, emergency calls, and include calls. These facilities are defined in Section 8.3.2. It must again be stressed that all these facilities and services are only possible thanks to the existence of a digital signaling system over a common control channel. Another important feature of trunking MPT-1327 systems is the possibility of setting up multisite networks to provide regional or even national coverage.

On the other hand, these systems have the drawback of using analog voice transmission techniques (FM voice modulation). New trunking systems, such as, for example, the European TETRA standard (Chapter 11), combine both the advantages of trunked systems and those of fully digital systems.

8.3.2 Types of Calls and Facilities

The MPT-1327 [4] standard defines the following call types and facilities:

1. Voice calls: Individual or selective voice calls may be set up with different priority levels. Also, voice group calls can be made either in conversational mode (i.e., where all terminals can speak to each other) or in announcement mode, where only the terminal originating the call is able to transmit.
2. Data calls: In the same way as for voice, individual and group calls may be set up to transmit data over traffic channels. The transmitted data stream first modulates an audio subcarrier. There is no limitation in length or format.
3. Emergency calls: Both individual and group emergency calls (both voice and data) may be set up in MPT trunked systems. If there is not a free traffic channel available, the emergency call request will preempt one of the current communications. No acknowledgment from the called party is needed to complete the call.

4. Include calls: This facility may be used to set up conference calls or call transfers.
5. Status messages: A maximum of 32 status messages whose meanings have to be agreed in advance between the calling parties can be transmitted through the control channel. Only two messages have a predefined meaning: "call-me-back request" and "cancel previous call-me-back request."
6. Short data messages: Messages of up to 184 bits of free format can be transmitted through the control channel without the need to seize one of the traffic channels.

Individual radio terminals may make or receive calls to/from other individual radio terminals or line-connected units, groups or all system terminals, PABX extensions, or PSTN numbers. Mobile terminals may belong to several groups. Group addresses are selected in a totally independent way from the individual terminal addresses.

8.3.3 Characteristics of MPT-13xx Systems

The addressing capabilities established in the MPT-1327 standard are the following:

- 1,036,800 addresses per system;
- 1,024 channel numbers;
- 32,768 system identity codes.

The signaling protocol uses a transmission binary rate of 1,200 bps using fast frequency shift keying (FFSK) modulation of an audio subcarrier which is, afterwards, taken to RF (VHF, UHF) by means of an angular modulation.

Radio terminals operate in two-frequency simplex mode while base stations operate in full duplex mode (half-duplex operation, Chapter 2). The required signaling for setting up calls is carried out over a common control channel. This *control channel* may be implemented using two strategies: dedicated or nondedicated. A dedicated control channel is permanently devoted to the interchange of signaling information only. On the other hand, a nondedicated control channel can transmit signaling and user information. The use of a dedicated channel is advisable in systems with a large number of traffic channels while nondedicated channels are more suitable for systems with a small number of traffic channels. MPT-1327 allows both operational modes.

The control channel transmits a continuous data stream (synchronous) in the base-to-mobile link (downlink), while the reverse-link (uplink) transmissions from mobile terminals will only take place when any of them needs to set up a communication. As a consequence, there is a contention between mobiles in the uplink with the risk of two mobiles transmitting at the same time, thus producing collisions between signaling messages, which are then lost. The signaling protocol must take care and sort out these problems so that the system does not get blocked. In case of system failure, base stations may revert to simple repeater operation mode or fall-back mode.

8.3.4 Call Processing

Radio trunked MPT-13xx networks are queuing systems. As indicated in Section 8.1, in these systems, if a call cannot be served immediately it is queued until a traffic channel becomes free. Also, call requests can be queued in case the called terminal is busy even though there are free traffic channels available.

The TSC must verify that the called terminal is within reach (within coverage). In this way, traffic channels are not unnecessarily seized.

The system also sends maintenance and release signals once the call has been set up. These signaling messages are transmitted over the traffic channel being used. Maintenance signals facilitate a swift release of the traffic channel at the end of a call or, in case of radio link outage, during the call. Each time the PTT is released, maintenance signaling will be sent through the traffic channel. In certain systems, signals will also be sent whenever the PTT is pushed. The TSC, in case it receives a DISCONNECT signal or a timer expires, will send RELEASE messages through the traffic channel.

8.3.5 Multisite Systems

The MPT-1327 standard allows the setting up of multisite systems for extended coverage. To accomplish this, several techniques may be used:

- Simulcast operation;
- Use of different control channels at each site;
- Use of a time-shared control channel.

The MPT-1327 protocol allows the establishment of a *registration protocol* to allow the setting up of multisite systems and networks with several TSCs. By means of the registration procedure, the terminal informs the TSC of its position as it travels from site to site within the network.

The TSC, via the control channel, may broadcast information to advise the terminals of the control channels of the different sites.

8.3.6 Access Protocol

This section briefly describes the basic aspects of the MPT signaling protocol [4]. Also, examples of different signaling message interchange sequences for setting up different call types are presented. The control channel is arranged in time slots of 106.7 ms (128 bits) each. Random access thus follows a slotted Aloha configuration. Messages can only be transmitted at the beginning of each time slot. System synchronization is guaranteed thanks to the continuous transmission in the downlink control channel as is shown in Figure 8.9(a).

Two 64-bit codewords are sent in each time slot:

1. Control channel system codeword (CCSC) that identifies the system to the radio terminals and provides the required synchronization for the reception of the next codeword.
2. Address codeword that is the first word in any message and defines its type.

For longer messages, data codewords will be sent in ensuing time slots. In this case, the next CCSC and address codewords will be delayed until the message is finished.

Mobile radio terminals shall be able to receive a signaling message from the TSC in one time slot, transmit their answers on the next time slot, and then listen to the control channel downlink to receive further messages from the TSC (Figure 8.9(a)). Each codeword is structured as shown in Figure 8.9(b). Of bits 49 to 64, the first 15 are check bits and are calculated by means of a (63,48) cyclic code. Finally, at the end of each 63-bit block an even parity bit is added.

When messages require the transmission of additional codewords, these cannot break the time slot structure. This is achieved by delaying the transmission of the next CCSC and address codewords until the beginning of a new time slot as shown in Figure 8.9(c). When a message consists of an odd number of data codewords, a filler codeword shall be used to maintain the slotted Aloha time structure.

8.3.7 Types of Signaling Messages

Signaling messages may be classified into different categories as indicated in Table 8.1. In Table 8.2 the most important messages used are summarized.

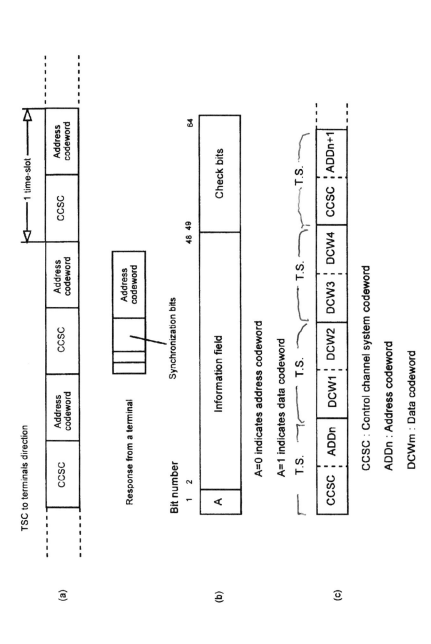

Figure 8.9 (a) Time structure in the control channel downlink and terminal response. (b) Codeword structure. (c) Delay in transmitting the CCSC+ADD codewords to transmit DCWs [4].

Table 8.1
Types of Signaling Messages [4]

Message Type	Direction	Function
Aloha	TSC-terminals	Used to invite the mobile terminals to transmit their call set up requests and regulate the random access to the control channel
Requests	Terminals-TSC	Communication requests/transactions
AHOY	TSC-terminals	To request an answer from an addressed terminal
Acknowledgments (ACK)	Both directions	Receipt acknowledgments
Go-to-channel (GTC)	TSC-terminals	Tune to a given traffic channel
Single address messages	Terminals-TSC	For extended addressing procedures
Short data messages	Both directions	Short data
Miscellaneous messages	TSC-terminals	For system control

Table 8.2
List of Main Signaling Messages [4]

ALH	General Aloha invitation	RQS	Requests: "simple"
ALHS	Standard data excluded	RQD	Standard data
ALHD	"Simple" calls excluded	RQX	Cancel/abort
ALHE	Emergency only	RQT	Divert
ALHR	Registration or emergency	RQE	Emergency
ALHX	Registration excluded	RQR	Registration
ALHF	Fall-back mode	RQQ	Status
ACK	Acknowledgments: general	RQC	Short data
ACKI	Intermediate	AHY	AHOYs: general availability check
ACKQ	Call queued	AHYX	Cancel alert/waiting state
ACKX	Message rejected	AHYQ	Status message
ACKV	Called unit unavailable	AHYC	Short data invitation
ACKE	Emergency	MARK	Miscellaneous: control channel marker
ACKT	Try on given address	MAINT	Call maintenance
ACKB	Call-back/negative acknowledge	CLEAR	Call clear-down
SAMO	Single address messages: outbound	MOVE	Move control channel
SAMIU	Inbound unsolicited	BCAST	Broadcast
SAMIS	Inbound solicited		
HEAD	Short data message		

8.3.8 Principles of the Random Access Protocol

A signaling protocol must have three relevant features:

1. Minimize the number of collisions/clashes between signaling messages;
2. Minimize call setup delays;
3. Assure good stability and throughput under heavy traffic conditions.

To achieve this, a dynamic frame length structure is built on top of the slotted Aloha time division. The operation fundamentals are illustrated with the aid of Figure 8.10(a), where the message flow in both directions (uplink and downlink) is shown. The TSC transmits an Aloha synchronization message ALH including the parameter (N) that indicates the number of ensuing time slots available for mobile terminal access. The set of (N) time slots forms a frame.

A mobile terminal always tries to transmit on the first time slot available without the need to wait for the start of a new frame. Only in case access is forbidden for some reason will the terminal wait until the current frame ends and will randomly select a time slot belonging to the next frame. The frame length (N) is optimized as a function of the traffic load on the control channel to avoid excessive collisions and minimize access delays. Towards this end, the TSC monitors the activity on the control channel. Figure 8.10(b) illustrates the frame length modification mechanism.

The TSC detects the collision between requests RQS1 and RQS2 and defines a longer frame by transmitting the message ALH(2). The terminals repeat their requests by randomly selecting any of the time intervals in the following frame. In the example, they select different slots. Each successful request is followed by an acknowledgment message from the TSC. When no collisions are detected, the frame length is reduced. In the example, message ALH(0) does not define a frame. However, message ACKQ(1) not only acknowledges a request but also defines a new one-time slot frame.

The signaling load on the control channel is reduced by allowing that acknowledgment and GTC messages include the parameter (N) (frame length). In this way, frames are defined without the need to explicitly send an Aloha message ALH.

During a frame, the TSC can transmit messages requesting an answer from a particular mobile terminal. These messages inhibit the random access from other terminals, thus reserving the next time slot for the addressed terminal to answer.

The TSC may reserve specific frames for a particular type of call by means of specific Aloha messages; for example, message ALHE prompts emer-

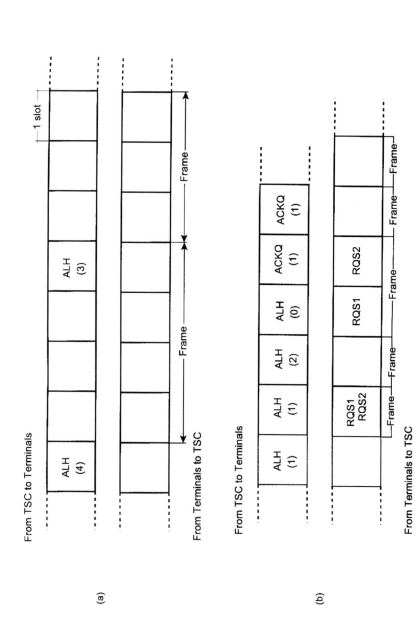

Figure 8.10 (a) Dynamic frame length structure. (b) Dynamic frame length modification as a function of the traffic load [4].

gency requests only. Also, frames may be reserved for specific subsets of terminals (i.e., address subdivision).

8.3.9 Terminal Addressing

MPT-1327 specifies that the address of a radio terminal is a 20-bit number divided into two fields:

- 7-bit PREFIX (PFIX);
- 13-bit IDENTITY (IDENT).

Normally, all terminals in the same fleet will have the same prefix. This numbering system made up of prefix and identity numbers allows most messages to include two addresses, that of the calling terminal and that of the called terminal, since the prefix will be included only once. For example, call requests or GTC messages contain two identities and only one prefix. To make a call to a terminal with the same prefix, the request message contains enough information. However, to set up a call to a terminal with a different prefix, the details of the call requested cannot fit into a single address codeword. This type of call requires the use of extended addressing procedures.

However, the numbering system defined in MPT-1327 is not flexible enough to setup public PAMR networks such as those in VHF-Band III. Standard MPT-1343 defines a new numbering system superimposed on that defined by MPT-1327. MPT-1343 defines an individual network number and network group numbers. The individual network number consists of the following:

- PN: Prefix number in the range 200 to 327;
- FIN: Fleet individual number in the range 2001 to 4999;
- UN: Unit number consisting of two digits in the range 20–89 for fleets of up to 70 terminals or three digits in the range 200–899 (700 terminal fleets).

Network group numbers consist of the following:

- PN: Prefix number in the range 200 to 327;
- FGN: Fleet group number in the range 5001 to 6050;
- UN: Group number that can have either two digits in the range 90 to 99 or three digits in the 900 to 998 range.

Each terminal can be assigned more than one group number so that it can receive calls within more than one group.

Fleets of terminals are set up in such a way that most calls are normally made between users sharing the same prefix. The identity range may thus be divided into blocks that the network operator will assign to each fleet or group (Figure 8.11). Each fleet should be assigned a reasonably large number of identities to allow for later expansions.

One of the basic objectives in a numbering plan is to allow the use of a small number of digits to address the most frequently used terminals instead of using long sequences of digits. To assign an individual network number for individual calls within the same fleet, the first number in the fleet (20 or 200 depending on the fleet size) will correspond to the lowest identity assigned to that fleet: base identity (IBI). The same procedure is followed to assign numbers for group calls. The lowest group number (90 or 900 depending on the fleet size) will be assigned to the lowest group identity in the fleet, called base identity (GBI).

Figure 8.12(a) shows numbering examples according to MPT-1343. Figure 8.12(b) illustrates the algorithm used to convert MPT-1327 to MPT-1343 addresses and vice versa. The MPT-1343 address dialed by the user is translated within the radio terminal itself to a MPT-1327, which is really the one transmitted over the air.

Figure 8.13(a) illustrates a scenario where fleets and groups are defined. In Figure 8.13(b), different possible call types with their corresponding numbering requirements are shown. The numbering system presented in Figure 8.13(b) is also reproduced in Tables 8.3 and 8.4.

8.3.10 Examples of Signaling Sequences

This section describes several examples of signaling sequences [4] in a simplified way. The messages interchanged between the TSC and the mobile terminals for setting up different call types are presented. In these examples, repeated messages due to collision, degradations or propagation channel fading/shadowing effects are not shown. It must also be borne in mind that once a call has been set up, maintenance signaling is interchanged through the traffic channel itself.

In general, for setting up a call some or all the following steps are followed:

- Call request;
- Instructions to use extended addressing;
- Checking availability of terminals;
- Assignment of a free traffic channel or queuing of the call request until a traffic channel becomes available.

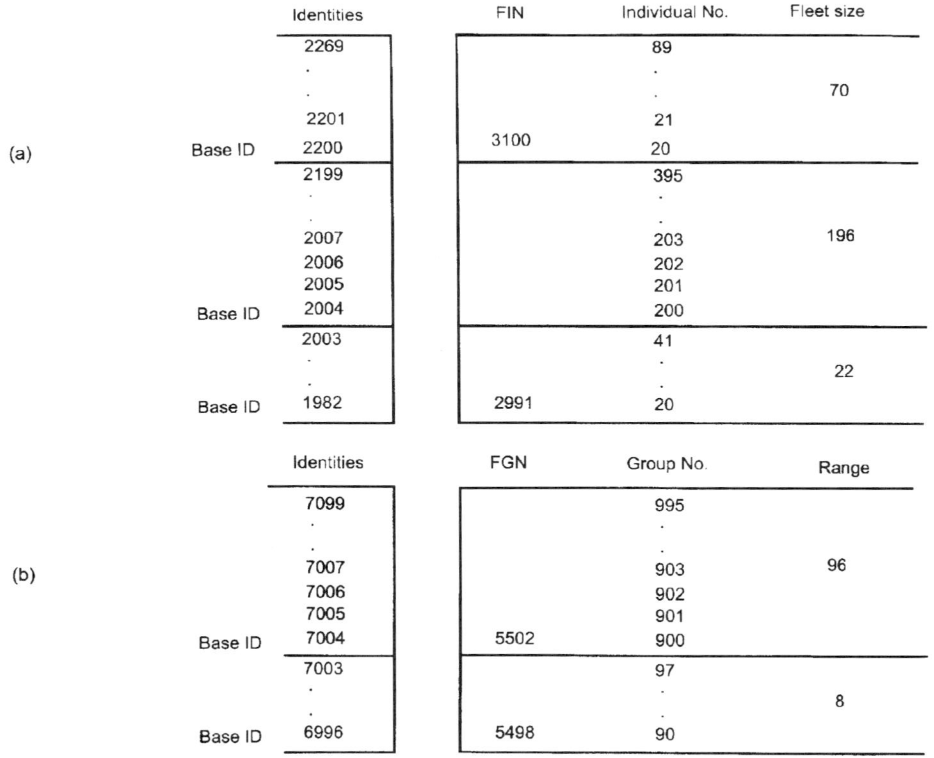

(a)

Identities		FIN	Individual No.	Fleet size
2269			89	
.			.	
2201			21	
2200	Base ID	3100	20	70
2199			395	
.			.	
2007			203	
2006			202	
2005			201	
2004	Base ID		200	196
2003			41	
.			.	
1982	Base ID	2991	20	22

(b)

Identities		FGN	Group No.	Range
7099			995	
.			.	
7007			903	
7006			902	
7005			901	
7004	Base ID	5502	900	96
7003			97	
.			.	
6996	Base ID	5498	90	8

Figure 8.11 (a) Address blocks for individual terminal addressing according to the MPT 1343 standard. (b) Address blocks for group calls according to the MPT 1343 standard.

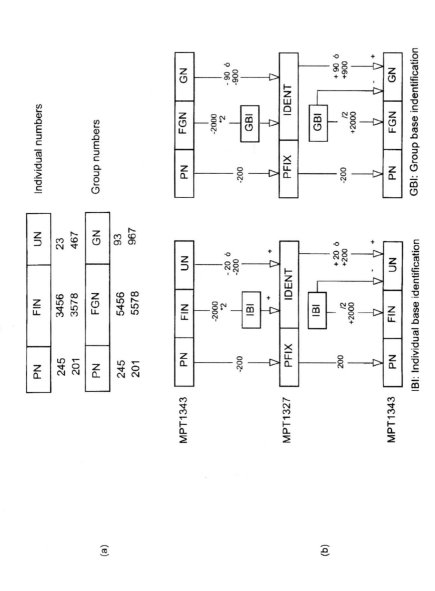

Figure 8.12 (a) MPT 1343 numbering examples. (b) Translation algorithms between MPT-1327 and MPT-1343 numbering systems.

Trunked Systems

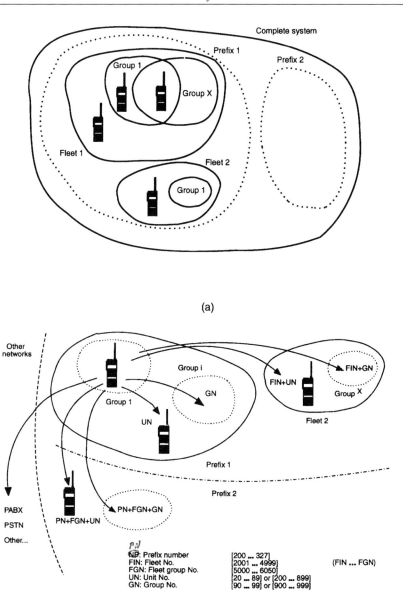

Figure 8.13 (a) Organization of terminals, groups and fleets in a MPT-1343 numbering system. (b) Types of calls and addressing used.

Table 8.3
Individual Call Numbering System

Called Unit	Type	NP	FIN	UN
Within the same fleet		Not used	Not used	20...89
In a different fleet	Interfleet call	Not used	2001...4999	or
With other prefix	Interprefix call	200 ... 327		200...899

Table 8.4
Group Call Numbering System

Called Group	Type	NP	FGN	GN
Within the same fleet		Not used	Not used	90...99
In a different fleet	Interfleet call	Not used	5001...6050	or
With other prefix	Interprefix call	200 ... 327		900...999

Group Call

Figure 8.14(a) illustrates the sequence of messages interchanged between the TSC and the mobile terminal to set up a group call between terminals with the same prefix [4]. For group calls, no terminal availability checking is carried out. In this example, all traffic channels are busy so that the call request is queued until a free channel is available.

1. ALH: General Aloha invitation and indication of the beginning of a one-slot slot frame.
2. RQS: Communication request.
3. ACKQ: The TSC acknowledges the RQS message and informs the calling terminal that its request has been queued until.
4. GTC: When a traffic channel is available, the TSC transmits the GTC command. This message orders the addressed terminals to switch to the assigned traffic channel. In the example the GTC message is sent twice to improve reliability.

Should any traffic channel be available at the time of the request, the TSC would send the GTC message immediately after the arrival of the RQS.

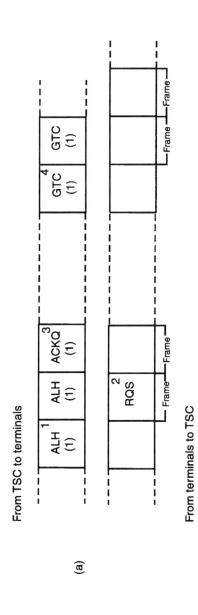

Figure 8.14 (a) Group call. (b) Individual call to a terminal with the same prefix. (c) Individual call to a terminal with a different prefix. (d) Short data message over the control channel. (e) Status message [4].

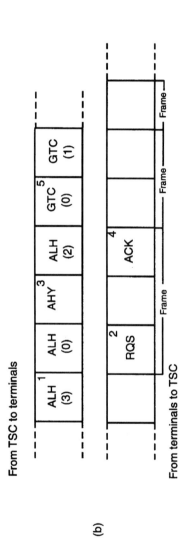

Figure 8.14 (continued).

Trunked Systems

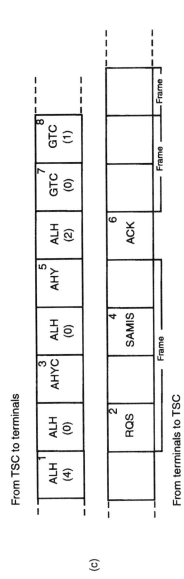

Figure 8.14 (continued).

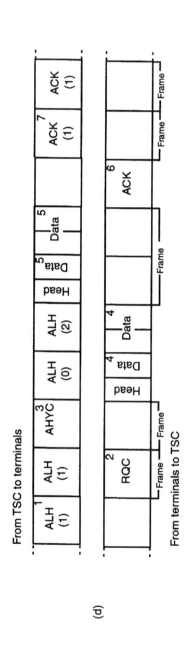

Figure 8.14 (continued).

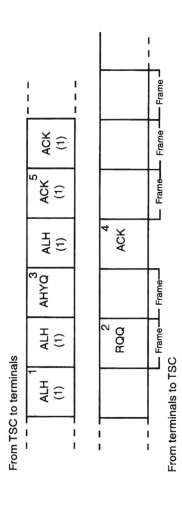

Figure 8.14 (continued).

Individual Call to a Terminal With the Same Prefix

Figure 8.14(b) [4] illustrates the sequence of messages interchanged between the TSC and mobile terminals to set up an individual call to a terminal with the same prefix. In this case, the following steps are followed: communication request, terminal availability check and channel allocation. They are detailed as follows:

1. ALH: Aloha message defining a three-slot frame;
2. RQS: Communication request;
3. AHY: Availability verification message with the following purposes:
   - RQS request acknowledgment;
   - Request to the called terminal to verify if it is within radio contact;
   - Inhibits random access during the next time slot, so that it can be used by the called terminal to answer.
4. ACK: Called terminal acknowledgment transmitted in the time slot specially reserved by the TSC for it to answer;
5. GTC: Go-to-channel message instructing both terminals to switch to the assigned traffic channel. Again, in the example the GTC message is sent twice for increased reliability.

In this example, the called terminal is within radio range, and thus, it answers with an ACK message. If the called terminal were not available (not within reach or disconnected) the TSC would indicate it to the calling terminal using the acknowledgment message ACKV (called unit unavailable).

Message ALH (0) in these examples is used as a dummy message in those time slots with no relevant signaling interchange. In practice, these slots carry signaling messages relative to other calls or broadcast messages relative to the system.

Individual Call to Terminal With a Different Prefix

Figure 8.14(c) illustrates the sequence of messages interchanged between the TSC and mobile terminals to set up an individual call to a terminal with a different prefix [4]. In this case, the signaling sequence includes, as in the previous example, the following steps: call request, terminal availability check, and channel allocation. However, an additional step is required where, upon reception of the RQS message, the TSC will transmit an AHYC message requesting the calling terminal to transmit the complete address of the called terminal (extended addressing). Also, separate GTC messages will be sent to the calling and called terminals to tune the assigned traffic channel since GTC messages can only contain one prefix. The steps in this process are detailed as follows:

1. ALH: General Aloha invitation to transmit indicating a four-slot frame.
2. RQS: Interprefix call request. This message contains the calling terminal's address (prefix + identity), but the identity of the called terminal is replaced with a specific number indicating that an extended addressing procedure is required.
3. AHYC: Message prompt to transmit a short data message which, in turn:
   - Serves as an acknowledgment of the RQS message;
   - Prompts the calling terminal to transmit the address of the called terminal;
   - Inhibits random access on the next slot.
4. SAMIS: Single address message sent by the calling terminal and containing the full address of the called terminal.
5. AHY: Called terminal availability check message requiring a response from the called terminal. In this example the message consists of a single codeword. This means that only the called terminal address is included in the message.
6. ACK: Called terminal acknowledgment.
7. GTC: Go-to-channel message ordering the called terminal to switch to the assigned traffic channel.
8. GTC: Go-to-channel message ordering the calling terminal to switch to the assigned traffic channel.

Short Data Message Over the Control Channel

Figure 8.14(d) illustrates the sequence of messages interchanged between the TSC and mobile terminals in a short data communication over the control channel. In this example, the message consists of an address codeword and three additional data codewords sent immediately after the address codeword. Each data codeword consists of 46 format free bits.

In this case, the following steps are followed: The calling terminal sends in a request; the TSC instructs the calling terminal to send the data; and, once the TSC receives the data massage, a retransmission occurs toward the called terminal. Finally, acknowledgment messages are exchanged. These steps are detailed as follows:

1. ALH: General Aloha invitation to transmit and definition of a one-slot frame.

2. RQC: Short data message request. The request indicates the necessary number of slots—two, in this case.
3. AHYC: Invitation to transmit the data, which, in turn, does the following:
   - Acknowledges the RQC message;
   - Instructs the calling terminal to send the short data message using the next two slots.
4. HEAD+data: The terminal sends its short data message to the TSC. In this example the message consists of an address codeword (HEAD) and three appended data codewords.
5. HEAD+data: The TSC forwards the message to the called terminal.
6. ACK: Acknowledgment from the called terminal indicating that the message has been received.
7. ACK: Acknowledgment forwarded by the TSC to the calling terminal to advise it that the message has arrived. In this example this message is repeated for added reliability.

In this case, it has been assumed that three data words are transmitted. It is also possible to transmit up to four data words. In case two or four data words are transmitted, a half time slot is empty and a filler word must be added.

Status Message

Figure 8.14(e) illustrates the sequence of messages interchanged between the TSC and a mobile terminal to transmit a status message to a terminal with the same prefix.

1. ALH: General Aloha message indicating a frame length of one slot.
2. RQQ: Random access request to send the status message to the addressed terminal.
3. AHYQ: AHOY status message to do the following:
   - Acknowledge the RQQ request;
   - Send the status information to the called terminal and request an acknowledgment.
4. ACK: Acknowledgment from the called terminal indicating that the information was received.
5. ACK: Acknowledgment from the TSC to the calling terminal advising it that the message has been delivered. In this example the ACK message is repeated for added reliability.

8.4 Trunked Network Dimensioning

Dimensioning or sizing a trunked radio network means calculating the number of traffic channels required to provide an adequate GoS to the terminals in the system or, conversely, to figure out the number of mobiles that can be served with a fixed number of channels while maintaining a given GoS.

In addition to the number of traffic channels calculated, a control channel must be added. The reader is reminded that channel, in this context, means a pair of frequencies for half-duplex operation.

Since trunked radio networks are of the waiting or queuing type, the Erlang-C formula shall be used in the computations. The dimensioning is linked to a previous specification of an objective from the point of view of the traffic quality or GoS. The GoS definition combines both the probability that a call request has to wait to be served (because all traffic channels are busy) and a waiting time limit W_o:

GoS (prob. of having to wait) & (prob. of waiting more than $W_o(s)$)

Another relevant parameter needed to dimension a trunk system is the average call duration, which is different in every application (dispatch, security, data transmission). Some networks implement a call timer that cuts off all calls exceeding a previously set call duration limit. The average call duration can be estimated before network operation starts, but, later, it can be obtained from operational network statistics.

In the dimensioning of a trunk radio network, the following parameters are involved:

1. The GoS which includes the following:
   - A reference waiting time W_o;
   - A probability of exceeding the waiting time W_o at least with two probability levels, a design objective $P_{wo}(\%)$ and an overload level $P_{wos}(\%)$.
2. An average channel occupation time, which includes the following:
   - The mean call duration, H, during the busy hour;
   - The average busy hour call attempts (bhca), L.

If the total number of mobile terminals is M, the total offered traffic A is

$$A = \frac{M \cdot L \cdot H}{3{,}600} \text{ Erlang} \qquad (8.5)$$

The probability that a call has to wait is given by the Erlang-C formula

$$P_D = C(A, N) = \frac{\dfrac{A^N}{N!} \dfrac{N}{N-A}}{\displaystyle\sum_{k=0}^{N-1} \dfrac{A^k}{k!} + \dfrac{A^N}{N!} \dfrac{N}{N-A}} \tag{8.6}$$

The Erlang-C formula can be expressed in terms of the Erlang-B formula as follows:

$$C(N, A) = \frac{NB(N, A)}{N - A[1 - B(N, A)]} \tag{8.7}$$

This expression allows the use of the Erlang-B tables in Appendix 9B for the computation of the Erlang-C formula.

The probability that the waiting time exceeds W_o seconds is

$$P(W > W_o) = P_D \exp\left[-(N - A)\frac{W_o}{H}\right] \tag{8.8}$$

and the average waiting time for any call is

$$W_a = P_D \frac{H}{N - A} \tag{8.9}$$

The above expressions, together with the GoS objective are used to dimension trunk radio networks as it is illustrated in the following two examples.

Example 8.1. Given the following data:

- Average call duration: $H = 20$s;
- BH call attempts: $L = 1$ calls and a GoS objective specified as;
- Reference waiting time $W_o = 20$s;
- Probability of exceeding W_o;
  - Design objective, $P_{wo} = 5\ \%$;
  - Overload value, $P_{wos} = 30\ \%$.

the following two cases, direct and inverse, are studied.

Direct case. Computation of the number of traffic channels required for $M = 1,000$ mobile terminals. The study starts with the computation of the total offered traffic applying (8.5)

$$A = \frac{1000 \cdot 1 \cdot 20}{3600} = 5.56 \text{ Erlang} \qquad (8.10)$$

By using (8.6) and (8.8), a table with the probabilities of exceeding the waiting time W_o for several values of N (number of traffic channels) is computed. The value of N sought is the one closest to the GoS objective

$$N = 6 \quad P(W > W_o) = 0.5190$$
$$N = 7 \quad P(W > W_o) = 0.1123$$
$$N = 8 \quad P(W > W_o) = 0.0229 \leftarrow$$

In this example, for $N = 8$, $P(W > W_o) = 0.0229 < 0.05$ (design objective). In this way, it has been found that 8 traffic channels are needed to handle the traffic demand and fulfill the GoS objective. Additionally, another channel for signaling functions (control channel) is needed.

The average waiting time will be, applying (8.9),

$$W_a = P_D \frac{H}{N - A} = 8.2 \text{ s} \qquad (8.11)$$

For $N = 8$ traffic channels it is also possible to compute the maximum number of mobiles to reach the saturation level. Again, constructing a table for increasing values of A, and consequently of M (number of mobiles), using (8.5), the following values are obtained:

$$A = 6 \quad P(W > W_o) = 0.0483$$
$$A = 7,0 \quad P(W > W_o) = 0.2337$$
$$A = 7,1 \quad P(W > W_o) = 0.2716 \leftarrow$$
$$A = 7,2 \quad P(W > W_o) = 0.3152$$

From the table, it results that for $A = 7.1$ Erlang, the probability of exceeding W_o, $P(W > Wo) = 0.2716 < 0.3$ (saturation level). A traffic of $A = 7.1$ Erlang corresponds to $M = 1,278$ mobiles.

Inverse case. Given a number of available traffic channels N, the maximum possible number of mobiles M will be computed. To carry out this study, firstly the offered traffic per mobile a is computed as

$$a = \frac{H \cdot L}{3,600} = 5.56 \ 10^{-3} \text{ Erlang/mobile} \qquad (8.12)$$

For a given value of N, in this case eight traffic channels, and again computing a table for different values of A using (8.8), the following values are obtained

$A = 5.9 \quad P(W > W_0) = 0.0409$

$A = 6.0 \quad P(W > W_0) = 0.0483 \leftarrow$

$A = 6.1 \quad P(W > W_0) = 0.0570$

In this example, the value sought is $A = 6$ Erlang which corresponds to $P(W > W_0) = 0.0483 < 0.05$ (objective). Then the maximum number of possible mobile terminals is $M = A/a = 1,071$.

The same study can also be carried out for the saturation level, i.e., for $P(W > W_0) = 0.3$. The following table can be calculated

$A = 6.9 \quad P(W > W_0) = 0.2086$

$A = 7.0 \quad P(W > W_0) = 0.2337$

$A = 7.1 \quad P(W > W_0) = 0.2716 \leftarrow$

$A = 7.2 \quad P(W > W_0) = 0.3152$

resulting that, for $A = 7.1$ Erlang, $P(W > W_0) = 0,2716 < 0,30$ (*saturation level*), which corresponds to a maximum number of mobile terminals $M = A/a = 1,267$.

Example 8.2. Assume the following dimensioning parameters for a trunked radio network:

- Average call duration: 30s;
- Average number of calls per terminal in the BH: 1 call;
- Reference waiting time: 20s.

The values in Table 8.5 are obtained for two GoS levels, 5% and 10%.

Table 8.5
Computed Offered Traffic Values and Number of Users for Example 8.2

No. of Channels	Traffic 5% GoS	No. of Users	Traffic 10% GoS	No. of Users
1	0.0917	10	0.1735	20
2	0.5770	69	0.7910	94
3	1.2634	151	1.5652	187
4	2.0426	245	2.4079	288
5	2.8755	359	3.2884	394
6	3.7420	449	4.1929	503
7	4.6315	555	5.1142	613
8	5.5380	664	6.0474	725
9	6.4577	774	6.9900	838
10	7.3872	886	7.9390	952
11	8.3245	998	8.8945	1067
12	9.2687	1112	9.8540	1182
13	10.2182	1226	10.8175	1298
14	11.1718	1340	11.7845	1414
15	12.1299	1455	12.7540	1530
16	13.0912	1570	13.7260	1647
17	14.0552	1686	14.7000	1764
18	15.0220	1802	15.6760	1881
19	15.9910	1918	16.6540	1998
20	16.9620	2035	17.6325	2115

References

[1] Bellamy, J., *Digital Telephony,* Second edition, New York: John Wiley & Sons, Inc., 1990.

[2] Macario, R. C. V., *Cellular Radio. Principles and Design,* Houndmills, Basingstoke, Hampshire and London, UK: MacMillan New Electronics, 1993.

[3] www.open.gov.uk/radiocom/mpt. Radiocommunication Agency, DTI, UK.

[4] MPT-1327, A signaling standard for trunked private land mobile radio systems DTI, UK (Revised 1991).

[5] User Access Definition Group, MAP 27. Mobile access protocol for MPT-1327 equipment. Version 1, UADG-September 1992.

9

The Cellular Concept

9.1 Introduction

The cellular concept [1] refers to the scheme adopted by the most important wireless systems. Using this concept it is possible to serve at a reasonable price entire countries or even continents using a limited portion of the RF spectrum. The cellular concept is based on two main key points:

- Frequency reuse;
- Cell splitting.

As an illustration of the concepts covered in this chapter, a brief description of the advanced mobile phone service (AMPS) system used in North America and its European version total access cellular system (TACS) will be given.

A cellular layout in a wireless communication network permits the adequate fulfillment of the following objectives [1]:

1. High user capacity;
2. Efficient use of the RF spectrum;
3. Extended coverages up to national or continental level;
4. Adaptability to different traffic density conditions;
5. Provision of service to mobile and hand-held terminals;
6. Telephone service plus special services including dispatching;
7. Service quality similar to that of the fixed telephone network;
8. Affordable price.

9.2 Keys to The Cellular Concept

To introduce the cellular concept, an illustrative example will be given relative to the so-called zero generation public mobile telephony (PMT) systems. In these systems a high-power multichannel control transmitter was employed for the downlink direction (from the base to the mobile stations). The mobile uplink range limitations were solved by using voting receivers (see Chapter 2) with a coverage that matched the lower transmission powers of the mobiles.

Suppose that a PMT system covering a town and its suburbs is set up. Assume that the coverage area is a circle of radius R = 10 km. The blocking probability objective is set to 2% and 40 two-frequency RF channels are available. Assuming that all the channels are used for voice calls (i.e., the need for signaling channels is neglected), the inverse Erlang-B formula (See Section 9.7 for a detailed explanation) gives the traffic intensity that can be handled by the system

$$A = B^{-1}(40; 0.02) = 31 \text{ Erlang}$$

The coverage area is $\pi \times 10^2$ = 314 km^2; thus, the traffic density is

$$\rho_t = \frac{31}{314} \approx 0.1 \text{ Erlang/km}^2$$

Assuming a traffic per mobile of 25 mErlang, this corresponds to a vehicular density, ρ_v, of

$$\rho_v = \frac{0.1}{0.025} = 4 \text{ mobiles/km}^2$$

As it can be seen, these values are very small. Traffic forecasts for PMT systems predicted densities in the range 0.5 Erlang/km^2 to 1 Erlang/km^2.

Suppose now a traffic density ρ_t = 1 Erlang/km^2. In this case, the offered traffic would be A = 314 Erlang. Maintaining the same blocking probability objective, the inverse Erlang-B formula gives the number of traffic channels required

$$N = B^{-1}(314; 0.02) = 328 \text{ channels} \quad (9.4)$$

This is an extremely large number of very difficult, if not impossible, implementations on any radio station.

Assuming a channel spacing of 25 kHz, a bandwidth of 0.025 × 238 = 8.2 MHz would be required. This is also an excessive requirement. On the other hand, if the initial number of 40 channels is maintained, the blocking probability would be

$$p_t = B(40; 314) = 0.873$$

that is, 87.3% of calls would find the system busy, unable to serve new call requests. This cannot be acceptable. As a consequence, the single base station implementation cannot handle the kind of expected traffic loads. This problem was detected early on in those zero generation mobile telephony systems. They reached a saturation state very rapidly, and the operators had to limit the demand by charging high tariffs.

Assume now that the coverage area is divided into smaller zones, called cells, circular in shape with radii of 1.5 km. The area of each cell is now $Area_{cell}$ = 7.1 km². A radio station is set up in each cell. Assuming a uniform traffic distribution, the offered traffic in each cell would be

$$a_c = \rho_t Area_{cell} = 7.1 \text{ Erlang}$$

The number of RF channels needed to handle this traffic, with the same blocking probability of 2% is

$$n_c = B^{-1}(7.1; 0.02) = 13 \text{ channels}$$

If these channels could be reused in all the cells, these 13 channels would suffice, instead of the 328 channels of the single base station case.

Nevertheless, with classical multi-access techniques, FDMA and TDMA, it is not possible to use the same frequencies on adjacent or nearby cells, due to unacceptable cochannel interference.

Frequency reuse is possible at a distance for which interference is not harmful. As a consequence, a group of cells must be established with different sets of frequencies in each cell. This group of cells is called a *cluster*. Several clusters reusing the same sets of frequencies will be set up to cover the target service area.

Continuing with the previous example, assuming a 12-cell cluster (in fact, this was the cluster size firstly used in cellular mobile telephony), the number of different RF channels needed is then 12 × 13 = 156. This is still half the number required for the single cell setup.

The number of clusters (i.e., the number of frequency reuses) is given by the ratio between the total coverage area and the cluster area

$$Q = \frac{314}{12 \cdot 7.1} = 3.7$$

So, in this example, three full clusters and one fractional cluster would be deployed. This number is usually called the *reuse index* of the network.

The previous example helps to understand two basic facts of the cellular concept. The first one can be represented by the following equation:

$$A = B^{-1}(N; p_t) = \rho_t \text{Area}_{\text{cell}} \quad (9.1)$$

where N is the number of channels, p_t is the blocking probability, ρ_t is the traffic density, and $\text{Area}_{\text{cell}}$ is the coverage area of an N channel cell site.

N is generally limited, and p_t is a fixed quality parameter; thus, A is also a fixed value. Then, if ρ_t increases, as is the usual case, $\text{Area}_{\text{cell}}$ must be reduced at the same rate. In fact, the cellular concept allows for this reduction of $\text{Area}_{\text{cell}}$. In a mature, consolidated cellular networks, further increases of ρ_t are handled by carrying out additional reductions of $\text{Area}_{\text{cell}}$ by means of the so-called cell splitting technique.

The second fact comes from the cluster concept. The number of RF channels needed in a cellular network is given by

$$C = NJ \quad (9.2)$$

where J is the cluster size. The reuse index is

$$Q = \frac{\text{Area}_{tot}}{J \cdot \text{Area}_{\text{cell}}} \quad (9.3)$$

Equation 9.3 indicates that any coverage area can be served with N channels, because the frequency reuse can be as intensive as needed. Observe that Q is also the number of simultaneous calls handled by each available RF channel (reuse index).

From (9.2) the convenience of selecting a value of J as low as possible in order to minimize the number of frequencies required in the network, is apparent. This value is related to the interference acceptance of the radio stations and depends on two conditions:

1. Cell deployment (network condition);
2. Protection ratio (radio condition).

The protection ratio is a function of the modulation and multi-access technologies used in the cellular network. Modern mobile TDMA digital telephony networks require lower protection ratios than the older analog networks. This enables the use of smaller cluster sizes.

The interference produced by several cochannel transmitters depends on the reuse distance and relative directions of the interference paths. These are geometrical issues that help to reduce harmful interference effects. An example of this is the use of sectorized cells leading to smaller values of J.

An example of cellular network layout is shown in Figure 9.1(a) where A, B, C, D, E, and F represent sets of two-frequency full duplex channels that are reused throughout the network.

The number of available channels C is divided into J sets, each of them having $N = C/J$ channels, which are the resources allocated to each cell. The Erlang-B formula gives, for this number and a blocking probability (or GoS, expressed as a percentage) the amount of offered traffic, A, that can be handled by a single cell. It may happen, however, that the traffic demand in some cells (mainly in cities or highways) is again larger than A.

To allow for a localized increase in traffic demand that cannot be handled appropriately with the N channels available in a cell, that cell can be split into smaller cells with N channels each. This procedure is called cell splitting, and the concept is illustrated in Figure 9.1(b). In Figure 9.1(a), the area corresponding to set F_1 is now handled by four cells H_3, I_3, B_3, C_3. Hence, cell splitting allows the handling of larger localized traffic demands (recall (9.1)). In this way, it is possible to handle large localized traffic demands without any increase in the RF spectrum used. To set up smaller cells while still limiting interference effects, smaller powers have to be used at base stations and mobiles (power control).

Mature systems will tend to use smaller and smaller cells. In their initial stages, up to 20-km cell radii may be common. In large cities, cell radii tend to be smaller than 3 km. Figure 9.1(c) illustrates the evolution of the network in Figures 9.1(a, b). Later on, areas with even larger traffic demands may be served by street microcells or even indoor picocells (for large airport or station halls, buildings, or other similar structures).

9.3 Cellular Geometry

Network planners must define a suitable geometrical structure that enable the implementation of the cellular concept. During the first stages of a cellular network, omni-directional antennas defining circular shaped cells are deployed. A suitable geometrical approximation to a circle (omni cell shape) is achieved using hexagons.

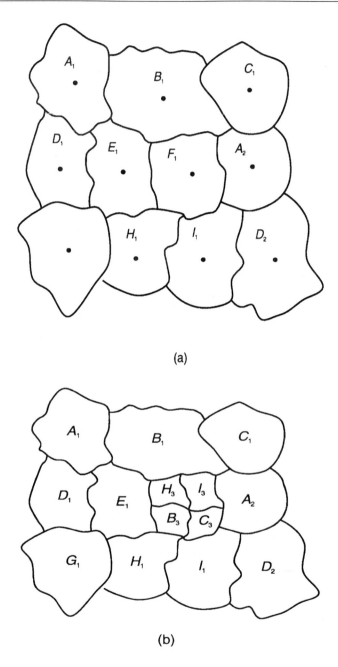

Figure 9.1 Three stages of a cellular network deployment.

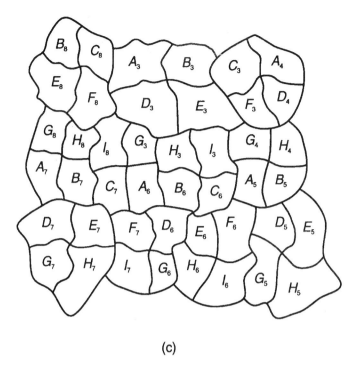

(c)

Figure 9.1 (continued).

The rules followed to assign channel sets to the different cells are based on the definition of two parameters, i and j (with $i \geq j$), called shift parameters. In Figure 9.2(a), channel set A is repeated six times around the origin cell. The channel assignment rule is the following:

- Move i cells along any of the six chains of hexagons around the original cell;
- Then move j cells along the cell chain that is 60 degrees in counterclockwise direction.

To continue the assignment of channel sets to cells, another channel set, B, is selected and the corresponding B cells are identified in the same way as the A cells. In this way, the same cell layout is replicated six times around the original cell A. In Figure 9.2(a) an example using $i = 3$ and $j = 2$ is presented. When the whole process is completed, *clusters* repeated throughput the service area are formed (Figure 9.2(b)). As indicated earlier, the number of cells per cluster is a fundamental system parameter since this number will condition

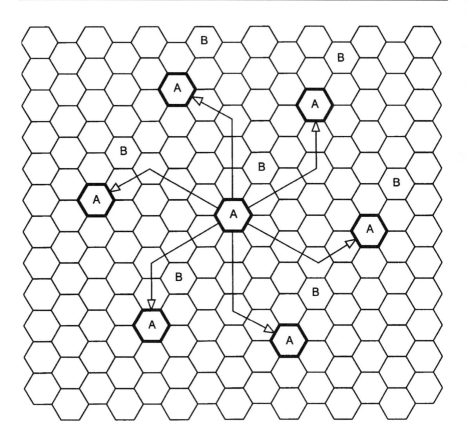

(a)

Figure 9.2 (a) Location of cochannel cells. (b) Cell clusters.

the number of available channels in each cell and thus, the traffic handling capability of the network.

The number of cells in a cluster, J, is directly related to the shift parameters through the expression

$$J = i^2 + ij + j^2 \qquad (9.4)$$

As i and j must be integers, the possible valid values of J are limited.

The ratio D/R is called normalized cochannel reuse distance, where D is the distance between cochannel cell centers, and R is the cell radius. This is related to J by the expression

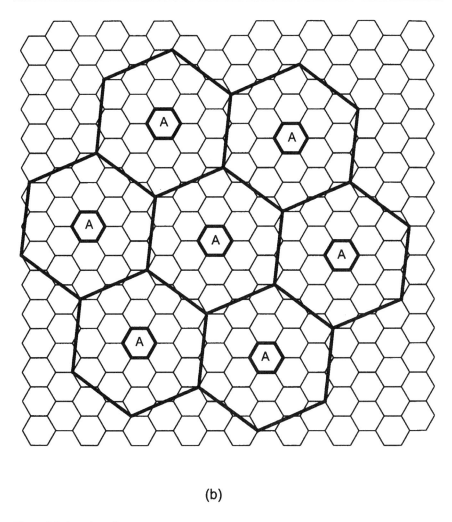

(b)

Figure 9.2 (continued).

$$\frac{D}{R} = \sqrt{3J} \qquad (9.5)$$

In real systems, the selection of J is based on cochannel interference criteria. Interference is quantified by means of the carrier-to-interference ratio, c/i. If J increases, D/R increases and thus c/i increases (i.e., the interference effect decreases). Conversely, if J decreases, the opposite effect is observed.

Base stations defining cells can be fitted with omni-directional or directive (sector) antennas. In Figure 9.3 [1] both cell types are illustrated. Also, Figure 9.3 shows the nominal cell locations on a hexagonal cellular grid. Another

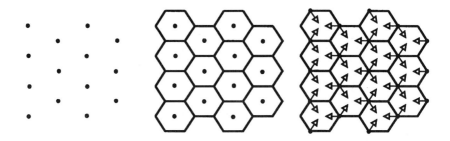

Figure 9.3 Base station positions in a cellular network.

approach would be to turn to nonhomogeneous cell pattern, especially when the terrain is very irregular and it is not possible to keep a regular network pattern.

At the first stages of cell rollout, omni-directional antennas are typically used. This is due to the fact that the main goal is to promptly achieve the required coverage area (service area). The corresponding base stations are located approximately at the center of the cell.

When the network has been in operation for a given period, directive antennas are more suitable to improve the carrier-to-interference ratio and thus reduce the cochannel reuse distance to increase network capacity. In these mature cellular networks the antenna directivity defines 120-degree cells (Figure 9.3). Sometimes six-sector cells (60 degrees) may be used.

9.4 Analog Cellular Systems

In this section, general aspects relative to analog, first-generation cellular networks are presented. A great number of systems exist, such as NMT, AMPS, TACS, Radiocom 2000, and C. The AMPS and its European version TACS have been selected here to illustrate the operation of analog cellular networks.

The main entities in any cellular network are (Figure 9.4) the following:

- Switching center (mobile switching center (MSC) or mobile telephone switching office (MTSO));
- Base stations or cell sites;
- Mobile stations.

Switching centers control the system. They also provide the interconnection to the fixed PSTN. MSCs control both base stations and mobiles. They are connected to base stations by means of the following:

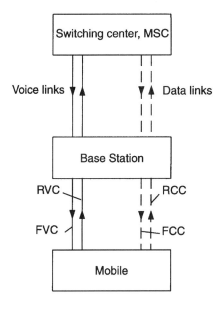

Figure 9.4 Introduction to cellular network operation fundamentals.

- *Voice links,* one for each traffic channel in the base station;
- *Data links* for signaling information interchange.

Base stations consist of the following components:

- Several transceivers—one for each voice or *traffic channel*;
- A special transceiver for the signaling channel (*control channel*).

In turn, mobiles will have a frequency synthesizer that allows them to tune to any of the RF channels allocated to the network. Figure 9.4 illustrates the connection between an MSC, a base station, and a mobile station.

When a mobile station is first turned on, it looks for a signaling channel and listens to the messages sent through it. If several signaling channels are received, possibly from neighboring cells, the mobile will measure their levels and select the one with the highest average level. The mobile tunes to this channel and will remain this way unless the received level, as the mobile moves, goes below an unacceptable value. Should this occur, the mobile will go again into a search mode to look for a more suitable signaling channel.

The signaling channel transmits the addresses of the mobiles being called. When a mobile detects that it is being called, it transmits its answer back to the network. The system, in turn, commands the called mobile to tune to an available traffic channel to carry out the actual communication.

When the mobile originates the call, a signaling interchange between the mobile station and the system will take place through the control channel to send the call request to the network.

In the course of a call, as the mobile station travels through the service area, it may happen that it may go from the original cell to another cell. The base station serving the call periodically checks the received signal level. If the received power is below a given threshold, the system looks for another base station with free channels from which the communication link is more favorable. The process of periodically monitoring the received signaling level is called *location* [1] and is carried out under control of the MSC. The process of changing the base station handling the call in progress is called *hand-off* or *hand-over*. Both mechanisms, location and hand-over, allow keeping the transmission link quality above a previously setup standard that should be the same or similar to that in fixed telephone networks. This was one of the objectives specified when the cellular concept was defined (see Section 9.1).

9.4.1 AMPS System Parameter Selection Criteria

In this section the criteria used to select several key parameters in AMPS cellular networks to fulfill the transmission quality, traffic capacity, and cost objectives are presented [1].

Base Station Location Tolerance

Generally, it will not be possible to select the base station sites at exactly their nominal positions on the cellular grid. There is, however, a tolerance. For AMPS this tolerance was set to a fourth of the cell radius ($1/4\ R$) taking into account transmission quality criteria based on acceptable c/i levels.

Maximum Cell Radius

The maximum cell radius is basically limited by the maximum mobile transmission power. This is especially true in today's networks where the most popular terminal is of the hand-held type with powers that may even be smaller than 1W. When AMPS was defined, service was mainly aimed at mobile terminals with a less drastic power limitation. It was thought that both mobiles and base stations would use EIRP levels in the order of 10W in the 800-MHz band. This meant that vehicle-mounted terminals would be fitted with transmitters with powers in the order of 12W, while base station transmitter powers would be in the order of 40W (accounting for feeders and other losses).

During the first deployment phases of the network, link ranges would be limited by noise (thermal and man-made) since frequency reuse would be carried out in a limited way. However, in subsequent evolution phases, interference would be the limiting factor. It was experimentally verified using transceivers and Rayleigh fading simulators that, with an RF c/n ratio of 18 dB a good voice quality was achieved (under noise-limiting conditions). Similarly, it was verified that, with an RF c/i ratio of 17 dB, an adequate quality was achieved. The transmission link quality level should be maintained for high percentages of locations within the service area. These percentage levels were set to 90% of locations.

To figure out the maximum attainable range under urban environment propagation conditions, experimental propagation study results were used where a propagation law of the type shown in the equation below was observed

$$L_b(\text{dB}) = K + 10n \log d \text{ (km) with } n \approx 4 \qquad (9.5)$$

with a standard deviation or locations variability, σ_L = 8 dB (Chapter 4). It was verified that the c/n threshold of 18 dB for 90% of locations was exceeded for cell radii up to approximately 12 km.

Minimum Cell Radius

The minimum cell radius is determined by totally different considerations. Each cell splitting stage places new base stations halfway between existing ones, and thus, their surfaces are a fourth of those of the original cells. This, in turn, allows an increase in the traffic handling capability of the network, as discussed earlier.

A practical value for the minimum cell radius of 1.5 km was set. This value allows for three cell splitting stages starting from 12-km cells down to 6, 3, and 1.5 km. Smaller cells would give rise to problems if the maximum value of base station location tolerance is used. On the other hand, smaller cells would give rise to a large number of cell hand-overs which might collapse the processing capabilities of the controlling switching centers, MSCs.

Cochannel Reuse Ratio

The D/R or cochannel reuse ratio greatly affects both the transmission and traffic handling capabilities of the network. Bearing in mind that the c/i objective of 17 dB must be exceeded for 90% of locations, simulations were carried out using the above-mentioned propagation model. The following D/R values were obtained:

- 4.6 for 120-degree sector cells;
- 6 for omni-directional cells.

These values correspond to cluster sizes of $J = 7$ and 12, respectively.

9.4.2 Operation of an Analog Cellular System: AMPS/TACS

As indicated in the introduction, cellular systems were conceived to achieve, among others, the following objectives: national coverage, efficient use of the assigned spectrum, voice quality close to that of the fixed telephone network, and technical and economical capability to operate both in small and extended areas with the ability to adapt to the offered traffic conditions.

The last of these objectives not only refers to the capability of handling large traffic loads but also doing it at a reasonable cost. Half or even more of the total cost of the network is in the terminals [2]. The possibility to manufacture mobile terminals at a reasonable price meant the adequate sharing of network operation functions between the mobiles and the fixed infrastructure. This was not technologically possible until the beginning of the 1980s when the AMPS system started its operation. At that particular moment the use of three important technologies was already possible [2]:

1. Stored control program (SCP) switching;
2. LSI custom circuits;
3. Microprocessors.

Without the widespread availability of these technologies the AMPS system would not have been feasible.

General System Description

Figure 9.5 shows a simplified diagram of the components of a cellular network and their connection to the PSTN. A bandwidth of 20 × 20 MHz was initially available, which, in turn, was split into 666 full duplex channels. Seven cell clusters are used, which means that 96 channels per site or 93/3 channels/sector are available. Base stations are located on three of the six vertices of the hexagonal pattern, and sector antennas are used. Each base station contains the following pieces of equipment:

- Radio transceivers;
- Control devices;

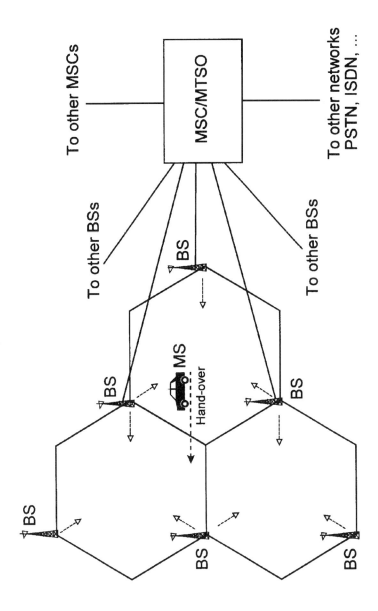

Figure 9.5 Schematic representation of an AMPS network.

- Audio processor (2:1 syllabic compander);
- Maintenance gear;
- Receive and transmit antennas with associated elements such as associated combiners, multicouplers, and feeders.

Base stations are connected to switching centers (MSC, MTSO in AMPS terminology) that control the system by means of line or radio links. Each traffic channel is directly connected to the MSC. Also, each base is connected to its corresponding MSC by means of 2,400-bps data links to interchange traffic and control information, such as call setup and hand-over.

During the first stages of network expansion, omni-directional cells were used, while, in mature stages 120-degree sector cells have been deployed. Each base station has, thus, three directive antennas for transmission and three pairs of receive antennas for space diversity. This means that full 360-degree coverage is achieved around a sectorized base station. Each directive antenna is assigned 32 channels (96/3). The use of directive antennas instead of omnidirectional antennas introduces a c/i improvement of 4–5 dB.

As the mobile travels across the service area, location measurements are carried out every few seconds. These measurements are carried out by the base currently serving the call. Also, neighboring cells carry out measurements on the frequency used for the call being monitored. These measurements determine if the antenna handling the communication with the mobile terminal must continue doing so or the call must be handed over to another antenna of the same base station or of another base station. In this way, transmission quality is guaranteed throughout the whole network service area.

9.4.3 Control System

In this section, several basic operational procedures as well as control system elements are described [2].

Call Setup

Each base station is fitted with, at least, one radio channel for setting up calls: control or signaling channel. Figure 9.6 [3] illustrates the call setup process, which is explained in more detail later. In this process, data messages are interchanged via the control channel and signaling tones over the traffic channels.

Control channels carry out the following functions:

- *Paging:* Paging messages, including the called user terminal's ID, are sent over the control channel downlink to warn of the existence of a call addressed to it.

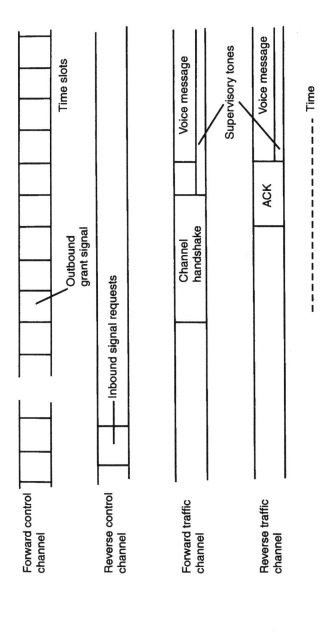

Figure 9.6 Signaling and conversation phases [3].

- *Access:* Messages are exchanged in both directions, base station-mobile station and mobile station-base station. In the uplink, mobile call setup messages and mobile location updating and registration messages are sent; while, in the downlink, registration acknowledgments as well as mobile traffic assignment messages are sent.
- *General information:* In the downlink, messages, including basic system parameters, location area IDs, and page channels used, are sent.

When a mobile is in the idle state, it automatically tunes to the control channel received with the highest average level. The downlink control channel transmits a continuous synchronous stream of data including paging messages containing the called terminal ID. When no information is transmitted via the forward control channel, *filler text* messages are sent to maintain continuous synchronous transmission. Also, *overhead words* will be transmitted periodically with the following information:

1. Location area identification;
2. Number of channels to be scanned;
3. Data necessary for the mobile to identify the access channels when the paging and access functions are not implemented on the same radio channels in systems with a high traffic load.

Fixed Network Originated Calls

When the mobile detects that it is being paged, it answers by trying to capture the reverse control channel [3]. This channel operates in a totally independent way with regard to the forward one. Through it, isolated data bursts are randomly transmitted by the mobiles. Since there is a contention between all mobiles in the same cell to seize the uplink, collisions may occur. Several techniques are implemented to minimize the number of collisions. To do this on the downlink binary data flow, every eleventh bit is set aside to inform the mobiles of the busy/idle state of the reverse control channel.

The mobile transmits a channel seizure message over the reverse control channel. This message is called *seizure precursor*. After sending this message the mobile waits a given time (time window) to see if the busy/idle bits on the downlink change from idle to busy. If this does not happen within a given time window, the channel capture attempt is immediately aborted. If the initial channel capture did not succeed, the mobile will try again to seize the channel a limited number of times.

When the mobile finally captures the uplink control channel, it sends a response to the received paging message. In its response, the mobile sends back to the base its ID or mobile identity number (MIN), its serial number or equipment serial number (ESN), and a paging acknowledgment message. This information is, in turn, sent from the base station to the MSC through the 2,400-bps data link. The MSC will assign a free traffic channel to complete the call. The control assignment message is sent through the downlink control channel.

As a consequence of the received instructions, the mobile tunes to the assigned traffic channel and informs the base station of this by retransmitting one of three audio tones at around 6 kHz (supervisory audio tone (SAT)) sent by the base through the forward traffic channel. When the base station receives back the SAT retransmission, it sends an *alert signal* through the traffic channel that makes the mobile terminal ring.

The mobile acknowledges receipt of the alert message by sending back a 10 kHz signaling tone (ST) over the reverse traffic channel until the mobile answers the call. Next, the communication itself can commence through the assigned traffic channel. In Figure 9.7, [3] a time diagram with the different events and messages interchanged between the mobile station, the base station, and the MSC is given.

Mobile-Originated Call (Figure 9.8)

In this case the process is very similar [3]. The mobile user keys the number of the called terminal (fixed or mobile) into the terminal's memory. No transmission on the control channel is carried out until the full called number is keyed and the "send" button is pressed. In this way, no useless seizure of the reverse control channel is made.

The mobile tries to seize the reverse control channel and transmit its own ESN as well as the called terminal number. This information is passed on to the MSC, which, in turn, assigns a free traffic channel to the call request. At the same time, the call must be completed to the other end by appropriate routing procedures.

Cell Hand-over (Figure 9.9)

One of the key features of mobile cellular networks is their capability to hand over one call being served by a base station to another base station or to another sector of the same base station. This also means that the mobile will be tuned to a different radio channel after the hand-over process is completed without the mobile user noticing it. In the AMPS system the location information is obtained by the base station serving the call, together with neighboring base stations, by periodically carrying out received field strength measurements and

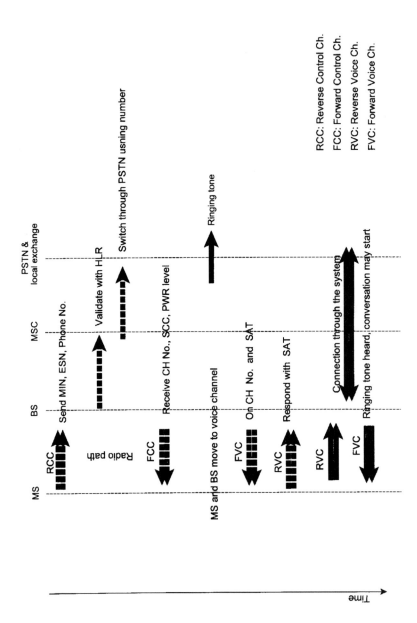

Figure 9.7 Mobile-terminated call signaling message interchanges [3].

The Cellular Concept

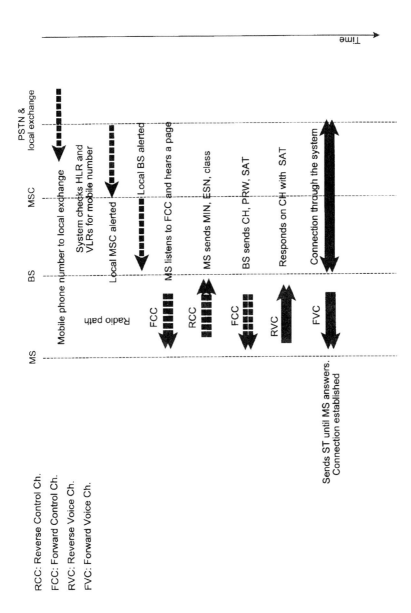

Figure 9.8 Mobile-originated call signaling message interchanges [3].

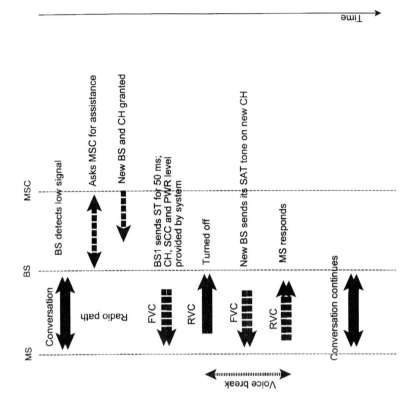

Figure 9.9 Hand-over signaling message interchanges [3].

RCC: Reverse Control Ch.
FCC: Forward Control Ch.
RVC: Reverse Voice Ch.
FVC: Forward Voice Ch.

by monitoring the reception quality of the SAT tone. Location information is transmitted by the base stations to the MSC by means of data links. The MSC will decide if a call hand-over is needed. If this is the case, a free traffic channel belonging to another sector in the serving base station or to another base station is assigned to the call. The MSC commands the candidate base station to activate this traffic channel and start sending its SAT tone. Moreover, the MSC commands the base station currently serving the call to send over the traffic channel a data message ordering the mobile to tune to the new traffic channel.

The data transmission through the traffic channel is carried out by means of a technique called blank-and-burst (i.e., the voice communication is momentarily interrupted for approximately 50 ms to send a burst of data). The mobile station, after receiving the hand-over command, sends back to the serving base station a short 10-kHz ST burst over the reverse traffic channel. After that, the mobile transmitter turns itself off, tunes the new traffic channel, and starts retransmitting the SAT tone received from the new base station.

Likewise, the MSC must reconfigure its switching matrix to adequately route the call to the new base station. The new base station, after receiving the SAT tone considers that the hand-over process has been carried out successfully. The full process takes approximately 0.2s including the blank-and-burst transmission and the radio channel retuning. The whole process does not introduce a significant degradation on a voice communication. However, a data transmission may be significantly affected unless appropriate correction procedures are implemented.

Signaling Tone (ST)

Some of the basic functions of the 10-kHz ST were explained in the preceding paragraphs. Here, these functions are revised [3]. The ST is used for the following purposes:

- *Hand-over acknowledgment:* When a hand-over command is received at the mobile station it stores the new traffic channel, the SAT tone, and the power level in the message and sends back to the base station a ST burst for 50 ms before going on to tune the new assigned channel.
- *Hook flash:* It is used to request supplementary services during a call in progress. The user keys in the required service and presses the "send" button. This will originate a 400-ms ST burst on the traffic channel uplink.
- *Call cleardown:* It is used to indicate to the base station that the call has finished (i.e., the user has hung up). A 1.8s ST burst is sent to the serving base station.

- *Alert confirmation:* When the mobile terminal starts ringing, it sends the ST back to the base station until the user answers the call.

Supervision Audio Tone (SAT)

The other audio tone used for signaling purposes over the traffic channel is the SAT, which has the following missions [3]. It enables the distinction between the same radio channels used in different cell clusters by means of a *color code* (Figure 9.10). Three SAT frequencies are available. In addition to the three SATs, there is a *digital color code* (DCC) used within digital messages.

The second major function of the SAT tone is to verify the continuity of the radio link between the mobile and the base. If SAT tone reception is lost, a timer is started. If the timer count exceeds a given value, the communication is aborted.

During the call setup phase, the cellular network indicates the mobile, which is the SAT tone that it should expect.

9.4.4 Signaling Message Formats

The transmission of signaling data messages is subject to fast fading and shadowing effects. This makes it necessary to include suitable protection in the transmitted codewords [2]. Fast fading effects cause errors that tend to

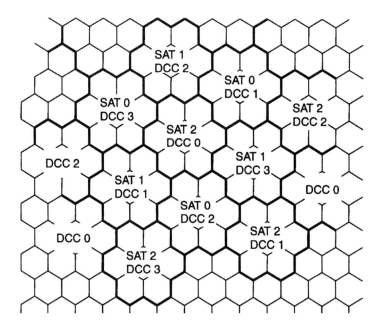

Figure 9.10 Cellular layout and DCC and SAT assignments.

group together in bursts. To protect the transmissions of signaling information messages, these are coded and retransmitted several times. At the receiving end a majority voting decision is carried out. In addition to the retransmission scheme, a BCH error control code is used. Table 9.1 indicates the coding schemes and the number of retransmissions for each of the channels and directions. Figure 9.11 shows the data formats used both on control and traffic channels.

With this protection, acceptable BERs and false alarm rates are achieved. Signaling information transmission is carried out at a 10-kbps gross binary rate using direct FSK modulation of the RF carrier. Due to the coding scheme used and the repetitions, a maximum net rate of 1,200 bps is available for signaling purposes. Before reaching the modulator, the data is Manchester-encoded to increment the number of transitions in the digital stream and thus facilitate the synchronization functions.

As regards the forward control channel, mobile identities are split into even and odd. Even parity ID mobiles only listen to "A" words in the forward traffic channel, while mobiles with an odd ID will only listen to "B" words (Figure 9.11).

The main messages sent over the forward control channel are the following [3]:

- Mobile paging: MIN;
- Power level: Mobile attenuation code (MAC);
- Transfer (voice) channel: CHAN;
- SAT code: DCC.

9.4.5 Specific Features of AMPS and TACS

Even though the AMPS and TACS systems are similar, they present some differences mainly due to the frequency bands and channel spacings used in

Table 9.1
Codes and Number of Repetitions Used on the Different Channels for Signaling

Channel	Data Bits	Parity Bits	Repeats
FCC	28	12	5
FVC	28	12	11
RCC	36	12	5
RVC	28	12	5

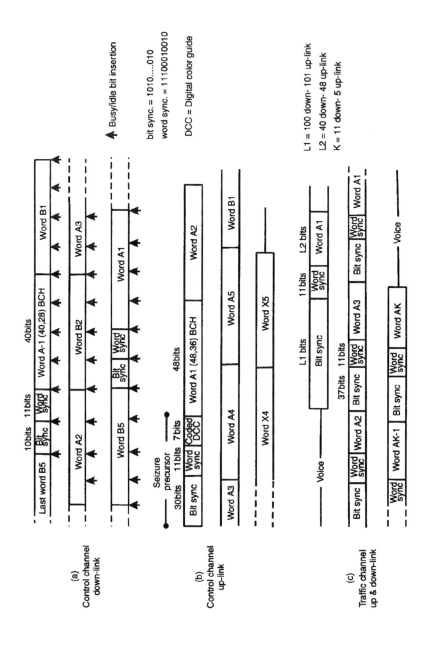

Figure 9.11 (a) Forward control channel data format. (b) Reverse control channel data format. (c) Traffic channel data format [2].

America and Europe. In this section, some additional technical details on the TACS system are also presented.

As for AMPS, in the United States, cellular systems have been allocated to the 800-MHz band, while, in Europe, the allocated band to TACS is 900 MHz. Another relevant difference between the United States and Europe is the channel spacing used: 30 kHz in America and 25 kHz in Europe. Figure 9.12(a) shows the channels allocated to the AMPS system. The available spectrum is split in two halves to be used by two different operators in each region or market. In addition to the original 666 channels, the band was expanded to 166 further channels (5 MHz).

The spectrum allocation in Europe is shown in Figure 9.12(b). 50 MHz (25 × 25) have been allocated in the 900-MHz band for mobile communications. This means that 1,000 full duplex 25 + 25 kHz channels are available. In the long term this band will be devoted to exclusive GSM use. However, during several years the band will be shared between TACS and GSM networks in the countries where both systems exist. Eventually, the TACS system will release the band to be used exclusively by GSM networks. The TACS band was also extended (ETACS) so that a total of 1,320 channels have been defined.

The peak frequency deviation in AMPS is 12 kHz, while in TACS, since the channel spacing is smaller, a value of 9.5 kHz is used. This deviation, according to Carson's rule, corresponds to a modulated signal bandwidth of BW = 2(9.5 + 2.5) = 24 kHz. The large frequency deviation causes the modulated signal to overflow to the two adjacent channels. This means that it will not be possible to use the adjacent channels in the adjacent cells.

Other differences between TACS and AMPS consist of the ST frequency used. In AMPS the ST is 10 kHz while in TACS it is 8 kHz. As for the binary data rate used, this is 10 kbps in AMPS while in TACS it is 8 kbps.

Other TACS System Features

The ERP of base stations is fixed and selected according to the required cell radius. The maximum allowed value is 100W. For mobile stations, several power levels, according to different mobile station categories have been established. Power levels go from a minimum of −22 dBW (6,3 mW) to the maximum indicated in Table 9.2. The power used varies according to the MAC assigned at call setup.

Class 1 terminals are vehicle mounted while other classes correspond to portable and hand-held terminals. The power control dynamic range is 32 dB for mobile stations and 20 dB for hand-held terminals.

Example 9.1 Typical TACS Link Budgets For the case of vehicular terminals, Table 9.3 shows a typical link budget for both the uplinks and downlinks.

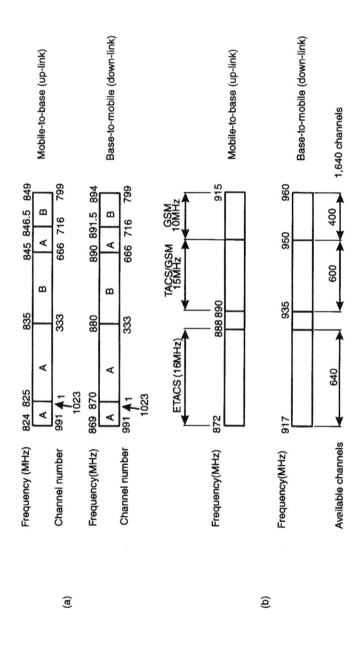

Figure 9.12 (a) AMPS system spectrum allocation. (b) Provisional sharing of the 900-MHz band between ETACS and GSM [3].

Table 9.2
Maximum ERPs for the Different TACS Terminals

Mobile Station Category	Nominal ERP	Transmitter Power (Assuming G_{MS} = 1.5 dBd)
Class 1	10 dBW (10W)	8.5 dBW (7W)
Class 2	6 dBW (4W)	4.5 dBW (2.8W)
Class 3	1.6 dBW (1.6W)	0.5 dBW (1.1W)
Class 4	−2 dBW (0.6W)	−3.5 dBW (0.45W)

Table 9.3
Link Budget for Vehicular TACS Terminals

Parameter	Base Station-Mobile Station Direction	Mobile Station-Base Station Direction
Transmitter power (dBm)	40	35
Tx antenna gain (dB$_i$)	10	3
Rx antenna gain (dB$_i$)	3	10
Diversity gain (dB)	0	6
Rx sensitivity (dBm)	−116	−116
Margin (dB)	(13)	(13)
Path gain (dB)	156	157

A 13-dB margin has been included to account for fading effects. A link budget for handheld terminals is presented in Table 9.4. In this case, the lower mobile station power (0.6W) with regard to that of vehicular terminals (3.5W) is compensated using 60-degree sector antennas which provide a 17-dB gain.

9.5 Rolling Out an AMPS Network

In this section the elements used to configure an AMPS network are presented so that the required objectives of transmission quality and localized traffic demand adaptation capability are achieved [1].

9.5.1 Adjacent Channel Interference Reduction

A channel spacing of 30 kHz is used to transmit voice using frequency modulation with a peak deviation of 12 kHz. Signaling data is transmitted at low

Table 9.4
Link Budget for Hand-held TACS Terminals

Parameter	Base Station-Mobile Station Direction	Mobile Station-Base Station Direction
Transmitter power (dBm)	40	28
Tx antenna gain (dB$_i$)	10	0
Rx antenna gain (dB$_i$)	0	17
Diversity gain (dB)	0	6
Rx sensitivity (dBm)	−116	−116
Margin (dB)	(13)	(13)
Path gain (dB)	153	154

data rates using FSK modulation. A 40-MHz band is available in the 800-MHz band. This band is split into two sub-bands to implement full duplex channels. The total number of channels is thus

$$C = \frac{40 \text{ MHz}}{2 \cdot 30 \text{ kHz}} = 666 \text{ channels} \qquad (9.15)$$

A set of full duplex channels is assigned to each cell. The transmit-receive separation must be large enough (45 MHz) for an adequate duplexer operation, and the separation between transmitted frequencies in the same sub-band must be enough to allow a good operation of the cell-site transmitter combining devices.

Once the parameter J (number of cells per cluster) has been selected, the available channels are split into J sets. If a given cell is assigned the n-th set, this means that the cell will use channels n, $n + J$, $n + 2J$, ... For example, if $J = 7$, set 4 will consist of channels 4, 11, 18, . . . The channel set distribution within the cluster must also avoid adjacent channel interference. Figure 9.13 illustrates this for a $J = 12$ cluster.

This $J = 12$ layout was used in the initial phases of expansion of AMPS networks. The purpose was to achieve a fast coverage of the service area. During evolved phases of network deployment, 120-degree sector cells were used (in the case of TACS networks, even 60-degree sector cells were used). When going from a $J = 12$ to a $J = 7$ network layout it is impossible to avoid using adjacent channels in neighboring cells. However, it is possible to benefit from the directivity properties of sector antennas by dividing each seven channel set into three subsets. In this way, adjacent channel interference may be prevented thanks to the front-back gain ratio of sector antennas. As an example of subset

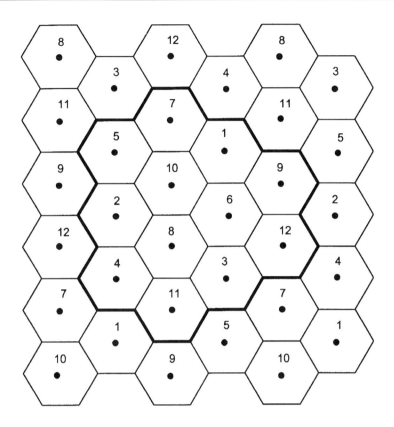

(a)

Figure 9.13 (a) Twelve-cell cluster of omni cells. (b) Seven-cell cluster with 120-degree sectors [1].

channel division, set 4 made up of channels 4, 11, 18, 25, 32, 39, 46, 53, 60, ... is split into the following three subsets [1]:

- Subset 4a consisting of channels 4, 25, 46, ...
- Subset 4b consisting of channels 11, 32, 53, . . .
- Subset 4c consisting of channels 18, 39, 60, . . .

Figure 9.13(b) illustrates how adjacent channel interference is rejected by the front-back ratio of sector antennas [1]. For set 6 serving point M,

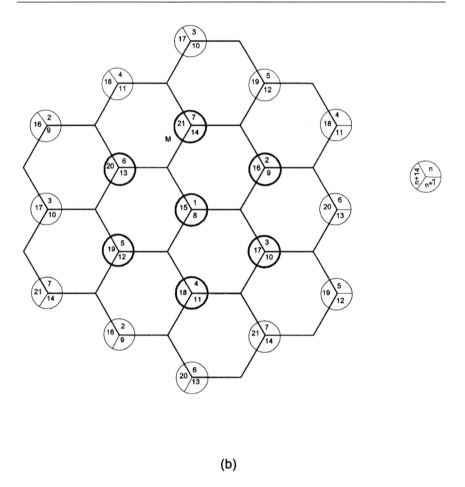

(b)

Figure 9.13 (continued).

adjacent channel 7 interference is sufficiently attenuated by the small gain in the backward direction.

9.5.2 Cell Splitting

By means of the cell-splitting approach a cellular network can accommodate nonuniform traffic densities. When cell splitting is carried out, the distance between adjacent cells is divided by half, so that the cell surfaces are divided by four, and thus, the traffic handling capability is increased by four as well. The nominal positions of the new cells are defined halfway between existing cells. However, a certain tolerance is admitted as indicated in Section 9.4.1. Both new and existing base stations form two hexagonal grids.

Figure 9.14 shows a network made up of large cells to which six smaller cells have been added. The channel set assigned to each new base station is determined by observing that the new station will be halfway between two existing cochannel cells. In Figure 9.14, this procedure is illustrated with a circle whose cells belong to set 2. The reuse patterns both in large and small cell clusters are kept in successive splitting phases but with a 120-degree rotation [1].

9.5.3 Overlaid Cell Concept

As the network evolves adapting itself to localized traffic demand, both large and small cells will co-exist. An example of this is shown in Figure 9.15 [1]. This co-existence leads to problems that must be known and solved.

To keep adequate transmission quality conditions throughout the network service area, an appropriate c/i level must be assured. This is equivalent to keeping the required cochannel reuse ratio, D/R. In Figure 9.15(a), a network with two cell sizes and a seven cell reuse pattern is shown. Considering a large cell A1 belonging to a seven-cell reuse pattern, the D/R ratio of 4.6 must be kept. This is fulfilled for cells A_2 and A_3 as well as with the small cells, A_4 and A_5.

Station A_1 is also correctly placed with respect to cochannel cells A_4 and A_5 given that the D/R requirement is accomplished, since the radius $R(small)$ is smaller for the small cells. The channels in cell A_1 will not cause harmful interference on calls handled by cells A_4 and A_5, given that their mobiles will be at a distance $R(small)$ from cells A_4 and A_5 in the worst case.

However, there is a problem for the calls handled by cell A_1. Its mobiles can be at a distance $R(large)$ from it. Consequently, cells A_4 and A_5 with regard to cell A_1 will not comply with the $D/R = 4.6$ requirement. This requirement would be fulfilled if cell A_1 would handle mobile stations only within an area defined by $R(small)$, but this, of course, would give rise to coverage gaps.

Overlaid cell networks are those where cells of different sizes co-exist. The best way to view these networks is to think of a cell grid of a given step on which another incomplete (fragmentary) cell grid with half the initial step is superposed. The larger cell pattern can not be removed in a given area until the smaller cell pattern is completed.

The practical implementation of the concept of overlaid cell networks requires that, in the region where cells with different sizes co-exist, each channel subset is further divided into two groups, one for the larger cells and the other for the smaller cells.

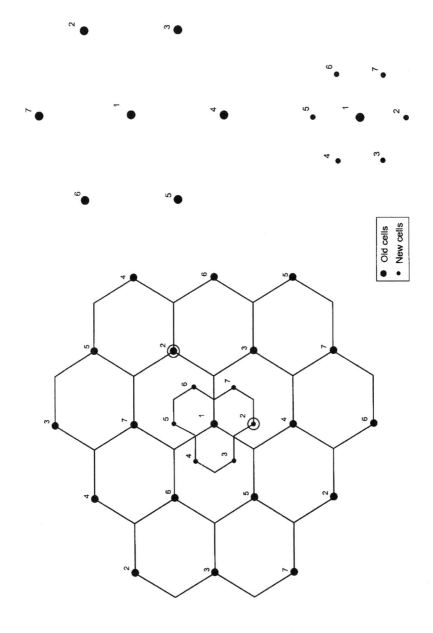

Figure 9.14 Cell splitting [1].

The Cellular Concept

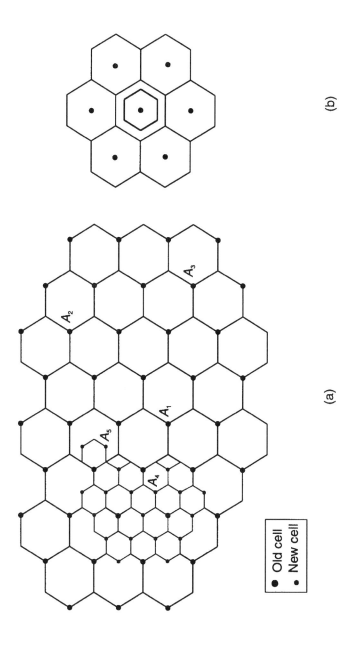

Figure 9.15 Co-existence of cells with different sizes.

Each sector in the older base stations will continue to use some of its assigned channels to provide a coverage area corresponding to a larger cell. The rest of the channels assigned to that sector will be restricted for smaller cell coverage. The proportion of large to small cell channels will depend on the channel needs of the neighboring smaller cells [1].

For example in Figure 9.15(a), any channel installed in base stations A4 and A5 cannot be used in cell A1 except to provide small cell coverage. These restrictions are implemented on the controlling MSC software.

In the case of a call being handled by a small cell channel that requires hand over, this could be made to a channel in a neighboring cell or to a channel belonging to the large cell group of the same base station.

9.5.4 Example of an AMPS Network Rollout Plan

In this section a possible AMPS network rollout plan is presented [4]. It will be assumed that a total of 666 channels are available, 21 of which are set aside for signaling purposes; the remaining 645 will be used as traffic channels.

When cells of different sizes or cells with omni and sector antennas co-exist, interference effects may occur in neighboring areas. To prevent this, the available channels are subdivided into two groups, *group A* and *group B*. Each of these frequency groups will be used on either side of the boundary between these regions.

Initially, each service area, e.g. a city, will be served by *large cells* of 12 km radius. Base stations will be located on the center of the cells, and *omni antennas* will be used. Twelve-cell clusters will be initially deployed as shown in Figure 9.13(a). Hence, $645/2 \times 12 = 26$ traffic channels per cell will be available for group A.

The next step, to accommodate larger traffic demands, will be to use sector cells. To this end, the cell hexagonal pattern is redefined by sliding the whole pattern until the existing base stations are placed on the corners of the hexagons in the grid. Now, directive antennas defining 120-degree sectors must be used (Figure 9.13(b)). In Figure 9.16 both hexagonal patterns are shown, one superposed on the other.

The use of directive antennas helps improve coverage conditions on poorly covered spots, as well as reduces interference effects. This step, then, consists of switching from a 12-cell cluster structure to a seven-cell cluster structure. Each sector antenna is assigned one-third of the channels corresponding to the base station: that is, $7 \times 3 = 21$ channel subsets are defined for the seven-cell cluster. Each cell will be fitted then with an average of $645/2 \times 7 = 46$ channels. These channels shall belong to *group B*.

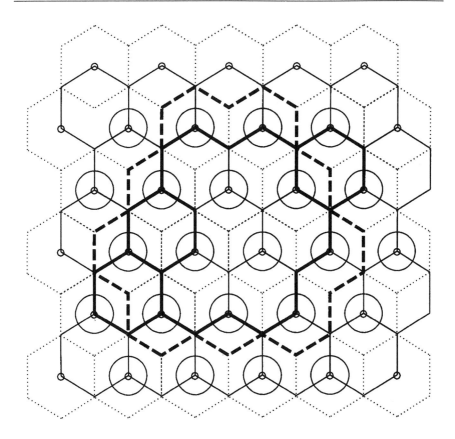

Figure 9.16 Superposition of the 12-cell and seven-cell patterns.

The next step is to define new sector cells with half the radius of the original *large cells* (*cell splitting*) (Figure 9.14). Each of these smaller cells will be fitted with 46 channels belonging to *group A*. However, since these cells have half the radius, there will be four times more smaller cells and, hence, four times more channels per km^2. Cell splitting can be carried out up to three times. The minimum cell radius depends on the following:

- The base station location tolerance (1/4 R);
- The capability of the MSC to handle large numbers of call handovers.

The minimum recommended cell size is 1.5 km. Some distance within the area served only by 1.5-km radius cells, channels from both *groups A and B* could be used. This means that an average of 92 channels per cell will be available. Table 9.5 illustrates and quantifies these steps. It has been assumed that the traffic density is uniform throughout the service area.

9.6 Structure and Basic Functions of a Cellular Network

Cellular networks are structured in a hierarchical way (Figure 9.17) in much the same way as fixed telephone networks. Base stations are on the lowest level and each of them define a base station area. Next are the location-paging areas made up of several base stations. A mobile can travel inside a location area without having its location register updated. The third level is comprised of the switching center areas, MSCAs, comprising several base station areas linked to the same MSC.

Several interconnected MSCs make up a public land mobile network (PLMN). Some of the MSCs will be responsible for the interconnection of the PLMN to fixed telephone networks, PSTNs. In turn, MSCs form a meshed network. If the network is large enough, higher hierarchical levels may be set up by introducing transit switching centers.

The ensemble of MSCAs defines the service area of the cellular network. Within this service area connections may be set up with other mobile or fixed users. The service area may partially or totally encompass one country or even expand over several countries.

Already, some of the basic functionalities of cellular networks have been analyzed:

1. Mobile paging;
2. Call hand-over.

Another basic feature must be implemented: the possibility to roam inside the service area. This last feature allows the location of a mobile wherever it may be within the network service area. In some cases, the mobile may even be located when it is traveling within other PLMN (international roaming)—for example, outside of the United States, in Canada or Mexico.

Each mobile is associated to a given MSC (home). When the mobile is first turned on it must perform a *registration* with the network, through the base station providing the highest signal level. This process may be repeated periodically (after a few minutes). Also a location update will be carried out whenever the mobile travels to another location area.

For the case when the mobile is not within its MSC area, databases are set up based on intelligent network (IN) concepts to implement the mobility

Table 9.5
Deployment Plan for an AMPS Network [4]

Cell Radius (Km)	Antenna	Cells/Cluster	Gr.A	Gr.B	Ch/Cell	Area (km²)	Ch/km²	TIR*	Traffic Density (E/km²)	Users/km²
12	Omni	12	x		26	452	0.057	1.0	0.035	0.7
12	120-degree	7		x	46	452	0.101	1.77	0.076	1.52
6	120-degree	7	x		46	113	0.407	7.14	0.303	6.06
3	120-degree	7		x	46	28	1.643	28.824	1.225	24.5
1.5	120-degree	7	x		46	7	6.571	115.28	4.9	98
1.5	120-degree	7	x	x	92	7	13.143	230.58	10.94	218.8

*Traffic load increment ratio.
†B(Blocking probability = 1%, 26 channels) = 15.8 Erlang; B(Blocking probability = 1%, 46 channels) = 34.3 Erlang; B(Blocking probability = 1%, 92 channels) = 76.6 Erlang.
‡Assuming one three-minute call per user during the BH = 0.05 Erlang/user.

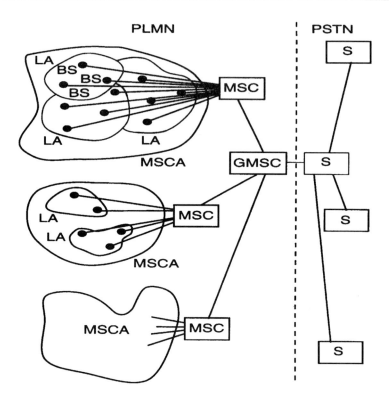

Figure 9.17 Structure of a cellular network.

management functions needed. These databases are called home location register (HLR) and visitor location register (VLR). When the mobile is not within its own location area, the MSC serving it will inform the mobile's HLR of where the mobile is located. The VLR of the new MSC will be supplied with all the required information to handle calls to or from the mobile (See Chapter 10 for more details). For mobile-terminated calls the HLR shall be consulted by the network to gather information on where the mobile is located to page it and to route the call to it.

9.7 Dimensioning of Cellular Networks

Cellular network dimensioning consists of the evaluation of the number of radio channels required at each base station to insure a given GoS. This quality parameter gives the probability of unsuccessful call attempts. These may happen due to traffic congestion or blocking (p_t) or due to a lack of radio coverage ($1 - p_c$). The GoS in percent units is given by

$$\text{GOS (\%)} = 100[1 - (1 - p_t) \cdot p_c] \quad (9.6)$$

In cellular telephone networks, the blocking probability is given by the Erlang-B formula

$$p_t = B(A, N) = \frac{\dfrac{A^N}{N!}}{\displaystyle\sum_{k=0}^{N} \dfrac{A^k}{k!}} \quad (9.7)$$

where A is the offered traffic and N is the number of traffic channels. In Appendix 9B a table with the traffic, A, for different values of N and p_t is available.

Example 9.2. If the overall probability objective is set to $p = 0.1$ (10%) and a typical coverage probability is $p_c = 0.95$ (95%), it results that

$$0.1 = 1 - (1 - p_t) \cdot 0.95$$

This section deals only with traffic issues. Previous chapters studied propagation effects and their influence on the coverage probability. It will be assumed that the available bandwidth $2B$ is divided in two halves of width B for the uplink and downlink. Assuming a channel spacing Δf, the total number of available channels is

$$C = \frac{B}{\Delta f} \quad (9.8)$$

A J-cell reuse pattern will also be assumed; thus, the number of channels per cell is

$$N = \frac{C}{J} \quad (9.9)$$

If the average call duration is H seconds and the number of calls that each mobile makes during the busy hour is L, the offered traffic per mobile will then be

$$a \text{ (Erlang/mobile)} = \frac{H \text{ (seconds)} \; L \text{ (calls/mobile)}}{3{,}600 \text{ seconds}} \quad (9.10)$$

If an overall lost call probability, p, is specified and the lost call probability due to lack of coverage is $1 - p_c$, then the blocking probability, p_t, is calculated using the expression

$$p_t = 1 - \left(\frac{1-p}{p_c}\right) \qquad (9.11)$$

Now, the maximum offered traffic per cell is computed as

$$A = B^{-1}(N-1, p_t) \qquad (9.12)$$

given that at least one of the N available channels is used for signaling purposes. B^{-1} is the inverse Erlang-B formula. If a maximum offered traffic A can be handled by a cell and each mobile generates a traffic a, then the maximum number of mobiles that can be handled by a base station for a given GoS is

$$m = \frac{A}{a} \qquad (9.13)$$

As seen in Section 9.2, traffic data is expressed as a density (Erlang/km^2). In this case, the maximum traffic density per cell is

$$\rho_a = \frac{A}{\text{Area}_{\text{Cell}}} \text{ (Erlang/km}^2\text{)} \qquad (9.14)$$

The total surface of a J-cell cluster is

$$\text{Area}_{\text{Cluster}} = J \cdot \text{Area}_{\text{Cell}} \qquad (9.15)$$

If the total coverage area is Area$_{\text{coverage}}$, the number of clusters in the network is

$$Q = \text{Integer part of } \left(\frac{\text{Area}_{\text{Coverage}}}{\text{Area}_{\text{Cluster}}}\right) + 1 \approx \left(\frac{\text{Area}_{\text{Coverage}}}{J \, \text{Area}_{\text{Cell}}}\right) \qquad (9.16)$$

This is also the network *reuse index*. As a consequence, the total number of traffic channels offered throughout the whole coverage area is

$$\text{Traffic channels offered} = QJ(N-1) \approx C\left(\frac{\text{Area}_{\text{Coverage}}}{J \, \text{Area}_{\text{Cell}}}\right) \qquad (9.17)$$

Taking a close look at this last expression, the following can be concluded:

- If Area$_{cell}$ is reduced then the total number of offered channels increases;
- If J is reduced also the total number of offered channels increases.

A problem arises, however, when the number of cells per cluster, J, is reduced since the carrier-to-interference ratio may go below the required protection ratio, R_p. This means that a network rollout criteria is to evaluate first the minimum number of cells per cluster, J and, once this is set, increase the network capacity by reducing Area$_{cell}$. Finally, the total number of mobiles that can be served with a GoS specified by a blocking probability p_t is

$$M = QJm \tag{9.18}$$

Example 9.3. Suppose a cellular system with omni cells of radius 2 km and clusters of $J = 7$ cells. The total number of available channels is $C = 287$. The blocking objective is set to 10%. Assume that the traffic per mobile is $a = 25$ mErlang and that the total area to be covered is Area$_{coverage} = 400$ km^2 (for example, a metropolitan area of 20×20 km^2). The results are as follows:

1. Number of channels per cell, N

$$N = \frac{C}{J} = \frac{287}{7} = 41 \text{ channels/cell} \tag{9.30}$$

That is, there are 40 traffic channels and 1 control channel.

2. Offered traffic per cell, A. It will be the maximum offered traffic that can be handled with 40 traffic channels for a blocking probability of 0.1.

$$A = B^{-1}(N-1, p_t) \tag{9.19}$$

To calculate this value the Erlang-B Table in Appendix 9B is used. An offered traffic value $A = 38.3$ Erlang/cell, can be read from the table.

3. Number of mobiles per cell, m

$$m = \frac{A}{a} = \frac{38.3}{0.025} = 1532 \text{ mobiles/cell}$$

4. Traffic density, ρ_a,

$$\rho_a = \frac{A}{\text{Area}_{\text{Cell}}} = \frac{38.3}{\pi 2^2} = 3.05 \text{ Erlang/km}^2$$

5. Reuse index, Q,

$$Q = \text{Integer part of} \left(\frac{\text{Area}_{\text{Coverage}}}{\text{Area}_{\text{Cluster}}}\right) + 1 = \text{Integer part of} \left(\frac{400}{7\pi 2^2}\right) + 1 = 5$$

6. Total number of offered channels:

Total number of traffic channels offered = $QJ(N-1) = 1365$

7. Total number of mobiles in the system, M:

$$M = QJm = 5 \cdot 7 \cdot 1532 = 53{,}620 \text{ mobiles}$$

Now, these results are compared with those obtained for a single-cell system using all available channels $C = 281$ (280 traffic channels and one control channel). The cell radius would be

$$\pi R^2 = 400 \Rightarrow R = 11.3 \text{ km}$$

and the offered traffic

$$A = B^{-1}(280, 0.1) = 302.8 \text{ Erlang}$$

This means that the number of mobiles in the system is

$$m = M = \frac{A}{a} = \frac{302.8}{0.025} = 12{,}112 \text{ mobiles}$$

while using a cellular system 53,620 mobiles were obtained. This means that a traffic capacity improvement factor of 4 is achieved—i.e.,

$$\text{Improvement factor} = \frac{53{,}620}{12{,}112} \approx 4$$

9.8 Cochannel Interference Issues

Cellular systems are interference-limited. This means that the usual case will be to receive a certain amount of controlled interference. Interference will thus determine the transmission quality throughout the service area. Adequate transmission quality will be achieved when a c/i threshold or protection ratio is exceeded. Moreover, cochannel interference will originate from multiple sources (several cochannel cells).

Suppose that a dominant interferer exists and assume an n power propagation law. Propagation losses can be expressed in linear units in the following way:

$$l_b = kr^n \qquad (9.20)$$

Consider now the carrier-to-interference ratio at the fringe of the interfered cell ($r = R$). Assuming that the cochannel cell is at a distance D, then the received powers of the wanted signal and of the single entry interference, are, respectively:

$$c = \frac{p_t}{kR^n} \qquad i = \frac{p_t}{k(D-R)^n} \qquad (9.21)$$

the carrier-to-interference ratio is

$$\frac{c}{i} = \left(\frac{D-R}{R}\right)^n \approx \left(\frac{D}{R}\right)^n \qquad (9.22)$$

If R is reduced at the same time as D, the c/i ratio will not change (i.e., the transmission quality will not be impaired). In interference-limited systems such as cellular networks it is the c/i value that sets the coverage limits and not the noise threshold (Figure 9.18(a))

In real cellular networks, each cell is surrounded by six cochannel cells at a distance D; this is called first cochannel tier. Sometimes, it is also important to take into consideration the effects of a second cochannel tier as illustrated in Figure 9.18(b).

Taking only into consideration the first tier of cochannel interfering stations, a simple estimate of the overall c/i ratio is

$$\frac{c}{i} = \frac{1}{6}\left(\frac{D}{R}\right)^n \qquad (9.23)$$

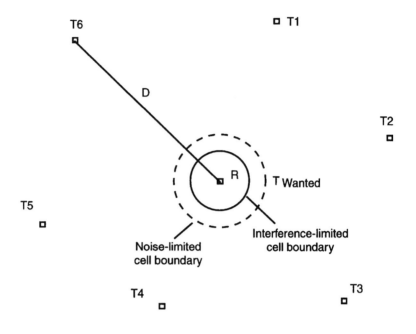

(a)

Figure 9.18 (a) Interference and noise limits in a cellular network. (b) First and second cochannel tiers.

that is, the c/i ratio is reduced, being one-sixth of that for the single interferer case.

Now, the influence of the c/i ratio on the cluster size J is assessed. First, J is expressed as a function of the c/i ratio

$$J = \frac{\text{Area}_{Cluster}}{\text{Area}_{Cell}} = \frac{1}{3}\left(\frac{D}{R}\right)^2 = \frac{1}{3}\left[6\left(\frac{c}{i}\right)\right]^{2/n} \quad (9.24)$$

For an adequate transmission quality, a given protection ratio r_p must be exceeded—i.e.,

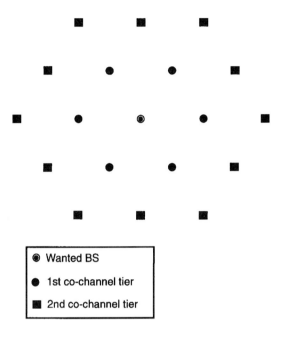

(b)

Figure 9.18 (continued).

$$\frac{c}{i} \geq r_p \tag{9.25}$$

This means, expressing J as a function of the protection ratio, that

$$J \geq \frac{1}{3}(6r_p)^{2/n} \tag{9.26}$$

This equation gives the lower bound value of J as a function of the protection ratio. As it is shown below, the value of J is extremely sensitive to the propagation law exponent n.

Example 9.4. For the AMPS/TACS analog system, the required protection ratio is R_p = 17 dB or, in linear units, r_p = 50.1 times. In Table 9.6 values of J are given for various propagation law exponents n.

On the other hand, as it is clear from previous sections, two conflicting objectives are sought. One is to achieve an adequate transmission quality that

Table 9.6
Cluster Size J for Different Propagation Exponents n

n	J
2	101
3	15
3.5	9
3.75	7
4	6

is conditioned by the received c/i ratio. The other one is to achieve the highest possible traffic capacity. However, if J is increased to improve transmission quality, the network traffic handling capability decreases since $N = C/J$. One of the major advantages of second-generation digital systems like, for example GSM, is that they require lower protection ratios. In the case of GSM the protection ratio is 9 dB (8 times) compared to the 17 dB requirement of AMPS/TACS.

Example 9.5. For $R_p = 17$ dB and $n = 3.9$, the required cluster size J is

$$J \geq \frac{1}{3}(6\ 50.1)^{2/3.9} \approx 6 \text{ cells/cluster}$$

Example 9.6. Now, the c/i values for clusters of $J = 12$ and 7 cells will be obtained for a propagation law $n = 3.9$. These cluster sizes were presented when the rollout of an AMPS network was studied. For $J = 12$, the c/i ratio is

$$\frac{c}{i} \approx \frac{1}{6}\left(\frac{D}{R}\right)^n = \frac{1}{6}(3J)^{n/2} = 180 \text{ times} = 22.6 \text{ dB}$$

On the other hand,

$$\frac{D}{R} = \sqrt{3J} = 6$$

For a typical large cell radius, $R = 15$ km, a cochannel reuse distance, $D = 90$ km is obtained. The c/i ratio of 22.6 dB is still above the specified protection ratio; this means that J can be further reduced. For $J = 7$, the c/i ratio is

$$\frac{c}{i} \approx \frac{1}{6}\left(\frac{D}{R}\right)^n = \frac{1}{6}(3J)^{n/2} = 63 \text{ times} = 18 \text{ dB}$$

On the other hand,

$$\frac{D}{R} = \sqrt{3J} \approx 4.6$$

For a typical large cell radius R = 15 km, a cochannel reuse distance D = 69 km is obtained. The c/i ratio is still 1 dB in excess of the required 17 dB. However, it must be borne in mind that the computations shown have been made for a static case (i.e., no statistical variations of both c and i have been accounted for). In real conditions, the c/i ratio must exceed the protection ratio for large percentages of locations: 90%, 95%. This means that a cluster size of J = 7 cannot be used with omni antennas. Only using sector antennas, operation with seven-cell clusters is possible in AMPS/TACS networks.

References

[1] McDonald, V. H., "The Cellular Concept," *Bell System Technical Journal,* Vol. 58, January 1979, pp. 15–41.

[2] Blecher, F. H., "Advanced Mobile Phone Service," *IEEE Trans. Vehicular Technology,* Vol. VT–29, May 1980, pp. 238–244.

[3] Macario, R. V. C., *Cellular Radio. Principles and Design,* Houndmills, Basingstoke, Hampshire, UK: Macmillan New Electronics, 1993.

[4] Stocker, A. C., "Small Call Mobile Phone Systems," *IEEE Trans. on Vehicular Technology,* Vol. VT-33, No. 3, Nov. 1994, pp. 269–275.

Appendix 9A: Summary of Parameters Used

C: Total number of available channels;
J: Number of cells in a cluster;
$N = C/J$: Number of channels in a cell;
R: Cell radius;
D: Cochannel reuse distance;
R_p: Protection ratio;
Δf: Channel spacing;
$2B$: Allocated bandwidth B+B (uplink + downlink);

Q: Reuse index;
$\text{Area}_{\text{Coverage}}$: Total coverage area;
$\text{Area}_{\text{Cluster}}$: Cluster area;
$\text{Area}_{\text{Cell}}$: Cell area;
A: Offered traffic in a cell;
a: Traffic offered by one mobile;
H: Mean call duration;
L: Number of calls in the BH;
m: Number of mobiles per cell;
M: Total number of mobiles in the system.

Appendix 9B: Erlang-B Table

Table 9.7
Blockage Probability P_t as a Function of the Number of Traffic Channels, N

N	1%	1.5%	2%	3%	5%	10%
1	0.0101	0.0152	0.0204	0.309	0.0526	0.111
2	0.153	0.190	0.223	0.282	0.381	0.595
3	0.455	0.535	0.602	0.715	0.899	1.27
4	0.869	0.992	1.09	1.26	1.62	2.05
5	1.36	1.52	1.66	1.88	2.22	2.88
6	1.91	2.11	2.28	2.54	2.96	3.76
7	2.50	2.74	2.94	3.25	3.74	4.67
8	3.13	3.40	3.63	3.99	4.54	5.60
9	3.78	4.09	4.34	4.75	5.37	6.55
10	4.46	4.81	5.08	5.53	6.22	7.51
11	5.16	5.54	5.84	6.33	7.08	8.49
12	5.88	6.29	6.61	7.14	7.95	9.47
13	6.61	7.05	7.40	7.97	8.83	10.5
14	7.35	7.82	8.20	8.80	9.73	11.5
15	8.11	8.61	9.01	9.65	10.6	12.5
16	8.88	9.41	9.83	10.5	11.5	13.5
17	9.65	10.2	10.7	11.4	12.5	14.5
18	10.4	11.0	11.5	12.2	13.4	15.5
19	11.2	11.8	12.3	13.1	14.3	16.6
20	12.0	12.7	13.2	14.0	15.2	17.6
21	12.8	13.5	14.0	14.9	16.2	18.7
22	13.7	14.3	14.9	15.8	17.1	19.7
23	14.5	15.2	15.8	16.7	18.1	20.7
24	15.3	16.0	16.6	17.6	19.0	21.8
25	16.1	16.9	17.5	18.5	20.0	22.8
26	17.0	17.8	18.4	19.4	20.9	23.9
27	17.8	18.6	19.3	20.3	21.9	24.9
28	18.6	19.5	20.2	21.2	22.9	26.0
29	19.5	20.4	21.0	22.1	23.8	27.1
30	20.3	21.2	21.9	23.1	24.8	28.1
40	29.0	30.1	30.9	32.4	34.6	38.3

ns
10

The GSM System

10.1 Introduction

In the European countries with more developed mobile communications, cellular systems since their outset have experienced a substantial increase in the number of users. This gave rise for the need to find more sophisticated and evolved technologies able to handle high volumes of traffic.

In the United Kingdom, for example, the analog TACS system was used. Cellular telephony service was provided by two operators with separate networks. Both networks have been in operation since January 1985, with such an increase in users that, in June 1989, figures rose to a total of 600,000. Already in the early 1990s subscriber figures were well over one million on both networks [1]. One of the common characteristics of all analog cellular systems has been their exponential grown in the number of users, to such an extent that some networks were on the brink of saturation. Hence, there was a real need for a more efficient system in terms of users per MHz.

A further relevant characteristic in Europe was the great diversity of existing systems varying from country to country with no operative compatibility. To sort out this problem the global system for mobile communications (GSM) was developed in Europe in the 1980s, with commercial operation starting in the early 1990s. GSM belongs to the so-called second-generation public cellular mobile communications systems.

The first generation, represented by analog systems such as NMT, C, AMPS, and TACS, was characterized by the use of frequency modulation for the voice signal and the solution of important technical problems such as efficient use of the RF spectrum, location of users on a national scale, and performance of call hand-overs between different base stations. One major drawback of these analog systems was the enormous difficulty in the provision of services other than basic telephony.

The second generation maintains the same features and advantages of the first generation. However, with the inclusion of digital transmission, both for voice, data, and signaling, allows a large number of additional (supplementary) services. Also, it has improved the spectral efficiency achieved by analog systems, producing, as a consequence of this, an increase in their traffic-handling capacity.

The complexity of the GSM system is high [2]; the official specifications take up over 8,000 pages. Table 10.1 shows how the specifications are divided up into a number of sections. This chapter briefly summarizes some of the most important aspects of the GSM system, paying particular attention to the radio interface.

Besides, GSM has been following an evolutionary approach from the initial Phase 1 with limited capabilities to the current Phase 2+ with a rich range of services and facilities. This approach intends to bridge the transition toward future third-generation mobile systems.

10.1.1 Origin and Evolution of the GSM System

In 1982, within the framework of CEPT (European Conference of Posts and Telecommunications) the so-called Groupe Spécial Mobile was set up, with the objective to define a new mobile communications system for the 1990s in Europe. The objectives initially defined for the new system included the following:

Table 10.1
Structure of the Specifications of the GSM System

00	Preamble
01	General vocabulary, abbreviations
02	Service aspects
03	Network
04	MS-BS interface and protocols
05	Physical layer on radio path
06	Audio aspects
07	Terminal adapters for mobiles
08	BTS/BSC and BSC/MSC interfaces
09	Network interworking
10	Service interworking
11	Network management, O&M
12	Equipment specifications and type approval

- The system should be pan-European (i.e., common for the entire continent) so that any subscriber could use the same terminal access the GSM network in any country.
- It should allow for a large number of services, within the context of the integrated services digital network (ISDN). In fact, it was intended that any service provided by the ISDN could also be available for the mobile user.
- There was a need to implement safety mechanisms both to safeguard the access and to guarantee privacy in communications in such a manner that they could not be picked up by unauthorized parties.
- The use of the spectrum needed to be more efficient, and the system had to have a greater traffic capacity than the previous systems.
- With time, the need to allow for increasingly lighter, low consumption terminals came to the forefront as the use of hand-held terminals became generalized.

In 1986 and 1987, evaluation tests were conducted on different methods of modulation and access. The main decisions were then made on these aspects. Although it was not one of the primary objectives for the system to use digital transmission, it was expected that such a decision should be adopted as a basic instrument for covering the objectives outlined above.

One of the first decisions was that the system would be digital, in view of its undoubted advantages over analog systems. On this basis, research programs were set up to develop candidate systems to the GSM system, all of them being digital with different access technologies:

- FDMA/SCPC (single channel per carrier);
- TDMA, either narrowband or wideband: NB-TDMA and WB-TDMA;
- CDMA.

After analyzing and evaluating the proposals, GSM tended toward the specification of a new system that took the most outstanding contributions from each proposal. In synthesis, the framework decisions that gave rise to the development of the GSM specifications were the following:

- Use of a common RF band reserved for GSM in all participant countries;
- Digital cellular structure;

- Narrowband TDMA multiple access technique (NB-TDMA);
- Low-speed binary source coding algorithm.

GSM is possibly the first telecommunications system that was specified at total system level. Its architecture and protocols are based on the layered OSI reference model. The objective of GSM integration in the ISDN has exerted an influence on the design of interfaces, taking on a logical structure closely resembling ISDN itself as regards the definition of traffic and signaling channels with a $B_m + D_m$ structure. The flexibility of this disposition works to the advantage of services that GSM will offer users in the future. GSM services will be dealt with in Section 10.7.

In 1989, the GSM group became a part of ETSI as a technical committee. The main novelty lay in the fact that manufacturers took part in this official body. Eighteen European countries agreed to start up the system from 1991.

In 1990, the GSM specifications were frozen and, in 1991, likewise for the twin DCS 1800 (GSM 1800) system. Specification of a series of additional and supplementary services was left for a second stage (GSM Phase 2), so that the system was implemented with a small set of services. At this point, it is obvious that, due to the tide of technological developments taking place and the use of these services by the public, the system had to have a certain evolutionary nature, allowing for the introduction of new services and facilities in the future. The first coverage limited systems were started up in 1991, but it was not until 1992 that the first operators started commercial operation of GSM networks.

The GSM system has also been adopted by a growing number of countries on all continents, thus changing the pan-European nature that was first envisaged. Due to its vocation towards a global system, the term Global System for Mobile Communications was adopted. This has become the official name of the system.

10.2 Basic Specifications of GSM

In brief here, follow the specifications and characteristics of GSM, which are developed in greater detail in subsequent sections:

- *Frequency bands:* Mobile station transmission of 890–915 MHz and base station transmission of 935–960 MHz.
- *Duplex separation:* 45 MHz.

- *Channel spacing:* 200 kHz, with a first adjacent channel selectivity of 18 dB (values corresponding to the second and third adjacent channels are, respectively, 50 dB and 58 dB as a minimum).
- *Modulation:* GMSK with a BT parameter of 0.3 and a modulation binary rate of 270.83 kbps.
- *Protection ratio:* For cochannel interference, R_{pc} = 9 dB and for first adjacent channels, R_{pa} = −9 dB.
- *Maximum BS EIRP:* 500W per carrier.
- *Nominal MS transmit power:* Five *MS classes* have been specified, according to their maximum transmitter powers (Table 10.2) i.e.: 0.8, 2, 5, 8 and 20W. Mobile stations may be of various types: vehicular, transportable, and hand-held.

  The market, however, has imposed the hand portable as the universal phone so, in fact, class 4 units have become the standard. There are also some class 5 products, mainly PC cards, designed for use in laptops.
- *Nominal BS transmit power:* For BSs, seven power classes were defined in 3 dB steps ranging from 2.5 W (4 dBW) to 320 W (25 dBW) (Table 10.2).
- *Cellular structure and frequency reuse pattern:* conventional cells (macrocells) may be used with radii between 1 km (urban areas) and 35 km (rural areas). In regions with a high traffic density, cells may be sectorized by the use of directive antennas. The usual reuse pattern for traffic channels is 4/12 (i.e., four trisectorized sites). With the aid of GSM functionalities such as frequency hopping, a more efficient 3/9 reuse pattern is achievable. Nevertheless, broadcast control channels

Table 10.2
MS and BS Classes

Class	MS Max. Power	BS Max. Power
1	20W (43 dBm)	320W (55 dBm)
2	8W (39 dBm)	160W (52 dBm)
3	5W (37 dBm)	80W (49 dBm)
4	2W (33 dBm)	40W (46 dBm)
5	0.8W (29 dBm)	10W (40 dBm)
6	—	5W (37 dBm)
7	—	2.5W (34 dBm)

(BCCH) follow a seven-cell reuse pattern. It is also possible to use *microcells* for traffic hot spots.
- *Multiple access technique:* TDMA with eight time slots per frame. The duration of each slot is 0.577 ms. Each frame comprises eight physical channels transporting both traffic and signaling (control) channels. Also multiframe structures has been defined.
- *Traffic channels:* Two traffic channels were established for voice and data, respectively.
- *Voice traffic channel.* A *full-rate channel* has been defined that makes use of a voice encoder giving a digital signal of 13 kbps. After channel encoding a gross binary rate of 22.8 kbps is obtained. Already from the beginning a *half-rate channel* was specified using a voice encoder with a speed of 6.5 kbps and a gross binary rate of 11.4 kbps.
- *Data traffic channel.* On full- and half-rate traffic channels, *transparent type* data channels have been defined with binary rates of 2.4, 4.8, and 9.6 kbps, with different speed adaptation procedures, channel encoding and interleaving. *Nontransparent type* data channels are also allowed, with a binary rate of 12.0 kbps.

  Transparency offers a guaranteed data throughput, but under poor radio channel conditions, the error rate can be high. On the other hand, nontransparent services adapt to poor propagation channel conditions by reducing the data throughput to ensure integrity of the delivered data.
- *Control channels* (signaling channels): Three categories of control channels have been established:
  - Broadcast;
  - Common;
  - Dedicated.
- *Interfaces.* The public land mobile network (PLMN-GSM) can set up connection interfaces with the public switched telephone network (PSTN) and the ISDN, among others.
- *Network signaling:* Signaling between base stations and MSC follows a structured procedure, similar to that used in ISDN networks. Between MSCs, CCITT's Signaling System No. 7 is used.
- *Interference reduction techniques:*
  - Transmission in slow frequency hopping (SFH) mode, under control of the network. The purpose is to provide an increased protection against multipath fading and cellular interference using frequency diversity. The FH technique also randomizes the effect of the different sources of interference.

- Discontinuous transmission (VAD/DTX), in order to economize the duration of the battery in hand-held terminals and to reduce interference.
- Transmit power control to minimize interference and battery drain.
- *Security:* Voice and data encryption capabilities and a complete authentication system for terminals accessing the system.
- *Maximum equalizable time dispersion:* 16 μs (4.8 km).
- *Maximum time-slot time advance:* 250 μs (37.5 km).
- *Maximum Doppler shift:* 200 Hz (250 km/h).
- *Hand-over and roaming capability.*

10.3 Elements and Architecture of GSM

The GSM network consists of three basic subsystems:

- Mobile stations;
- Base station subsystem (BSS);
- Switching subsystem (SSS).

The BSS is further split into base transceiver stations (BTSs) and base station controllers (BSCs). Also in the SSS, two main elements can be identified, mobile switching centers (MSCs) and location registers (HLRs and VLRs).

From the architectural point of view and according to the ISDN principles, a GSM network is structured into the following categories:

- Functional entities;
- Interfaces.

The functional architecture defines *entities* that carry out the various functions in the system. The *interfaces* establish functional separation boundaries. The functional architecture of the GSM network is illustrated in Figure 10.1 (Table 10.3) and refers to the BSs, MSs, MSCs, and a subscriber location database system (HLR/VLR).

The interfaces between the main elements of the system are the following:

- BSS-MS (U_m);
- BSS-MSC (A);
- BSC-BTS (A_{bis}).

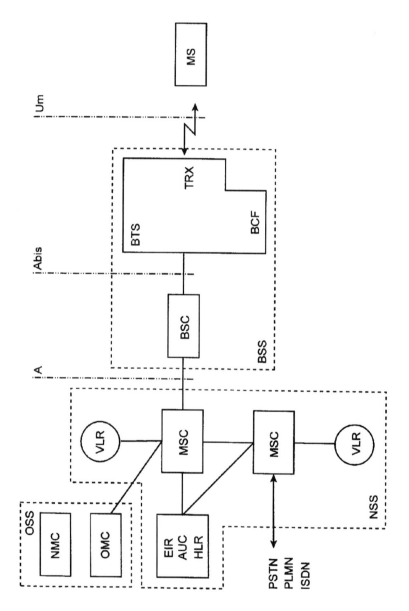

Figure 10.1 Main entities and interfaces in GSM.

Table 10.3
Elements in the GSM Network Structure

AUC	Authentication center
BSS	Base station subsystem (BTS+BSC)
BSC	Base station controller
BTS	Base transceiver station
BCF	Base station control functions
EIR	Equipment identity register
HLR	Home location register
MS	Mobile station
MSC	Mobile switching center
NMC	Network management center
NSS	Network subsystem
OMC	O&M center
ISDN	Integrated services digital network
PLMN	Public land mobile network
PSTN	Public switched telephone network
VLR	Visitors location register
TRX	Transceivers
SIM	Subscriber identity module
OSS	Operation subsystem

Connections are also defined with the other elements of the GSM network, such as HLR, VLR, OMC, etc. In turn, interfaces are defined with different telecommunications networks:

- MSC-ISDN;
- MSC-PLMN;
- MSC-PSTN.

The most important functional partition located at the A-interface where the functions dealing with network and switching aspects (associated with MSC, VLR, HLR) are separated from the *radio aspects*. As a consequence, two main parts are defined, termed network subsystem (NSS) and BSS. For network management there is also a further subsystem, the operation subsystem (OSS).

The following *network and switching* related functions may be highlighted:

- MS attach/detach;
- MS authentication;
- MS location;
- MS paging;

- Interworking with other networks (ISDN, PSTN, . . .);
- Communication control;
- Encryption of user and signaling data.

Some of the more outstanding *radio functions* include:

- Radio channel management: assignment of carriers and link supervision;
- Distribution of messages on broadcasting channels;
- Power control;
- Frequency hopping control;
- Transcoding of the digital voice signal;
- Rate adaptation for data signals.

A brief description of the functions carried out by the different elements in the GSM network is presented.

- BTS: This unit comprises the radio transceivers that make the link with mobile terminals possible. It also includes a minimum amount of equipment for generating TDMA frames and the communication channel with the BSC. In the specifications, it was decided to reduce the complexity of BTSs as much as possible so they could have a low size and reduced maintenance requirements. This helps their installation in unattended sites. A BTS includes a number of transceivers (TRX) which serve a single cell (omni-directional coverage) or several cells (sectorized coverage). The number of TRXs per sector typically varies between one and six.
- BSC: This element is responsible for performing most of the control functions related to a set of base stations. First, it is responsible for the configuration of each of its associated BTSs, so that when one comes back to work after a service interruption, it obtains its configuration parameters from a BSC. Also, the BSC manages the radio resources of the system, assigning channels, deciding on hand-overs, etc. The complexity involved in the functions performed by the BSC is considerable. For this reason, centralization of the functions corresponding to a large number of BTSs makes increased efficiency possible.
- MSC: Among others, it performs functions similar to those of a conventional telephone exchange, such as call routing. It also acts as an

interface with the PSTN and as a bridge between BSCs for call handovers. In addition, the MSC is the main element involved in the location process of mobile subscribers (mobility management). It also gives access to/from fixed telephone networks. To carry out mobility management and routing functions, MSCs are complemented with the two databases (registers) described below.

- HLR: It contains information about the network subscribers and the region in which they are located. This location is updated when the user is identified in the coverage area of a different MSC. The authentication center (AUC) associated with the HLR contains the required security data for the user authentication process.

- VLR: It contains information on all subscribers located within its associated location area (LA), so that these mobiles may be paged there and calls may be routed towards their actual position in the network. The information on a visiting MS is deleted from the database when the user registers on a base station belonging to a different LA.

- A further element is the EIR, which contains data on mobile equipment making it possible to discover if they have been stolen or are faulty.

Completely separate from the two basic subsystems described: BSS and NSS with its associated VLR and HLR databases, the GSM has a series of NMCs known as OMCs that take information from the above subsystems and databases. With this information, network operation statistics or call charging data can be compiled.

Some of the functions of the network are described below to further illustrate how these entities operate.

10.3.1 Identifiers and Contents of the Different Network Databases

Before moving on to describe some of the typical operational procedures in the GSM network, it would be appropriate to summarize some of the identifiers and information contained in the different elements of the GSM network: HLR, VLR, SIM.

First, there is a *personal number,* MSISDN, that is the telephone directory number that identifies the subscriber. Also, there is an equivalent to this for internal use only within the GSM network known as international mobile subscriber identity (IMSI). At the HLR, there is a table which relates these two identifiers. Furthermore, there is a temporary mobile subscriber identity (TMSI) used instead of the IMSI, as noted later, to protect it from eavesdroppers.

Another identifier of interest is the international mobile equipment identity (IMEI), which includes the equipment serial number and identifies the terminal for theft or faulty operation (inadequate radiation characteristics or terminal not type-approved, for example). This number may also be validated in the authentication process explained below. Finally, there is the mobile subscriber roaming number (MSRN), which is used for call routing purposes as is also discussed below.

The contents of the HLR referring to each subscriber are the following:

- The IMSI;
- The MSISDN;
- Information on the type of subscription;
- Service restrictions;
- Supplementary services;
- Information on the subscriber location and MSRN for call routing.

The VLR contains the following pieces of information:

- IMSI;
- TMSI;
- MSISDN;
- MSRN;
- Location area where the mobile is currently roaming.

The SIM is inserted into the terminal as a smart card or in the form of a removable chip. The fixed contents of the SIM are the following:

- IMSI;
- Identification code, K_i;
- Personal identification number, PIN;
- Subscriber information;
- Access control class;
- Nonauthorized PLMNs.

Also, during normal operation, the network updates the following SIM parameters:

- Encryption key, K_c;
- TMSI;
- Supplementary GSM services;
- Identity of the LA.

The SIM can be supplied as a full-sized card from which the mini-SIM can be snapped out if required by small hand-sets that cannot accommodate an SIM card. Call confidentiality and security is achieved in the following ways:

- By using the TMSI instead of the IMSI;
- By carrying out an authentication for each registration on the network;
- By encrypting the information on the radio path;
- By performing equipment validations.

Supranational GSM coverage is divided up, in the first place, into national coverages attended by the national networks of each country. Each national coverage area is, in turn, subdivided into a series of LAs. Figure 10.2 gives an idea of how LAs are structured together with their associated elements.

The VLR, HLR, and AUC are the elements in the GSM network that allow setting up calls and protect them from eavesdroppers. The functions that permit achieving these objectives are: registration, authentication, roaming and hand-over. A schematic explanation of these functions is given later.

Each mobile station connected to the system requires the achievement of registration functions in real time any time a new LA is visited and the authentication of the calling (and called) mobile station, in order to avoid fraud.

10.3.2 Registration and Location Update

This procedure will be initiated whenever one of the following events occur:

1. A change of LA;
2. A periodic update is requested;
3. At the time of connection/disconnection (attach/detach).

This procedure is as follows:

1. The BSS will, at regular intervals, broadcast the ID of its LA. The MS will verify it periodically. In the event of a change, a location update will be initiated.

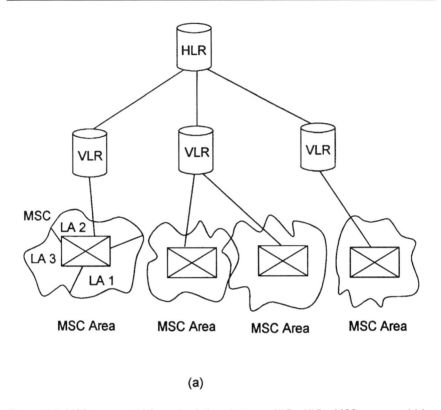

Figure 10.2 MSC areas and LAs and relations between HLRs, VLRs, MSC areas, and LAs.

2. The BSS broadcasts *update intervals*. The MS will initiate a registration when its timer count expires. The exception will be when the mobile terminal has had recent activity.
3. If the terminal is turned off, the terminal will send a "DETACH" message. When it is switched on again, it should send an "ATTACH" message.

The registration will be made with the BTS from which the highest signal level is received. To do this, the mobile scans the available broadcast channels, selecting the most appropriate.

10.3.3 Authentication

Depending on operator decisions/policies, authentication may take place at one of the following times:

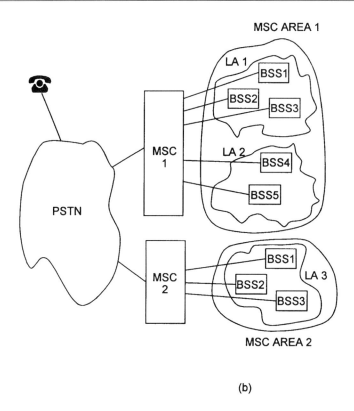

(b)

Figure 10.2 (continued).

- When a registration due to a change of LA or of MSRN has occurred;
- Prior to each call;
- Prior to providing supplementary services.

Authentication is based on a challenge-response method as described below. Each mobile station has a secret individual code, K_i, unknown by the user and residing in the SIM, which makes it possible for it to perform the authentication function. The AUC associated with the HLR to which the mobile belongs also has a copy of K_i.

The authentication process is performed under control of the visited VLR. The GSM network requests the mobile to send its IMSI that is passed on to the visited VLR. The VLR, in turn, sends the received IMSI to the HLR together with the identity of the current BSS where the mobile is located.

The HLR, in turn, requests its associated AUC to perform the authentication. To do this, it has a copy of the MS code K_i, which, together with a random number (RAND), are the parameters used in an algorithm (A3). This algorithm is used to calculate a signed response (SRES), which is sent back to the network. Also key K_i and the random number are the inputs to calculate another key, K_c, by means of algorithm A8. This new key, K_c, is used to encrypt the information on the traffic channel. The triad, random number, signed response and code K_c comprise the data set that identifies each communication.

The RAND is also sent to the MS so that it also calculates the SRES and K_c using the same algorithms (A3 and A8). These results are sent back to the network, which verifies the coincidence of the SRESs calculated by the network and the MS. In case they match, a positive authentication has been achieved. (Figure 10.3).

As seen earlier, the authentication process uses various elements to avoid the classical "password" approach used in analog networks which can be easily monitored. Instead of that, the network carries out the process described, using a secret key contained in the SIM. The SIM does have a password or PIN, which is not transmitted over the air but rather is used internally at terminal turn on.

Once the MS is authenticated, the VLR assigns it a TMSI, which will be valid for all calls made when in the same LA. The MS stores the TMSI in the SIM. While the mobile remains in the same LA, all transactions will use the same TMSI. Transmission of the TMSI to the MS is protected by encryption in such a manner that it is not possible to relate the TMSI with the actual mobile subscriber number.

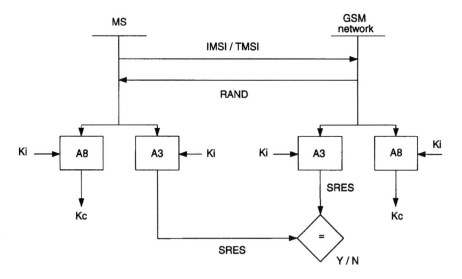

Figure 10.3 Authentication process.

10.3.4 Equipment Identification

The GSM network may have an additional protection provided by the EIR where all the serial numbers of the terminals reported as stolen or where any fault in transmission have been detected are stored. This protection is performed by merely consulting the database before allowing a communication to be set up.

10.3.5 Roaming

The area covered by the mobile radio network is divided into zones, which, in turn, are broken down into cells, each supported by a BTS. Several BTSs form what is termed an LA. The sizes of cells and LAs depend on network operation criteria (geographical, commercial, or traffic-density related). An MSC with its associated VLR may attend to one or more LAs.

When a mobile roams from one LA to another, the visited VLR is warned, even in the case the same VLR handles both LAs, by means of a location updating. A new TMSI is assigned every time there is a change in LA. At the same time, location information in the VLR is updated.

If the change of LA also involves a change of VLR, the HLR and the previously visited VLR are informed of this. The HLR updates the information on the mobile location and the previously visited VLR deletes all the information related to this particular MS. The new VLR stores the required MS data.

Paging messages for this mobile will only go on air via the BTSs in the LA where the mobile is registered. Routing information (MSRN) will also be available on the HLR, so that incoming calls may be routed to the MSC associated to the LA where the mobile is roaming.

10.3.6 Call Hand-Over

This is the function that makes it possible to transfer a mobile from one cell to another during a communication without it being noticed. In the GSM system, at the start of an incoming or outgoing call, the mobile is attended by a base station and the communication controlled by the MSC to which this belongs. Let BTS_1 be the base station through which the communication with MS_1 in cell 1 is set up, as shown in Figure 10.4. Assume that a voice channel has been established and that the links A and L_1, are used.

MS_1 and BSC_1 continually monitor the levels of reception and quality of the signal. Assuming that MS_1 starts to detect a low level of received signal, it then advises BSC_1 of this fact, then BSC_1 selects an appropriate adjacent cell and selects a traffic channel in the new cell advising the mobile of these

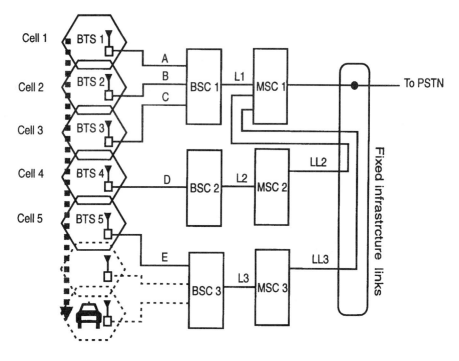

Figure 10.4 Call hand-over.

selections. All this is performed with the intervention of the BTS (BTS_1) that is going to be left. When everything has been set up, MS_1 is requested to switch to the new traffic channel belonging to BTS_2 and the corresponding fixed infrastructure link B-L_1 is set up, thus connecting BTS_2 with MSC_1, via the same BSC_1. Link A is released. The same procedure is used to hand over the call to cell 3, associated to the same BSC_1.

Now, assuming that, on its way, MS_1 passes from one cell to another associated to a different base station controller. As the mobile is handed over from cell 3 to cell 4, the same procedure is followed, except that now the BSC_2 controller prepares a traffic channel and a link D-L_2, this being carried out under instructions from MSC_1. The old traffic channel and its corresponding fixed infrastructure link C-L_1 are released.

In fact, MSC_1 acts as a transit switching center (anchor MSC [2]), since now the link used is D-L_2-LL_2. Furthermore, if the mobile station still reaches another cell depending on another MSC_3, the first MSC_1 will set up a new link E-L_3-LL_3 with the same operational philosophy and the previous link would be released. The process continues until the communication is released. In the last cases, MSC_2 and MSC_3 are called relay MSCs [2].

10.3.7 Stages in a Call

This section briefly summarizes the steps taken to set up a call with a fixed network terminal both originating in a mobile or ending in a mobile, or between two GSM mobiles. The steps followed in the different cases are listed in Tables 10.4, 10.5 and 10.6.

10.3.8 Call Setup and Routing Protocol

A difference is made between a call originating in a mobile and a call ending in a mobile.

Mobile-Originated Call

The process starts when the MS dials a mobile or fixed terminal number and presses the send button. The call request is directed to the visited VLR, which, in turn, starts the authentication processes and prepares the encryption key to be used on the assigned traffic channel. Then, the MSC reserves a circuit with

Table 10.4
Fixed Network Terminal to Mobile Terminated Call

Routing analysis
Alert
Authentication
Ciphering
Equipment validation
Call setup
Hand-overs (if necessary)
Release

Table 10.5
MS to Fixed Network Call

Registration
Service request
Authentication
Ciphering
Equipment validation
Call setup
Hand-overs (if necessary)
Release

Table 10.6
MS to MS Call

Registration	MS originating the call
Service request	
Authentication	
Ciphering	
Equipment validation	
Call setup	
Call routing	Network
Registration	MS receiving the call
Alert	
Authentication	
Ciphering	
Equipment validation	
Call setup	
Hand-overs (if necessary)	
Release	Network

the BSC and the process continues in much the same way as that followed in fixed telephone networks.

Incoming Call

Routing an incoming call (i.e., a call ending at a mobile from a PSTN/ISDN subscriber) requires additional functions to those demanded by fixed networks. Figures 10.5 and 10.6 show the sequence of events in the connection protocol, covering two commonly used procedures that basically depend on whether the telephone network switch has a sufficient signaling capability to dialogue directly with the HLR.

The procedure shown in Figure 10.5 is followed when the local or transit telephone switch does have for a direct access to the HLR. In this case, the MS number dialed allows the telephone network switch to route the call to the MSC that supports the MS's HLR. The HLR is interrogated, and this answers with a location number MSRN. The MSRN appears as a telephone number, indicating the MSC address where the mobile can be found.

The VLR belonging to the destination MSC obtains the TMSI of the station being called. If the mobile being called is not busy, a paging message using the TMSI is sent to it from all the BTSs in the corresponding LA where that MS is registered. The mobile called answers and is identified using the procedure described in above sections. Figure 10.5 also shows the signaling interchange (between the different elements of the network) when the call is made to a mobile located in the home MSC.

The GSM System   395

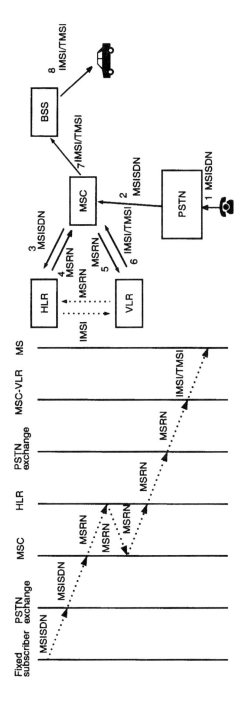

Figure 10.5 Routing protocol when direct dialog with the HLR is not possible.

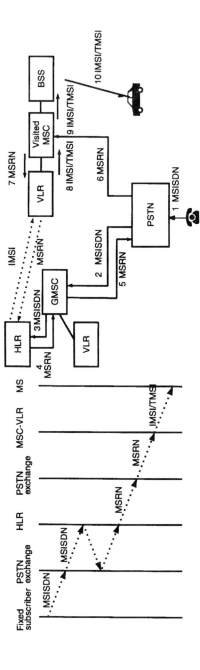

Figure 10.6 Routing protocol when it is possible to talk directly with the HLR.

Figure 10.6 shows the case when the local or transit telephone switch is able to directly interrogate the HLR to which the mobile belongs. The telephone switch identifies the HLR as a result of the subscriber number of the MS's MSISDN dialed by the fixed telephone network subscriber. The interrogated HLR responds to the ISDN or PSTN, by supplying the MSRN. This number is then used to route the call directly to the MSC where the mobile is located. By way of this procedure, the "fixed subscriber to mobile subscriber" connection is established with the *minimum number of switching sections* and, therefore, in a minimum period of time. Figure 10.6 also illustrates the dialog between network elements for carrying out the call routing.

10.4 The Radio Interface

As in any radio system, the resource that exerts a decisive influence on its design is the available RF spectrum. Cellular radio systems attempt to maximize the use of this resource, limiting the coverage area of the BSs and reusing the same frequencies at distances as close as possible. As in the previous generation systems, digital cellular systems define cells, cell clusters, and sectors. If the number of cells per cluster is reduced, the possibility of the system to make available a greater number of channels for each cell will increase. However, also, as a consequence of this, there will be an increase in cochannel interference. The limit on the quality of the received voice signal will condition the maximum level of acceptable cochannel interference, which, in its turn, will condition the system's capacity.

Needless to say, there should be a call *hand-over* mechanism. Another mechanism available is *power control*. This mechanism gives a greater control over interference and enables battery saving. The signal transmitted by the MS will be dynamically controlled under BSC command, maintaining the signal slightly above the limit required to ensure a transmission of acceptable quality.

GSM operates on the following RF frequency bands:

- 890–915 MHz in the MS-BS direction (up-link or reverse link);
- 935–960 MHz in the BS-MS direction (downlink or forward link).

It does so using a duplex separation of 45 MHz. The carrier spacing is of 200 kHz. A TDMA time structure with eight time slots is defined for each RF carrier. Due to the international regulations on out-of-band radiation, a guard band of 200 kHz is taken at the both ends of the band. So, there are 124 carriers available

$$\frac{25000 - 400}{200} + 1 = 124 \ channels$$

For the uplink:

$$F_u(n) = 890.2 + 0.2 \times (n - 1) \text{ MHz } (1 \leq n \leq 124)$$

For the downlink:

$$F_d(n) = F_u(n) + 45 \text{ MHz}$$

There will not necessarily be an exact mapping between the channel assigned to a mobile and the RF carrier frequency since the channel may go through an SFH process among the available carriers. This is a protection mechanism against interference and fading as explained later in Section 10.5.5.

10.4.1 TDMA Structures

Each RF carrier is divided into time slots (Figures 10.7 and 10.8) of approximately 577-μs duration (15000/26 μs). Eight consecutive time slots form a TDMA *frame*. These frames are grouped together in two different ways to form multiframes.

- *Multiframes of 26 frames* with a duration of 120 ms comprising 26 TDMA frames. These are used for traffic channels and their associated control channels.
- *Multiframes of 51 frames* with a duration of 235.5 ms (3060/13 ms). These are exclusively used by control channels.

A physical channel is defined by the combination of an RF channel and a TDMA time slot. This means that each RF carrier in the system has eight physical channels.

It is interesting to note that the multiframe structure used in each physical channel is independent of that of the other channels. More specifically, for channels on the same RF carrier and a different time slot. This means that the physical channels using the same RF frequency (8 physical channels with different time slots) do not have to use the same type of multiframe.

The multiframe structure is completed with a *super-frame* structure with a duration of 6.12 s. A super-frame may comprise the following:

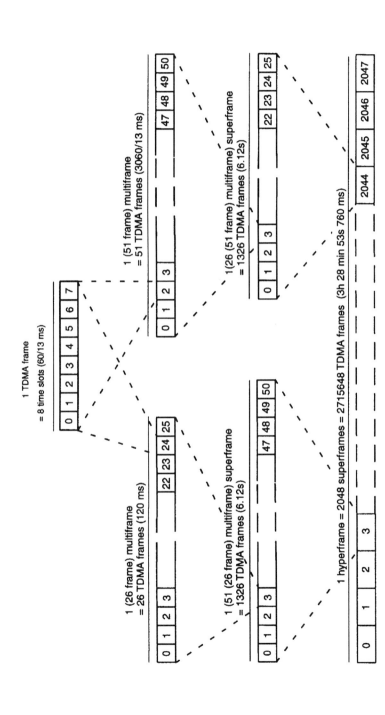

Figure 10.7 Frame and multiframe structure in GSM [1].

400   Introduction to Mobile Communications Engineering

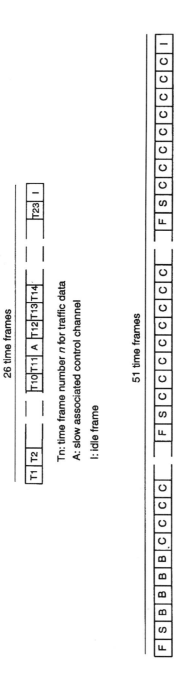

Figure 10.8 Traffic and control channel multiframes [1].

- 51 multiframes of 26 frames;
- 26 multiframes of 51 frames.

It is only allowed for the multiframe type used on a physical channel to change at the end of a super-frame.

Finally, there is a higher hierarchical level, known as *hyper-frame,* comprising 2048 super-frames (2715648 TDMA frames) and with a duration of 12533.76 s (3 h 28 m 53 s 760 ms). TDMA frames are numbered with reference to this hyper-frame structure, i.e., FN = 0 to 2715647. Such a long hyper-frame period is needed to support the GSM ciphering system.

Each channel is full duplex, i.e., it permits simultaneous transmission and reception both by the BTS and the MS. In order to avoid the mobile station actually transmitting and receiving simultaneously, it uses a *TDMA frame staggering* mechanism, which will be dealt with in Section 10.5.

10.4.2 Channel Types

A description follows of the types of channels specified in GSM and how they fit in the TDMA structure.

Types of channels. There are two main types of channels in GSM:

- Traffic channels (TCH);
- Control channels (CCH) (signaling channels).

Some of these channels are bidirectional and others are unidirectional.

Traffic channels. These are used to transmit encrypted voice or data; there are two types of them:

- *Full-rate* traffic channel (TCH/F) with a gross binary rate of 22.8 kb/s;
- *Half-rate* traffic channel (TCH/H) with a gross binary rate of 11.4 kb/s.

Control channels. Through the eight physical channels in each RF carrier, not only voice or data traffic is transmitted but also signaling information has to be sent. A brief description follows of the different control channels, indicating how they are inserted in the physical channels. The signaling channels may be of three types:

- *General information.* These are unidirectional channels through which general information is broadcast from a BTS to the mobiles, to be used by all of them;

- *Common.* These may be used by all mobiles, but the information transmitted at a given point in time only refers to one of them;
- *Dedicated.* They are devoted to the setting up of communications or for interchanging signaling data between the MS and the BTS during the course of a communication, so they are used only by an MS.

Following, the *General Information* unidirectional control channels are listed.

- FCCH. *Frequency Correction Channel.* This is a channel used by the MSs for frequency correction. A fixed frequency carrier is transmitted (simply by zeroing all bits), so that it may be used by the mobiles for frequency synchronization;
- SCH. *Synchronization Channel.* Its purpose is to permit frame synchronization in all mobiles. It also includes BTS identification data;
- BCCH. *Broadcast Control Channel.* Transmits general information on the system.

The group FCCH + SCH + BCCH is the minimum transmitted by any BTS in the absence of traffic. The frequency used by these channels must be transmitted continuously. To achieve this, the bit stream is padded with dummy data in case no signaling or traffic data is available.

The *Common Control Channels* (CCCH) are also unidirectional and are used at the beginning of each communication. They are the following three:

- PCH. *Paging Channel.* Used to page a mobile terminal, for example, if an incoming call is received from the network;
- RACH. *Random Access Channel.* Mobiles access this channel when they wish to address the network. A slotted ALOHA type protocol is used. This is a unidirectional channel towards the base station;
- AGCH. *Access Grant Channel.* Used by the base station to assign a dedicated signaling channel to the mobile to complete the call set up process.

Finally, there are three bidirectional *Dedicated Signaling Channels.*

- SDCCH. *Stand Alone Dedicated Control Channel.* This signaling channel is used before a traffic channel and has been assigned to carry out the actual communication. It is used to complete the call set up procedure;

- SACCH and FACCH. *Slow* and *Fast Associated Control Channels*. They are signaling channels associated with a traffic channel during a call. The difference between them lies in the speed at which information is transferred. Low priority information, such as the result of the measurements taken by the mobile of the received signal level from different BTSs, is transmitted by means of the SACCH, with a low binary flow. High priority information, such as that exchanged during a call hand-over, is transmitted through the FACCH.

The manner in which the logical channels are inserted in the physical channels is somewhat complex. The basics of this process are described briefly. The signaling data associated with a call in progress is transmitted on the same RF carrier and time slot as the traffic information (Figure 10.8). To do so, the intervals corresponding to 26 consecutive frames are multiplexed forming a multiframe; 24 frames are dedicated to traffic information and one frame to SACCH. The remaining interval is idle. The FACCH channel is inserted into the traffic channel by "stealing" traffic data time slots.

The first interval (interval 0) of one of the BTS carriers (always the same) is assigned to carry out signaling tasks. This carrier is named *beacon* or BCCH carrier. In the up-link it contains the random access channel (RACH). In down-link it contains the general information channels (FCCCH + SCH + BCCH) and the common channels (PCH + AGCH).

In cells with little traffic, four SDCCH channels are also multiplexed in this RF channel and time slots. The multiplexing of these channels is performed at multiframe level, grouping together the intervals corresponding to 51 consecutive frames (Figure 10.8). In cells with larger traffic loads, additional time slots are allocated to signaling tasks, these being used basically by SDCCH channels.

10.4.3 Setting Up Calls

To illustrate the functions of the different channels described above, the procedures for setting up calls are explained. Only the exchanges of information between the BTS and the MS during call set up stage are presented; the channels used are also indicated. The process will vary depending on whether the call originates or terminates in the mobile. In the first case (*mobile originated call*), the steps followed are (Table 10.7):

- The MS makes a call establishment request through the RACCH channel;

Table 10.7
Mobile Originated Calls

Action	Origin	Channel
Channel request	MS	RACCH
Channel assignment	BTS	AGCH
Service request	MS	SDCCH
Authentication process		SDCCH
Communication encryption	BTS	SDCCH
Call terminal number	MS	SDCCH
Channel assignment	BTS	SDCCH
Connection	BTS	TCH

- The BSC answers, via an AGCH channel, allocating a SDCCH channel, through which all the required signaling information is interchanged;
- The mobile makes a request for service. This specifies the type of service required;
- The BSC commands an MS authentication process. To do this, the network sends a random word with which the mobile computes a response in accordance with its internal identification data (as explained in Section 10.3.3);
- The MS sends back to the BSC the results of the authentication. The mobile transmits its result to the base: signed response. This response is compared for validation with results of the same operation performed within the network;
- Establishment of encrypted mode. In the authentication process, an encryption key is generated at the two ends. The base commands the mobile to pass to encrypted mode. This must be acknowledged by the mobile;
- The MS transmits the number called. The base confirms reception;
- Traffic channel allocation. The base transmits to the mobile the channel allocation on which the actual call will be carried out. When reception is confirmed by the mobile, the SDCCH is released and all the signaling information is then transmitted through the associated channels;
- Call alert and connection. The base transmits this information using the FACCH. When the mobile confirms the connection, a bidirectional communication is established on a traffic channel, TCH.

If the *call terminates in the mobile* (Table 10.8), the process is slightly different:

- The mobile is paged via the PCH channel is used to page the mobile on all the base stations belonging to the same LA;
- The mobile requests access to a signaling channel using the RACH. As before, the dedicated control channel assignment is received via the AGCH. The assigned channel is a SDCCH;
- The mobile informs that the reason for accessing the network is that it has received a page message on the PCH channel;
- Authentication processes and pass over to ciphered communication mode in a similar way to the previous case;
- Call data transmission. The base transmits, in encrypted mode, data on the incoming call, including the type of service and the identification of the subscriber who initiated the call;
- Mobile acknowledges this transmission;
- The base assigns a traffic channel and, once the mobile confirms the assignment, deactivates the SDCCH;
- Alert and connection. The mobile indicates the activation of the alert tone and subscriber connection, using the FACCH associated channel. Once the base confirms connection, the communication is set up through a traffic channel, TCH.

Burst Types

The RF carrier is modulated at a binary rate of approximately 270.838 kbps (1625/6 kbps). This means that if a time slot lasts approximately 577 μs, a total of 148 bits are transmitted during one time slot.

Table 10.8
Establishing Calls to a Mobile [2]

Action	Origin	Channel
Mobile paging	BS	PCH
Channel request	MS	RACH
Channel assignment	BS	AGCH
Message acknowledgment	MS	SDCCH
Authentication process		SDCCH
Communication encryption	BS	SDCCH
Call alert	BS	SDCCH
Channel assignment	BS	SDCCH
Connection	BS	TCH

$$270{,}838 \cdot 10^3 \cdot 577 \cdot 10^{-6} \approx 156.25 \text{ bits}$$

During a time slot, the carrier is modulated by a binary stream known as *burst*. A burst comprises an *active* or *useful part* and a *guard time*. The useful part contains the data to be transmitted, an intermediate part (*training sequence*) and tail bits. The guard time is a period in which no transmission takes place, thus reducing the overlapping of transmissions on adjacent time slots.

If any overlapping occurs between the useful parts of two consecutive bursts, there would be a loss of information. In order to avoid this, an *adaptive time alignment system* is used. Time alignment is achieved by the mobiles performing their burst transmissions at different times (time advance) depending on their distances from the BTS so that the bursts arrive at the BTS on their nominal time slots.

There are five types of bursts in GSM, four long and one short (Figure 10.9):

- Normal burst;
- Frequency correction burst;
- Synchronization burst;
- Dummy burst;
- Access burst.

The *normal burst* includes the following:

- 8.25 bits of guard period;
- 116 ciphered bits (data bits);
- 3 + 3 tail bits;
- 26 bit training sequence for equalization.

The *frequency correction burst* has the same structure as the previous one but, instead of the ciphered data sequence and the training sequence (142 bits in total), a fixed sequence of zeros is sent. This burst is used by the MS for RF frequency synchronization. It also allows the mobile to identify the *broadcast channel* since this type of burst is easily identified.

The *synchronization burst*. This differs from the above because it contains an extended training sequence of 64 bits which, as a result, the number of data bits is smaller. It is used by the mobile for time synchronization.

The *dummy burst*. This has the same structure as the normal burst, although encrypted bits are replaced by a given sequence of bits. This burst

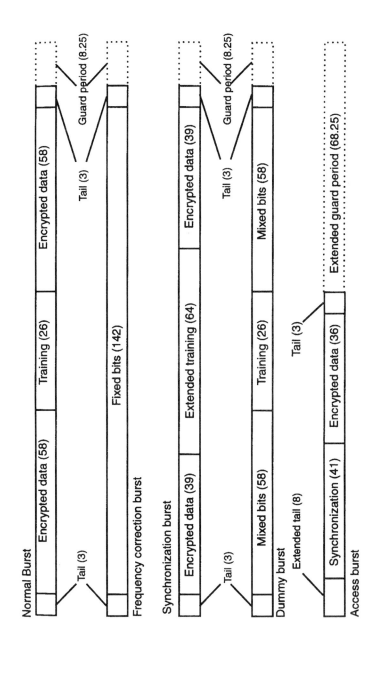

Figure 10.9 Types of bursts in GSM.

is used instead of the normal burst when there is no information to be transmitted, but the carrier must be on the air because it is the BCCH carrier.

The *access burst*. This type of burst is used by the MS to access the system. It is characterized by an extended guard period lasting 68.25 bits. This helps to avoid transmission problems from the mobile when the "timing advance" information is not yet known.

10.4.4 Voice Link

Figure 10.10 shows the main elements the GSM voice link. Below, these elements are briefly described.

Voice coder

An RPE-LTP voice coder (Regular Pulse Excited-Linear Predictive Coder with a Long Term Prediction loop) is used. This coder produces a binary rate of 13 kbps. It analyzes input voice blocks of 20 ms and produces 260 bits (13,000 × 0.02) at the output. It has been proven by subjective analyses that 182 of these bits (Class I bits) are highly sensitive to errors, i.e., they cause a severe quality degradation if they are received in error. The remaining 78 bits are more error resistant and are known as Class II bits. A *half-rate coder* has also been defined so that GSM system capacity may be doubled by using half-rate traffic channels (TCH/HS).

In order to enable the use of the *discontinuous transmission* feature described in Section 10.5.4, a *voice activity detector* (VAD) is included in the coder.

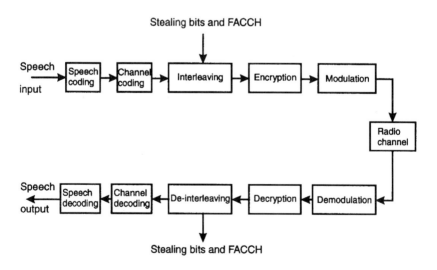

Figure 10.10 GSM voice link [1].

Channel Coding

Mobile systems, unlike fixed systems, are characterized by highly changing (fading) received signal levels as the mobile runs. In a large sized rural cell where the MS may be operating near its sensitivity limit, errors will occur when the received level decreases too much. In urban environments, where the system may be operating at a point near the cochannel interference limit, fading will give rise to errors when this limit is exceeded.

In order to protect transmissions, error control techniques are used (*Forward Error Correction*). To achieve this, additional bits are added to the actual messages. As described above, the voice coder (source coder) produces 260 bits each 20 ms, 182 of these bits belong to Class I and the remaining 78 are of Class II. The 260 bits in a voice frame are delivered to the channel coder in order of diminishing importance. The 182 Class I bits enter first. This ordering scheme was obtained through experimental studies where the effect of transmission errors on different bits in the voice frame were evaluated. Redundancy bits (cyclic redundancy code) are added to the first 50 Class I bits (Class I.a) to detect errors. The purpose of these 3 parity bits is to detect the errors present in Class I.a bits prior to delivery to the source decoder. Class I.a bits are the most important as voice quality is concerned and it has been shown that the voice signal which they produce when errors are present is of a very poor quality. In the event of having errors, it is preferable for the voice decoder not to produce any output (erased frames) but rather regenerating the voice signal by extrapolation, from previously received voice frames.

Afterwards, the Class I bits plus the 3 parity bits are reordered, placing the even bits at the beginning, the parity bits in the middle and the odd bits at the end. Four tail bits are set to zero, added to the Class I bits, giving rise to a stream of 189 bits (182 + 3 + 4 = 189).

The purpose of these 4 bits is discussed below. The 189 bits are coded using a rate 1/2 convolutional code with a constraint length of 4 producing 378 bits (189 × 2) at the output. Finally, the 78 Class II bits which have not been protected are simply added to the remaining 378 bits, giving an output of 456 bits (378 + 78) (Figure 10.11). Taking into account that this number of bits will be produced every 20 ms, a gross binary rate of 22.8 kbps (456 bits/20 ms) is generated.

A convolutional code was chosen in view of its error correction performance compared to that of the more common block codes. However, the disadvantage is that it is more difficult to decode. The Viterbi algorithm is used for this purpose and, as an aid to decoding, the 4 tail bits mentioned earlier were added which are known to be zero. This always leads the coder to a known final state. The decoder knows, a priori, which are to be the last

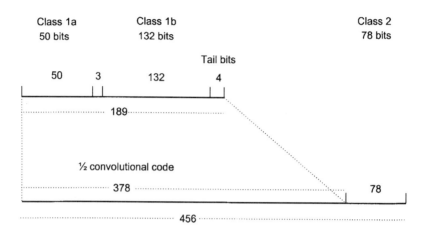

Figure 10.11 Channel coding for voice communications.

bits to be received, this helps in the decision process and, therefore, reduces the error rate at the decoder output.

Interleaving

Errors in mobile environments are due to fading. Fades will occur at a much lower rate than the binary transmission rate ($\approx$ 270 kbps). Hence, errors will occur in bursts which coincide with signal fades. Generally, error bursts lasting throughout the TDMA frame are not very likely. Observing the error behavior, there will be periods of high error rates followed by longer intervals with small errors. For error correction codes to be able to operate efficiently, errors should be distributed uniformly over time. In order to achieve this, reordering and interleaving techniques are used: the 456 coded bits are reordered and interleaved over a period of 8 TDMA frames, as shown in Figure 10.12. Also, flags are added to advise the receiving end in case voice communication data bits

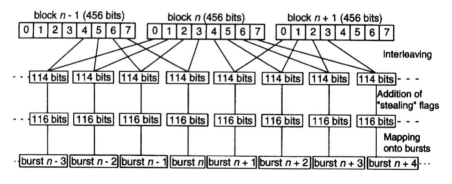

Figure 10.12 Voice frame interleaving process in TDMA frames [1].

have been "stolen" and replaced by signaling data corresponding to the FACCH channel (e.g., to send urgent call hand-over related messages).

For transmission, the 456 bits are divided into 8 blocks of 57 bits which are transmitted, interleaved, in 8 semi-intervals ("half-slots") of consecutive frames, as shown in Figure 10.12. When an error burst escapes the correcting action of the convolutional code, the block code detects these errors, if it is within its detecting possibilities. Since the system operates in real time, when an error is detected, the voice frame is lost, the lost information being replaced by extrapolated voice sample based on previous samples.

Encryption

GSM may encrypt both transmitted voice and data using a bit level algorithm. The encryption is performed prior to transmission, as shown in Figure 10.13. The encrypted information is obtained by executing an exclusive or logical operation between the actual data and a pseudo-random sequence of 114 bits obtained by the algorithm as a function of the cipher key K_c and the frame number FN. Figure 10.13 indicates the encryption/de-encryption process.

Modulation

The GSM modulation system was required to have the following characteristics:

- Conversion to the appropriate RF band;
- A relatively narrow band to permit a good spectral efficiency expressed in bits/s/Hz;

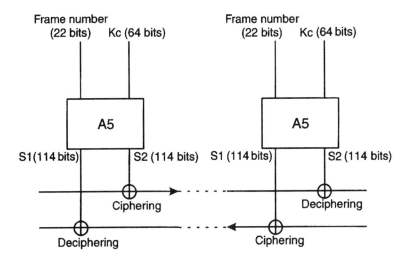

Figure 10.13 Encryption process.

- Constant carrier envelope to enable the use of simple, efficient amplifiers, e.g., Class C;
- Low out-of-band radiation to reduce interference on adjacent channels.

MSK (binary FSK with a modulation index of 0.5) complies with the first 3 specifications, but does not achieve the fourth, as it produces excessive out-of-band radiation. By carrying out a filtering process prior to modulation, it is possible to reduce the out-of-band radiation while maintaining the constant envelope property. It has been shown that with a Gaussian pre-modulation filter the desired characteristic is obtained. The resulting modulation is known as GMSK (Gaussian Minimum Shift Keying). A BT parameter value of 0.3 was chosen, where B is the filter bandwidth and T is the bit period.

Multipath propagation produces a time spreading of the transmitted pulses, generally on the order of some microseconds, thus producing intersymbol interference, ISI. The effect of ISI entails a severe impairment of the BER characteristics making it necessary to introduce some kind of countermeasure to avoid this problem which would render the GSM system unusable.

Each burst contains a 26-bit *training sequence* which the receiver uses to evaluate the channel response affecting each burst. This information may be used by an equalizer to compensate for the time spreading caused by multipath. A Viterbi equalizer is employed that makes use of the burst tail bits to assist in this process.

Generally, the multipath environment is dynamic and, therefore, changing from one frame to the next, particularly if the speed of the mobile is high. For this reason, each burst should contain its own training sequence. This is also the reason why it is placed in the center of the burst. In this manner, the measured channel impulse response will vary less with respect to each end of the burst.

10.4.5 Coding the Signaling Bits

In order to code the signaling channel bits, a different scheme is used to that employed in the voice link. The signaling bits are arranged in groups of 23 octets, i.e. 184 bits, to which a 40 bit redundancy Fire code is applied. This code is very powerful in detecting error bursts. Four tail bits are added to the 224 Fire coder output bits and the set of 228 bits is input to a rate 1/2 convolutional coder, producing 456 output bits, as in the voice link. These are transmitted after being interleaved. Figure 10.14 illustrates this process. It will be observed that the protection of signaling bits is even greater than that used for voice signal bits. The reason for this is the high sensitivity of the system to signaling errors.

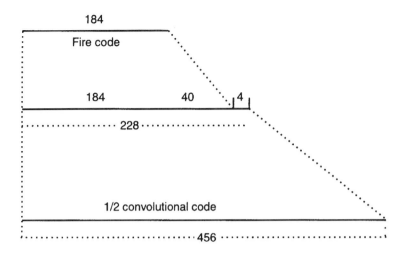

Figure 10.14 Channel coding for signaling channels.

Also, there are different channel coding schemes for the traffic channel when transmitting data at different rates. This topic is not discussed in this book.

10.5 Other Features of the GSM Radio Interface

10.5.1 Adaptive Time Alignment

In GSM, MSs obtain their transmit time reference from BTSs. In particular, MSs delay transmission of their bursts, sending them 3 time slots after the bursts from the serving BTS have been received (Figure 10.15) [1].

This technique avoids transmitting and receiving at the same time (still, the link is full-duplex). This mode of operation renders the use of a duplexer in the mobiles unnecessary, thus making them more economical and less bulky.

Furthermore, there is the problem that propagation time depends on the distance between the BTS and the MS. Thus, it is necessary to set up a control mechanism to prevent adjacent bursts from different MSs from overlapping. For example, if the BTS-MS distance were 10 km (or, equivalently, 33 μs), then the bursts would be received in the BTS some 66 μs later (BTS-MS-BTS link) relative to the case in which the MS were beside the BTS.

In order to solve this problem, GSM uses an *adaptive time alignment system* by which the BTS calculates the *timing advance* necessary for each MS

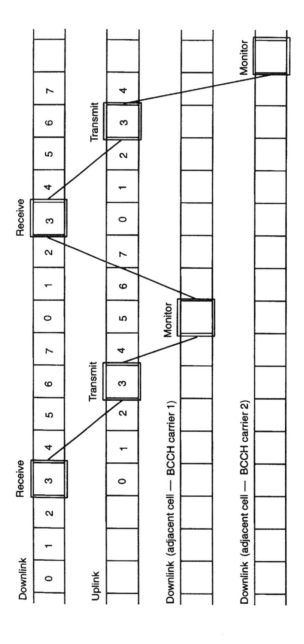

Figure 10.15 Intervals of transmission, reception and monitoring [1].

and, in this way, achieves that the bursts from different MSs reach the serving BTS in the correct time slot with no overlapping.

An alternative approach might be the use of greater guard times in all bursts. This option would render the system inefficient. Timing advance is initially calculated by the serving BTS on the basis of the access burst in the RACH (Random Access Channel). As shown in Figure 10.9, this burst (RACH) has an extended guard period of 68.25 bits or 252 μs (68.25 bits/270.838 kbps). The BS-MS separation may then reach a maximum of approximately 37 km ($3 \times 10^8 \times 252 \times 10^{-6}/2$) before overlapping of time slots occurs at the BTS.

At the BTS, the transmission timing advance necessary for the MS, which is expressed in terms of bit periods, is sent to the MS as a 6 bit binary number ($2^6 = 64$). In this way, timing advances may be programmed from 0 to 63 bit periods, allowing a maximum BTS-MS separation of 35 km.

$$63 \frac{1}{270,838} 3 \cdot 10^8 \frac{1}{2}$$

During normal operation, once a TCH traffic channel is already set up, the BTS will continually monitor the delay in the MS bursts and, if required, would correct the timing advance and send a change instruction over the down-link via the SACCH channel.

10.5.2 Power Control

Cellular systems are interference limited. First generation cellular systems were already fitted with a very simple RF power control mechanism to reduce the power transmitted by the MS when visiting small cells. GSM goes much further and is equipped with an adaptive mechanism.

Generally, if an MS is near the BTS, the power transmitted will be low and increase gradually as it moves away from the base. MSs are able to vary their transmitted power from a maximum to a minimum of 20 mW in 2 dB steps under BS control.

For initial access to the system using the RACH, the MS will use the maximum allowed power in the cell broadcast over the BCCH or, alternatively, its own maximum power, depending on which of the two is the smallest.

After the initial access, the BTS calculates the RF power level required by the MS and sends instructions to the MS for it to transmit with this power, making use of a 4 bit parameter field which is sent in the down-link through the SACCH every 480 ms. The MS acknowledges the change of transmitted power instructions via the SACCH in the up-link direction. Also, at the BTS,

15 2-dB steps are specified. The BTS uses MSs reports sent back to the base on the power level they receive.

10.5.3 Call Hand-Over

The hand-over mechanism conditions the allocation of MSs to BTSs. By adjusting the decision threshold used in the process, cell sizes will increase or decrease. Hence, the hand-over mechanism is used by the cellular network operators, not only to control the quality of links but also to adapt cell sizes to the local traffic densities (Erlang/km^2).

In order to control the hand-over process, it is necessary to know not only the quality of the link with the BTS serving the call, but also the possible quality with other, alternative BTSs. In analog systems such as, for example, AMPS, this knowledge was gathered by using scanning receivers in the BTSs in such a manner that received field strengths from mobiles being served by nearby cells could be known.

In GSM, however, the procedure to carry out call hand-overs is substantially different: a mobile is only active, i.e., transmitting or receiving, in two out of eight time slots in a TDMA frame. This allows an MS to monitor the transmissions of nearby BTSs in the remaining time slots. In this way, MSs are able to report on the results of these measurements. Measurements of received powers from nearby cells together with received power and link quality with its serving BTS are reported back to the network. These reports are, in turn, used by the network to decide on the need to perform a call hand-over. This scheme is known as mobile assisted hand-over (MAHO).

Figure 10.15 shows how the MS passes through the transmitting, receiving and monitoring states. It can be seen that there are approximately 1 or 2 time slots available to give sufficient time for the MS synthesizer to switch to the required frequency.

During the monitoring time slot, the receiver will tune to the BCCH carrier of its own BTS or of surrounding cells and will measure the received field strength. The BCCH carriers will be measured sequentially and averaged out over a SACCH block (480 ms, i.e., four 26-frame multiframes). In this period, there will be 100 active frames during which measurements are carried out and four idle frames during which the receiver is in a different state described below. The number of samples taken from each BCCH carrier will depend on the number of BTSs explored. For example, if there are eight carriers, 12 samples will be taken (100/8 = 12).

It is essential at the measuring and reporting stages to know which is the BTS actually being measured. It is possible, for systems with clusters of a small number of cells, not to identify cells with BCCHs using the same carrier.

During the idle frames, the MS carries out a demodulation of the BCCHs and extracts the identities of the measured base stations (base station identity codes (BSIC)). The BCCH should always be transmitted with the same power so that mobiles are able to measure propagation losses. This means that FH, DTX, and power control cannot be used in the BCCH. Also, during idle time intervals, dummy bursts must be transmitted for transmission continuity.

Once the BSIC and results of the level measurements are obtained, values for the six stations with the highest levels are reported back over the uplink SACCH.

For the base station giving service to an MS, not only is it possible to measure signal strength, but also the quality of the received signal. With a link quality measurement and a signal level measurement, it is possible to determine if the impairment is due to cochannel interference (e.g., high-level and low-quality) or to an insufficient signal level. In this way, it is possible to exert an efficient control of call hand-over and power control.

A further function of the MS is to measure the received signal in unused time slots. This is equivalent to a direct measurement of the interference level in these time slots. These measurements may be used by the BTS to hand over the communication with an MS to another of its time slots with a better quality (intracell hand-over).

10.5.4 Discontinuous Transmission and Reception

Provision was made for the fact that most of the bursts transmitted in the GSM system will carry voice traffic. In almost all cases, voice traffic channels transport signals in one direction at a time, whereas the other direction will not contain any conversation signal. So, on average, the channel in each direction will be transmitting an "active" voice signal for less than half of the total time. In GSM, this characteristic is exploited by using a discontinuous transmission (DTX) mechanism.

Only voice signal frames with active samples are transmitted (i.e., the TCH will not be transmitted during silence periods). This offers two clear advantages:

- The cochannel interference level will, on average, be 3 dB lower;
- In hand-held terminals where most of the consumed energy goes to the RF amplifier, battery life will be increased significantly.

In order to put the DTX mechanism in place, a voice activity detector (VAD) is required. The VAD distinguishes between voice and silence periods.

During the periods of silence, however, the receiving end would not hear anything at all, not even the background noise associated with the other communication end. It has been shown that switching between periods of voice activity with its associated background noise and periods of total silence is unpleasant to the listener. In order to circumvent this problem, silence frames are replaced in the receiver by background noise generated from information coming from the other end ("comfort noise"), which is adapted both in amplitude and in spectral characteristics to the actual background noise.

There are algorithms associated with the VAD to extract information to generate the comfort noise. This information is sent to the receiver during silence periods, in silence identification frames (SIDs), at a low rate so that the receiver end generates the background noise artificially.

Also, the paging channel (PCH) in the downlink of the common control channel (CCCH) will be organized in such a way that an MS will only need to listen to a small subset of all PCH frames. MSs are therefore designed to have their receivers open only when necessary. This is known as discontinuous reception (DRX). The main aim of this is towards prolonging the battery life in hand-helds when in standby mode.

10.5.5 Slow Frequency Hopping

One of the basic problems caused by multipath propagation is that a quasi-stationary wave pattern with deep nulls separated by approximately half a wavelength ($\lambda/2$) is formed. When the mobile runs at relatively high speeds and, for a reasonably high average received signal level, Rayleigh fading will not give rise to many errors. These errors may be compensated for by FEC channel coding.

For low speed or stationary mobiles, these nulls will affect consecutive transmission bursts and the number of bits in error reaching the channel decoder will increase. When this occurs, error bursts in the Class I output bits will be produced, and they will impair the received voice quality. This error generating mechanism is particularly troublesome for stationary or quasi-stationary MSs. If the MS is in one of these nulls, it is possible that the entire communication will be lost.

To avoid this problem, a slow frequency hopping (SFH) mechanism may be used (Fig. 10.16). The sequence of bursts comprising the TCH is assigned cyclically to different frequencies under control of the serving BTS. Synchronization of frequency hopping is achieved by making use of the timing signals available in the BTS and MS.

When using SFH, the received signal nulls will change their physical positions from one burst to the next one. In this way, TDMA bursts with

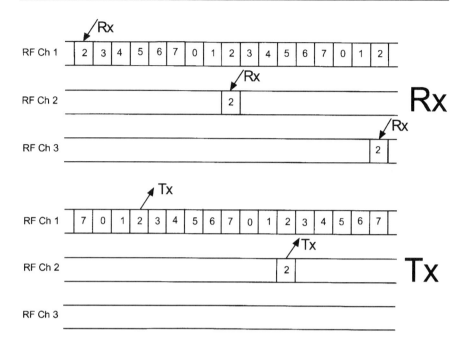

Figure 10.16 SFH.

large BERs will be spaced out in time, as occurs when the mobile runs at great speed.

Also, another important advantage of SFH is that cochannel interference is evenly distributed among all the MSs. This is known as *interference diversity*. In a system with no SFH, a given MS in a given cell may experience a high level of cochannel interference due to a particular mobile transmitting with a high level. All the bursts from this MS will have a high BER, producing a very poor end quality in the voice signal. If SFH is used, however, the interfering frequency will be distributed between various MSs. Each MS will then endure a small fraction of the bursts received with high BER, allowing the channel decoder to work properly.

10.6 Link Budget Data

Here, typical GSM link budgets are presented. The data and the methodology used are based essentially on GSM Recommendation 03.30 [3]. Nevertheless, a generalization is made, taking into account the antenna noise figure of base and mobile stations and the existence of coupling elements (combiners, antenna multicouplers) between the antenna and the radio equipment. First, system noise factor concepts are briefly reviewed, assuming a generic receiver installation comprising the antenna, A, feeder, F, preamplifier, PA, antenna multicoupler, AM, and receiver, Rx. The relevant parameters of these components are the following:

- Feeder loss: L_f (dB) or, in linear units, $l_f = 10^{L_f/10}$
- Preamplifier gain: G_{pa} (dB) or, in linear units, $g_{pa} = 10^{G_{pa}/10}$
- Preamplifier noise figure: F_{pa} (dB) or, in linear units, $f_{pa} = 10^{F_{pa}/10}$
- Multicoupler loss: L_{mc} (dB) or, in linear units, $l_{mc} = 10^{L_{mc}/10}$
- Receiver noise figure: F_r (dB) or, in linear units, $f_r = 10^{F_r/10}$

The system noise figure is given by

$$f_{sys} = f_a - 1 + f_{pa} l_f + \frac{l_f}{g_{pa}}(f_r l_{mc} - 1) \qquad (10.1)$$

and, in logarithmic units, F_{sys} (dB) = 10 log f_{sys}

The ratio

$$d = \frac{f_{sys}}{f_r} \quad \text{or} \quad D \text{ (dB)} = F_{sys} \text{ (dB)} - F_r \text{ (dB)} \qquad (10.2)$$

is called *noise degradation* of the receiving system. It gives the receiver sensitivity impairment due to noise contributions of the elements before the receiver front-end.

The receiver sensitivity, S, is defined assuming that the only disturbance is caused by the receiver's noise, and is given by

$$S \text{ (dBm)} = F_r \text{ (dB)} + E_b/N_o \text{ (dB)} + 10 \log R \text{ (bps)} - 174 \qquad (10.3)$$

where E_b/N_o is the bit energy-to-noise spectral density ratio and R is the channel bit rate. The isotropic power at the receiving antenna output is

$$P_{iso} \text{ (dBm)} = S \text{ (dBm)} + D \text{ (dB)} + M_I \text{ (dB)} + L \text{ (dB)} - G_r \text{ (dB)}$$
$$(10.4)$$

where

- M_I (dB) is the interference margin. This parameter is introduced to account for marginal cochannel interference effects in adequately frequency planned networks.
- L (dB) is the log-normal margin for the wanted coverage locations probability.
- G_r (dB) is the receive antenna isotropic gain.

In the link budgets which follow, it has been assumed that the BS's receiving system is made up of a preamplifier with a gain G_{pa} = 10 dB and a noise factor F_{pa} = 2.5 dB. Obviously, on the MS side there is neither an antenna multicoupler nor a preamplifier, so that $L_{mc} = F_{pa} = G_{pa}$ = 0 dB. The following assumptions were also made:

1. Noise figures: BTS antenna, F_a = 10 dB; MS antenna, F_a = 12 dB; BTS receiver, F_r = 8 dB; MS receiver, F_r = 10 dB.
2. Sensitivities: BTS receiver, S = −104 dBm; MS receiver, S = −102 dBm.
3. Interference degradation margin, M_I = 3 dB.
4. Log-normal margin corresponding to a 75% coverage percentage at the fringe of the cell. For a locations variability σ_L = 7 dB, $L = 0.67 \times 7 \approx 5$ dB (see Chapter 5, Section 5.2).
5. Representative midband frequencies: Uplink, f = 903 MHz; downlink, f = 948 MHz.
6. Antenna gains: BTS: 10 dB$_i$ for transmission and reception (it is assumed that there is a receive diversity gain of 3 dB); MS: −3dB$_i$ for transmission and reception, taking into account the effect of the user's body.
7. Terminal losses: BTS transmit: 2.5 dB combiner loss and 1.5 dB antenna feeder; BTS receive: 1.5 dB antenna feeder and 3.5 dB antenna multicoupler (two-channel BS).
8. Transmission powers: BTS: 6W (38 dBm); MS: 2W (33 dBm).
9. Antenna heights: BTS: h_b = 30m; MS: h_m = 1.5m.
10. Building penetration losses, L_p = 15 dB, are assumed.
11. Two coverage ranges were calculated: (a) when the MS is at street level ("on-street"); (b) when the MS is within a building ("in-building").

The Hata propagation formula that follows (see Chapter 4) was used to calculate the coverage range in an urban area, equating L_b to the maximum allowed loss. For the indoor coverage case, penetration losses were also added.

$$L_b \text{ (dB)} = 69.55 + 26.16 \log(f) - 13.82 \log(h_t) - a(h_m) \quad (10.5)$$
$$+ (44.9 - 6.55 \log(h_t)) \log(d)$$

In Table 10.9 the resulting uplink and downlink budgets are presented.

10.7 GSM Services

In GSM as well as in other communication networks, three types of services are provided: teleservices, bearer services, and supplementary services. They are defined as follows:

- *Teleservices* [4] are telecommunications services that provide complete capability, including terminal equipment functions, for communication between users according to established protocols (e.g., mobile telephony service).

Table 10.9
Typical GSM Uplink and Downlink Budgets

	Parameter	Downlink BS-MS	Uplink MS-BS	Relationship
A	Transmitter power P_t (dBm)	38	33	
B	Transmitter combiner loss L_f (dB) L_{comb}	2.5	0	
C	Feeder cable losses L_{at} (dB)	1.5	0	
D	Transmitter antenna gain G_{ti} (dBi)	10	−3	
E	Transmitter EIRP (dBm)	44	30	$D = A − B − C + D$
F	Receiver sensitivity S (dBm)	−102	−104	
G	Noise degradation D (dB)	3	4	Eq. (10.2)
H	Interference degradation margin M_I (dB)	3	3	
I	Log-normal margin L (dB)	5	5	
J	Receiver antenna gain G_{ri} (dB)	−3	13	
K	Isotropic power (dBm)	−88	−105	$k = F + G + H + J − J$
L	Allowed loss L_b (dB)	132	135	$L = E − K$
M	Street level coverage d (km) (h_t = 30 m, h_m = 1.5m)	1.4	1.7	Eq. (10.5)
N	Indoor coverage d (km)	0.5	0.6	

- *Bearer services* [4] are services that provide the capability for the transmission of signals between standard user-network interfaces (e.g., data services).

- *Supplementary services* [4] are defined as add-ons (i.e., additional features to both teleservices and bearer services). In a way they are value-added services on top of teleservices and bearer services.

Unlike the PSTN, the GSM network does not allow direct voice-band data transmission. Voice coders are adapted to the structure and characteristics of the voice signal and cannot operate with any other type of signal. For this reason, the modems developed for use on the conventional telephone channel for data transmission cannot be used on the GSM network. In fact, since GSM is a digital network, there is no need for the complex PSTN voice-band modem modulations to be used, but rather a digital rate adaptation must be carried out. It is necessary, therefore, to use an *adapter* to integrate the data to be transmitted in the TDMA frames and an interworking unit at the other end of the network to permit access to the PSTN from an MSC.

As regards the ISDN, the main difference with GSM lies in the binary rate limitation. GSM may transmit binary rates up to 9,600 bps, whereas the conventional ISDN can transmit up to 64 kbps (B-channel). In other aspects, many parallelisms are found between both networks.

Sections 10.7.1–10.7.3 describe the main services provided by GSM networks.

10.7.1 Teleservices

In this section the main GSM teleservices are listed [5].

- *Telephony.* This service enables bidirectional speech calls to be placed between GSM users and any telephone subscriber who is reachable through the general telephone network.
- *Emergency calls.* This service allows the user of an MS to reach a nearby emergency service (such as police or fire) through a simple and unified procedure by dialing 112.
- *Voice messaging.* This service enables a voice message to be stored for later retrieval by the mobile recipient.
- *Point-to-point short messages.* Alphanumerical messages may be sent to a mobile terminal or by the mobile to another subscriber. This latter modality was scheduled for Phase 2. The first case is directly competing

with other radio services although it includes some additional characteristics particular to the mobile network, such as confirmation of receipt of message or storage until the user connects to the network.

- *Message broadcasting.* Messages of all types may be sent to be received by all subscribers.
- *Fax, telex, videotex, etc.* All these services require *adaptation* elements. The purpose is that any general use commercial terminal may be used in the GSM network, with the appropriate adapter.

10.7.2 Carrier/Bearer Services

A series of services [6] has been specified for data transmission, summarized as follows:

- *Synchronous duplex data transmission services.* Different transmission speeds have been specified, up to 9,600 bps.

- *Asynchronous duplex data transmission services.* Different transmission speeds have been specified, up to 9,600 bps.

- *Gross duplex data transmission up to 12,000 bps.* This service may only be provided between GSM terminals, so that its importance is expected to be secondary.

- *Asynchronous access to a packet assembler disassembler (PAD)* for connection to packet-switched networks.

10.7.3 Supplementary Services

These are not independent services but are provided in association with tele- or carrier-services. These services include [4]:

- Advise of change;
- Call barring;
- Call forwarding;
- Call hold;
- Call waiting;
- Call transfer;
- Completion of calls to busy subscribers;

- Closed user group;
- Calling number identification presentation;
- Malicious call identification;
- Three-party service.

The number of services provided is very wide, but the possibilities of the system are so large it is likely that, in the next few years, further new services will be included that, to date, have not been considered in the current specifications. These services will, occasionally, be offered by different operators to those exploiting GSM networks so that new specialized trade niches will appear making it possible to broaden the use of the system.

References

[1] Hodges, M. R. L., "The GSM Radio Interface," *British Telecom Tech. Journal,* Vol. 8, No. 1, January 1990, pp. 31–43.

[2] Scourias, J., *Overview of the Global System for Mobile Communications,* http://ccnga.uwaterloo.ca/jscouria/GSM/gsmreport.html. Revised, October 1997.

[3] GSM Recommendation 03.30, ETSI.

[4] Mehrotra, A., *GSM System Engineering,* Norwood, MA: Artech House, 1997.

[5] GSM Recommendation 02.03, ETSI.

[6] GSM Recommendation 02.02, ETSI.

Selected Bibliography

Balston, D. M., "The Pan-European System: GSM," in *Cellular Radio Systems* (edited by D. M. Balston and R. C. V. Macario), Norwood, MA: Artech House, 1993.

Cheeseman, D., "The Pan-European Cellular Mobile Radio System," Chapter 13 in *Personal and Mobile Radio Systems* (edited by R. C. V. Macario), IEE Telecommunications Series 35, London: Peter Peregrinus, 1991, pp. 270–289.

Mouly, M., and M. B. Pautet, *The GSM System for Mobile Communications,* Palaiseau, France, published by the authors, 1992.

Redl, S. M., M. K. Weber, and M. W. Oliphant, *GSM and Personal Communications Handbook,* Norwood, MA: Artech House, 1998.

Redl, S. M., M. K. Weber, and M. W. Oliphant, *An Introduction to GSM,* Norwood, MA: Artech House, 1995.

Watson, C., "Radio Equipment for GSM," in *Cellular Radio Systems* (edited by D. M. Balston and R. C. V. Macario), Norwood, MA: Artech House, 1993.

11

Other Mobile Radio Systems

11.1 Introduction

This chapter presents a miscellany of mobile radio communication systems of particular interest due both to their field of application and their degree of development. The second-generation digital cellular telephone system used in the United States and Latin America, known as NADC (IS-54), is described. The modern cellular spread spectrum CDMA system (IS-95) is also studied. This standard is used both in the United States, Latin America, and Asia and may be considered as a bridge between second- and third-generation cellular systems. Also, the modern European digital PAMR and cordless systems TETRA and DECT are presented in this chapter.

11.2 The North American Cellular Digital System (NADC)

11.2.1 Overview

AMPS, which is a first-generation analog standard, has been used in the United States for cellular mobile telephony from the early 1980s. This system provides roaming capabilities between the various operators both in the United States and its neighboring countries, Canada and Mexico.

The FCC allocated a bandwidth of 40 MHz for mobile telephone systems. The specific frequency bands are 825–845 MHz (uplink) and 870–890 MHz (downlink). These bands are subdivided into two, called A and B, to be assigned to the wireline operator and to a non-wireline operator in each market in the United States. The channel spacing is 30 kHz, and channels are numbered from 1 to 666. The duplex separation is 45 MHz. The following frequencies correspond to the generic channel, n, for the uplink and downlink:

$$f_u(n) \text{ MHz} = 825 + 0{,}03 \cdot n \quad \text{(uplink)}$$
$$f_d(n) \text{ MHz} = f_u(n) + 45 \quad \text{(downlink)}$$

Frequency Group A comprises channels 1 to 333 inclusive, and Group B ranges from channel 334 to 666 inclusive. Later, the allocation was broadened by 5 MHz, as follows (see Figure 9.12(a)):

- 1 MHz from 824 to 825 MHz;
- 4 MHz from 845 to 849 MHz.

This new bandwidth was also divided in two segments of 2.5 MHz each, as follows:

- Subband 824–825 MHz, comprising 33 channels numbered from 991 to 1023 inclusive, was assigned to operator A.
- Subband 845–846.5 MHz, comprising 50 channels numbered from 667 to 716 inclusive, was also assigned to operator A.
- Last, sub-band 846.5–849 MHz, comprising 83 channels numbered from 717 to 799, was assigned to operator B.

Figure 9.12(a) shows the resulting channel distribution known as extended AMPS (EAMPS), with a total of 832 channels.

Each operator was assigned a total of 416 EAMPS channels. In AMPS, 21 control channels are used for communication set up and signaling purposes. These channels are located in the central portion of sub-bands A and B, as follows:

- Group A setup channels: 313–333;
- Group B setup channels: 334–354.

The rest are used as traffic channels. When AMPS started to show signs of saturation, the need to increase its capacity was discussed. In order to take maximum advantage of the existing base station site infrastructure and roaming services, while taking into account that it was not possible to make new frequency bands available, it was decided to develop a modified system capable of supplying the required increased capacity while maintaining the existing channel spacing. The new system should present backwards compatibility with the original AMPS system, allowing for a simple migration from analog to digital technology.

Three standards were established to improve capacity. These solutions are not compatible with each other, although they can co-exist in the available frequency spectrum. Each local carrier may replace analog AMPS channels with channels corresponding to the new standard adopted. Consequently, dual mode user terminals appeared that were able to operate both on AMPS analog channels and on channels operating with the new standard. In this manner, all networks, despite having introduced the new system, could serve the old AMPS terminals, using analog channels. Terminal roaming between different systems is also permitted.

The three standards developed for the new generation of AMPS were the following:

1. Narrow AMPS (NAMPS) where the original 30-kHz spaced traffic channels are divided into three analog channels using an FDMA multi-access technique.
2. NADC, which provides three TDMA traffic channels per carrier.
3. CDMA, which is a spread spectrum, high-capacity, wideband system.

In the first two systems, 21 control channels are used.

Figure 11.1 shows how EAMPS and NADC, are superimposed through the use of common control channels. Control channels include codes in the information transmitted. These codes indicate if the system is a dual one. If an analog EAMPS terminal roams to an NADC network, the right path shown in Figure 11.1 is followed. When turned on, the mobile explores the frequency band in search of a downlink control channel (FCC). When registering with the network via the uplink control channel (RCC), the mobile is identified by sending an MIN and its ESN. After suitable checks, the network assigns it an analog traffic channel (forward voice channel (FVC) and reverse voice channel (RVC)). Communications may be established when requested by the mobile or by the network. The network will carry out such procedures as hand-overs in the same manner as in analog networks.

If a dual mode terminal tries to access an NADC system, the left hand branch in Figure 11.1 is followed. The initial access to the network is carried out in exactly the same way as in the previous case. In the downlink control channel (FCH) information is included, such as the enhanced protocol indicator (EPI), that advises the user terminal on the type of system it is accessing.

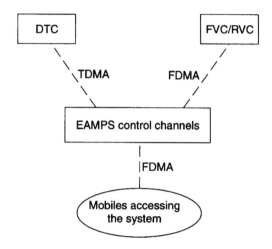

Figure 11.1 Dual terminal access to an NADC/AMPS network.

Once access to the network is achieved via the control channel, the network acknowledges the type of terminal, in this case dual, and assigns a digital traffic channel (DTC). The digital channels are set up on the original 30-kHz spaced channels using a TDMA access technique, giving rise to three time-multiplexed DTCs. If no DTC is available, the network may assign the dual terminal an FVC/RVC analog channel.

In Section 11.2.2, some of the main features of the NADC system, also known as "Digital-AMPS" (DAMPS), or IS-54 are summarized.

11.2.2 Description of the NADC System

The NADC system has many similarities with GSM. Its architecture comprises several functional entities separated by interfaces as shown in Figure 11.2 and in Table 11.1.

Unlike GSM, not all interfaces between entities have been standardized. The exceptions are the U_m air-interface, also known as "common radio interface," and the E interface between MSCs. The evolution of NADC networks takes place in two phases:

1. In the first phase, 21 control channels used by analog AMPS are shared. Only voice services are provided. MSs are dual mode, analog AMPS and NADC, and the operators continue to provide analog AMPS service as NADC is being introduced.

2. In the second phase, new digital control channels are established for accessing the digital voice TDMA channels, so that operation is possi-

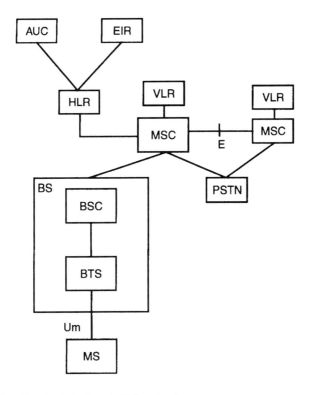

Figure 11.2 Functional entities in a NADC network.

Table 11.1
Elements in a NADC Network

AUC	Authentication center	MS	Mobile station
BS	Base station	MSC	Mobile switching center
BSC	Base station controller	PSTN	Public switched telephone network
BTS	Base transceiver station	U_m	Air interface
EIR	Equipment identity register	VLR	Visitor location register
HLR	Home location register		

ble with digital-only units. Also, a specific protocol for data transmission is introduced.

11.2.3 Logical Channels

Logical channels are data structures designed to support signaling and user information transmission. In the first version of standard IS-54, MSs used the same signaling channels as in the original AMPS system but traffic channels were already digital (digital traffic channels) and could transport both voice and data.

Call-associated signaling information is transferred via associated control channels, which share bursts with the traffic channels. There are two types of associated channels:

- *Slow associated control channel* (SACCH) for transmission of recurring, nonurgent information, associated with a call, as the result of measurements of neighboring cells carried out to assist with call hand-overs.
- *Fast associated control channel* (FACCH) for transmission of specific, urgent information, mainly for hand-over related messages. This channel overrides traffic channel information by "stealing" information bits.

In the IS-54+ specification, digital control channels are provided separated from the existing AMPS control channels. These new channels support enhanced features such as short message service (SMS), intelligent cell selection, and the use of MS sleeping mode to achieve longer mobile terminal battery lives.

In the downlink, the digital control channel (DCC) includes the broadcast control channel (BCCH), and the point-to-point SMS, paging and access response channel (SPACH).

At least one digital traffic channel is used at each cell sector to support DCC services. Mobile phones capable of utilizing DCCs will scan the RF spectrum in search of one DCC, gain synchronization and begin to decode the information transmitted over the BCCH. BCCH information includes system identification, neighbor lists of other DCCs, the DCC frame structure on the current cell, and other miscellaneous information. In the uplink direction control channels are contention based and are referred to as the random access channels (RACH).

11.2.4 NADC Radio Interface

The radio interface is based on a TDMA multi-access structure set up on each RF channel. A time slot (TS) and frame structure are defined. Each frame comprises 6 TSs and has a duration of 40 ms. Consequently, 25 frames per second are transmitted; the duration of each TS is $40/6 = 6.6$ ms. A physical channel comprises a TS and an RF carrier with a bandwidth of 30 kHz. The multi-access TDMA structure in NADC provides thus, six physical channels on each RF carrier.

Figure 11.3 illustrates the NADC system frames for digital traffic channels. It may be recalled that in Phase I, control channels are the same as those in the analog AMPS system.

Two types of traffic channels are defined:

1. *Full-rate traffic channels.* These channels use two time slots in each frame. Consequently, 3 channels per RF carrier are offered:
   - Channel 1 uses slots 1 and 4
   - Channel 2 uses slots 2 and 5
   - Channel 3 uses slots 3 and 6
2. *Half-rate channels.* Each channel uses a single time slot per frame, so that there are six possible TDMA channels per carrier. Channel n uses time slot n.

Each transmitted burst comprises 324 bits. Since NADC uses a quaternary digital modulation, two bits make up a symbol or modulation state, so that any one burst has 162 symbols. A frame contains $6 \times 324 = 1944$ bits (972 symbols). Since the frame duration is 40 ms, the channel bit rate is 1944/40 = 48.6 kbps.

Since the channel bandwidth is 30 kHz, the achieved spectral efficiency is $\eta = 48.6/30 = 1.62$ bits/s/Hz. In GSM, the channel bit rate and bandwidth are respectively 270.833 kbps and 200 kHz, hence its spectral efficiency is $\eta = 270 \times 833/200 = 1.35$ bits/Hz. It is evident that the modulation scheme used in NADC is 20% more efficient than that in GSM. However, this piece of information alone may be misleading since the overall spectral efficiency is also dependent on the minimum required protection ratio.

In the NADC system, there are several types of bursts, depending on whether a control channel or a traffic channel is transmitted. All traffic channel bursts consist of several bit fields, as described below (Figure 11.4(a)).

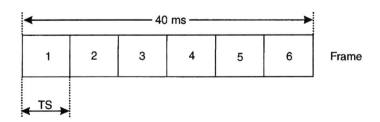

Figure 11.3 NADC frames for digital traffic channels.

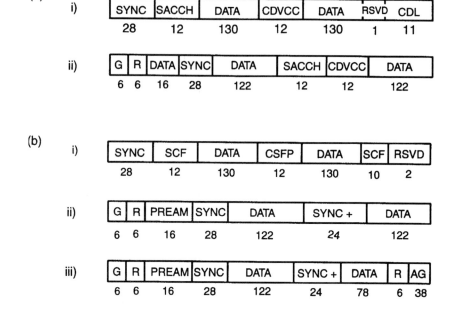

Figure 11.4 (a) NADC traffic channel bursts. (b) NADC control channel bursts.

a) Downlink burst (Figure 11.4.a.i) with the following bit fields:

- SYNC (28 bits), for MS synchronization;
- SACCH (12 bits), for mapping the slow associated control channel;
- DATA (130 + 130 bits), two 130 bit fields carrying the digital traffic (voice or data) information;
- CDVCC (12 bits), coded digital verification color code, similar to the SAT tone in analog AMPS;
- RSVD (1 bit), reserved field;
- CDL (11 bits), coded digital control channel locator;

b) Uplink burst (Figure 11.4.a.ii), structured as follows:

- G, guard time, with a duration equivalent to 6 bits. During this time, the MS has the transmitter disconnected. The purpose is to minimize any collisions which may occur between bursts transmitted by different MSs in their uplinks;
- R, ramp time, with a duration equivalent to 6 bits. Time for MS powering up;

- SYNC (28 bits), for BS synchronization;
- DATA (16 + 122 + 122 bits), one 16-bit field plus two 122-bit fields for digital traffic (voice or data) information;
- SACCH (12 bits), for mapping the slow associated control channel;
- CDVCC (12 bits), for the coded digital verification color code, as in the down-link.

Figure 11.4(b) illustrates the bursts used on the control channel. In this case, there are two types of bursts for the uplink: the normal and the abbreviated bursts.

a) Downlink burst (Figure 11.4(g)(i)):

- SYNC (28 bits), for MS synchronization;
- SCF (120 + 10 bits), for shared control feedback information indicating the status of reverse control channel;
- DATA (130 + 130 bits), two 130 bit fields of signaling data;
- CSFP (12 bits), for coded super frame phase information;
- RSVD (2 bits), reserved field.

b) Uplink burst (Figure 11.4(b)ii normal burst and iii abbreviated burst):

- G (6 bits), guard time;
- R (6 bits), ramp time;
- PREAM (16 bits), preamble information sent by the MS to facilitate synchronization of symbols at the BS;
- DATA: two fields of 122 bits for a normal burst or one field of 122 bits plus another of 78 bits for an abbreviated burst. This can be traffic or control data;
- SYNC+ (24 bits), additional synchronization;
- R (6 bits), ramp time for MS powering down;
- AG (38 bits), guard time for the abbreviated burst.

Abbreviated bursts are used with the RACH access channel and require a longer guard time. To avoid simultaneous transmission and reception, time slots are offset in time. For MSs the transit-receive offset is as follows:

forward time slot = reverse time slot + (1 time slot + 45 symbols) (i.e., MS time slot 1 of frame *n* occurs 207 symbol periods before time slot 1 of the same BS frame).

A super-frame structure is used for the mapping of the BCCH, SCF and SPACH channels onto the physical DCC downlink channel. Since a predetermined multiplexing structure is created, each MS will have knowledge of when to expect that a paging message may arrive addressed to it. Each super-frame is made up of 16 TDMA frames.

11.2.5 Modulation

NADC uses a linear $\pi/4$-shifted differential quadrature-phase shift keying ($\pi/4$-DQPSK) modulation at a 24.3 ksymbol/s rate. This modulation scheme uses the phase constellation shown in Figure 11.5. The information is differentially encoded (i.e., symbols (dibits) are transmitted as phase changes rather than absolute phases). A Gray code is used in the mapping: Dibit symbols corresponding to adjacent signal phases differ only in a single bit. Since the most probable errors due to noise result in the erroneous selection of an adjacent

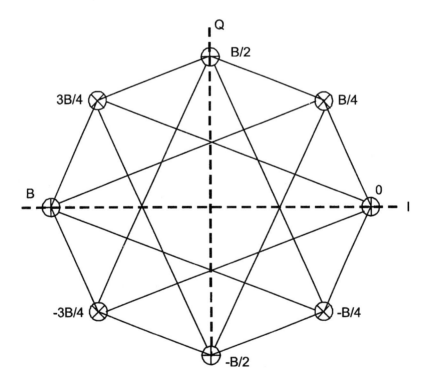

Figure 11.5 NADC $\pi/4$-DQPSK modulation system.

phase, most dibit symbol errors contain only a single bit error. Also note the $\pi/4$ rotation of the basic QPSK constellation for odd ($\oplus$) and even ($\otimes$) symbols.

The binary data stream is converted to two binary streams, X_k and Y_k, formed from the odd and even numbered bits, respectively. The quadrature symbols, I_k and Q_k, are obtained according to the expressions

$$I_k = I_{k-1} \cos[\Delta\phi(X_k, Y_k)] - Q_{k-1} \sin[\Delta\phi(X_k, Y_k)] \quad (11.1)$$

$$Q_k = I_{k-1} \sin[\Delta\phi(X_k, Y_k)] + Q_{k-1} \cos[\Delta\phi(X_k, Y_k)] \quad (11.2)$$

where I_{k-1} and Q_{k-1} are the amplitudes at the previous pulse interval. The phase change $\Delta\phi$ takes values according to Table 11.2.

As a consequence, symbols I_k and Q_k at the output of the differential phase modulator, can take on one of the following five values: $0, \pm 1/2, \pm 1/\sqrt{2}$

Symbols I_k and Q_k are then applied to the inputs of the I and Q baseband filters, which have linear phase and square root raised cosine frequency responses following the expression

$$|H(f)| = \begin{cases} 1 & 0 \leq f \leq (1-\alpha)/2T \\ \sqrt{\frac{1}{2}\left\{1 - \sin\left[\frac{\pi(2fT-1)}{2\alpha}\right]\right\}} & (1-\alpha)/2T \leq f \leq (1+\alpha)/2T \\ 0 & f > (1+\alpha)/2T \end{cases} \quad (11.3)$$

where T is the symbol period and α is the rolloff factor which determines the width of the transmission band. A value of 0.35 was selected for α. The resulting transmitted signal is given by

Table 11.2
Phases in the NADC $\pi/4$-DQPSK Modulation System

X_k	Y_k	$\Delta\phi$
1	1	$-3\pi/4$
0	1	$3\pi/4$
0	0	$\pi/4$
1	0	$-\pi/4$

$$s(t) = \sum_n h(t - nT) \cos\phi_n \cos\omega_c t - \sum_n h(t - nT) \sin\phi_n \sin\omega_c t$$

(11.4)

where $h(t)$ is the pulse shaping function (inverse Fourier transform of $H(f)$), ω_c is the radian carrier frequency, T is the symbol period and ϕ_n is the absolute phase corresponding to the *n-th* symbol interval, whose value results from a differential encoding process as follows:

$$\phi_n = \phi_{n-1} + \Delta\phi_n \qquad (11.5)$$

11.2.6 Speech Processing (Full-Rate Channel)

Speech processing covers the operations necessary for the digital representation of speech including channel coding for error protection. These operations are:

- Source coding;
- Channel coding;
- Interleaving.

The source coder uses a class of code excited linear predictive coder (CELP) called vector sum excited linear predictive (VSELP). The source coding is made using 20 ms-long speech blocks at the rate of 7.95 kbps, so that for each block 159 bits are generated.

These bits are applied to the channel coder for error control. The 159 bits in each speech block are divided into two classes: 77 *Class I bits* to which a convolutional code for forward error correction is applied and 82 *Class II bits* which are transmitted with no protection. Among the 77 Class I bits, 12 are the most significant and sensitive and, thus, they require further protection. To accomplish this, an error detection block code is used. This code adds seven cyclic redundancy bits (CRC) to the information bits. Five tail bits (five zeroes) are also included to form a 89-bit codeword, as shown in Figure 11.6(a).

This 89-bit codeword is input to a rate 1/2 convolutional coder producing $2 \times 89 = 178$ bits. Then, the 82 unprotected bits are aggregated, yielding a final result of a block of 260 bits for each 20 ms speech block, so that the speech gross binary rate is 260/20 = 13 kbps.

In order to improve the efficiency of the error correction codes, transmission is carried out with bit *interleaving* on two consecutive bursts, as follows: the bits of a speech block are stored in a 26 row 10 column matrix, writing them column-wise, as shown in Figure 11.6(b).

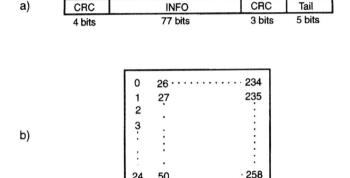

Figure 11.6 (a) NADC channel coding scheme. (b) Interleaving scheme in NADC.

For transmission, the bits are extracted from the matrix, reading them row-wise. The 130 bits of the even rows (0, 26, ... 234, 2, 28, ... 236, 24, 50, ... 258) are transmitted in a 130-bit burst and the 130 bits of the odd rows (1, 27, ... 235, ... 25, 51, ... 259) are transmitted on the next burst.

11.3 The Mobile Telephone CDMA System

11.3.1 Introduction to CDMA

Different narrowband FDMA and TDMA multi-access systems for mobile communications have been described in this and in previous chapters. These systems are very susceptible to cochannel interference which is also narrowband. Their correct operation requires that the wanted-to-interference signal ratio be higher than given threshold levels (protection ratio), which are relatively high. This limits their spectral efficiency and traffic handling capacity as they require large cluster sizes.

It is well known that digital communications provide a relatively good quality in the presence of white noise-like interference (i.e., with a wideband and a reduced power spectral density). Indeed, Shannon theory foresees an interchange between s/n or s/i and bandwidth, so that the transmission of digital information in an environment with high interference levels is possible. An application of these ideas is the spread spectrum technology.

Assuming an information source with a rate R bps, in these systems, a spreading code is applied to the source information transforming the original signal into another with a rate $W \gg R$, and a large bandwidth and reduced power spectral density so that the transmitted signal presents white noise-like characteristics. At the receiving end the same code, properly synchronized, is applied to the spread signal to despread it and recover the original signal. Any interference affecting the original signal is spread in the receiving process, thus converting the interfering signal into a noise-like, low spectral density perturbation, so that its effect on the wanted signal is minimum. In this way, it is possible to achieve a good baseband quality in an RF environment with high interference levels.

The spread spectrum transmission bandwidth is determined by the selected spreading code and not by the transmitted information rate. The improvement factor provided by the spread spectrum technique is proportional to the W/R ratio, called *processing gain, PG,* i.e., the signal-to-interference ratio at the receiver front-end is increased by PG (dB) at the demodulator input. In addition to this increased protection against interference effects, spread spectrum techniques present the following advantages [1, 2]:

1. Privacy. The use of a spreading code assures that an eavesdropper will not be able to retrieve the transmitted information.
2. Jamming protection. Any jamming perturbation is spread in frequency at the receiver in the same way as any other interfering signals.
3. Low probability of interception. Because of its low power spectral density, the SS signal is difficult to detect and intercept by a hostile listener.
4. Protection against multipath. Since the SS receiver is wideband, it has a greater time resolution of the received echoes that can be constructively combined with a RAKE receiver structure. In this way, multipath no longer will cause such destructive effects as in the case of narrowband receivers.
5. Multi-access capability. Different signals may be transmitted/multiplexed on the same RF carrier by assigning them different spreading codes. This is precisely the property on which the CDMA multi-access technique is based. SS techniques are briefly discussed below.

Given several source signals, a different spreading code is applied to each of them, and these signals are multiplexed on the same carrier. At the receiving end a bank of signal processors is available. In each processor in the bank, the code corresponding to each one of the users is applied to the multiplex signal. This despreads the wanted signal while the rest remain spread acting as low level interference sources on the wanted signal.

The processor is basically a *correlator*. This means that, in principle, the codes to be used in CDMA must be orthogonal (i.e., their cross correlation values must be zero), and thus, in each correlator, the wanted signal can be extracted alone without any trace of interference from the other signals. However, the selection of orthogonal codes presents two problems:

1. The need for a strict synchronization between the transmitter and the receiver. If synchronization is not complete the advantages of orthogonality are lost.
2. The number of orthogonal codes is small thus limiting the number of possible simultaneous users. Hence, nonorthogonal codes whose cross-correlation values are small but not zero are employed. The number of these codes is high, allowing the provision of service to a large number of users.

There exist three basic SS schemes:

- Direct sequence SS;
- Frequency hopping SS;
- Time hopping SS.

In these techniques the spectral spreading is accomplished in different ways. In DS-SS, spectrum is spread by multiplying the information signal directly by a high-rate spreading code. This technique is generally known as direct sequence-CDMA (DS-CDMA). In frequency hopping-SS (FH-SS), the carrier frequency at which the information is transmitted is changed according to the spreading code. In time hopping-SS (TH-SS), the information signal is transmitted in bursts, the time of these bursts being decided by the spreading code.

During the remainder of this section, DS-CDMA systems are discussed. In this type of system, CDMA signals are generated by means of a linear modulation process using pseudo-random binary sequences assigned to the users as *signatures* or *address codes*. The process consists on the multiplication of the original signal to be transmitted, $S_i(t)$, from user i by its address code $c_i(t)$ (spreading). At the other end, the received signal is multiplied again by $c_i(t)$ (de-spreading) thus recovering $s_i(t)$. Any other modulated signal $s_j(t)c_j(t)$ when multiplied by $c_i(t)$ remains spread and, at the receiver of user i, it behaves as white noise.

The number of simultaneous transmissions that can take place on one RF carrier is only limited by the number of unique codes available in the orthogonal case and by the interference level that each receiver can withstand in the non-orthogonal case. A code is made up of many elements (called "chips") transmitted for each information bit.

The coexistence of CDMA carriers and FDMA channels in CDMA-FM mixed cells requires the provision of guard bands to mitigate the out-of-band interference, so that acceptable performance of both systems can be ensured. Also, due to the fact that processing gain of practical CDMA systems is limited, the compatibility with other transmissions requires a strict power level control relative to the wanted signal.

The main advantage of CDMA technology is that capacity is interference limited: the more active channels allowed in the system, the higher the interference level will be. Thus, capacity is determined by the maximum level of interference that can be tolerated while maintaining the quality requirements of the system. Conversely, conventional systems are dimension-limited: there is a fixed number of resources (RF carriers in FDMA or time slots in TDMA) that determine the system capacity.

This interference-limited feature proves much more adequate in cellular environments, giving rise to a higher capacity in CDMA than in the other multi-access, narrowband techniques. Furthermore, in CDMA, adjacent cells can use the same RF frequency, so the need for network frequency planning disappears.

Capacity in CDMA cellular systems is maximized when all signals are received with the same level, which is the minimum necessary to achieve the desired quality. Hence, the need for a strict *power control* mechanism that tends to equalize all the received levels, overcoming the near-far effect (i.e., a nearby transmitter being received with a much higher level than a distant transmitter). The need for power control constitutes the most significant technical disadvantage of CDMA. The power control mechanism must ensure that all transmissions arrive with the same power, within a stringent tolerance (±0.5 dB). As a consequence, a method has to be implemented allowing the base station to control the transmitted power of the mobile terminals (reverse-link power control) and also permitting the mobile station to request different levels from its serving base station (forward-link power control).

In CDMA, for full-duplex operation it is convenient to use paired frequency allocations with separated frequency sub-bands for the uplink and downlink with a fixed duplex frequency spacing. This is appropriate for symmetrical traffic flow in both directions. Nevertheless, for asymmetrical traffic applications (e.g., Internet downloading) unpaired frequency allocations have been proposed using time division duplex (TDD), operating on the same RF carrier for both transmission directions, as discussed in Chapter 12.

A description is given in Section 11.4.7 on the capacity improvement obtained by using CDMA over other access technologies in cellular telephone systems. For a conventional (AMPS) analog system, the received signal must be at least 18 dB above the noise level or interference level in the channel in order to obtain an acceptable quality. This means that only part of the available spectrum may be used in any given cell. It is not possible to use the same channel in an adjacent cell. If a seven-cell reuse pattern is employed, only one seventh of the available spectrum may be used in each cell site (Figure 11.7(a)).

Even with this frequency distribution, it is necessary to use sectorized antennas to obtain the 18-dB signal-to-interference protection ratio (R_p) required. In the event of using three sectors per base station, in each sector, only one in each 21 channels available in the assigned spectrum may be used.

In CDMA, due to the bandwidth expansion, the signals may be received in the presence of high levels of interference compared to non-CDMA systems.

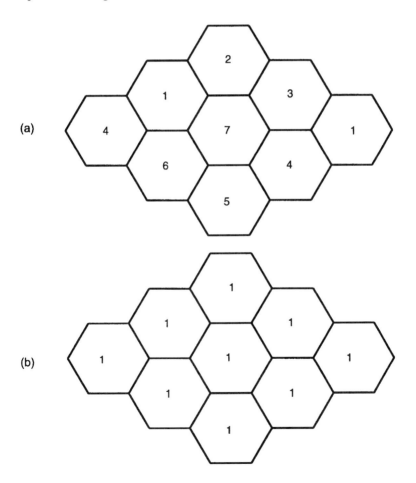

Figure 11.7 (a) Frequency reuse in FDMA and TDMA networks. (b) Frequency reuse in CDMA networks [4].

Although the practical limit depends on channel conditions, reception may take place in the presence of relatively strong interfering signals.

Due to this feature, the available channels may be reused in all sectors of all the cell sites (Figure 11.7(b)). Assuming uniform loading, approximately 60% of the total interference will originate at the cell itself and the other 40% will originate from adjacent cells, all operating on the same RF channel (Figure 11.8).

In order to explain CDMA system operation, the example is usually drawn of a packed meeting where everyone is talking in pairs and in different languages (different codes). Despite the fact that the noise level (interference) is high, it is possible to hold a conversation. The greater the difference between the different languages (the lower the correlation between them), the easier it is to operate with very high levels of interference.

The CDMA system, like the NADC system, is one of the second-generation evolutionary options of the American AMPS system. So, dual AMPS-CDMA terminals also exist. In such networks, analog AMPS transmissions may coexist with CDMA transmissions. Figure 11.9 illustrates a CDMA RF channel spectrum together with several analog carriers.

11.3.2 Diversity Mechanisms in CDMA

CDMA systems use various diversity techniques: space, frequency, and time. The normal form of space diversity is based on the use of more than one antenna at the base station. In parallel with this form of diversity (microdiversity), soft hand-over is also used (Figure 11.10), whereby, prior to call hand-over from one cell to another, both keep the link with the mobile simultaneously (macrodiversity). To do this, the mobile receiver is equipped with at least four correlators, three of which may be assigned to the current link; the fourth may be used to search for alternative routes for the signal.

Frequency diversity takes place within the bandwidth of the transmitted signal itself. In a multipath environment, time spreading effects occur which, in the frequency domain, appear as notch filters (Figure 11.11). The width of this eliminated band may vary although, as a general rule, it will be less than 300 kHz [4]. This bandwidth is sufficient to affect ten analog channels (of 30 kHz), but only eliminates 25% of the CDMA signal in the case of the IS-95 standard described in Section 11.4.

CDMA systems take advantage of the presence of multipath. Its different components (echoes with different delays) provide a form of time diversity. The different correlators in the receiver may be assigned to different delayed replicas of the same signal. These may be combined into what is known as a RAKE receiver (Figure 11.12), which comprises various branches or fingers.

Other Mobile Radio Systems 445

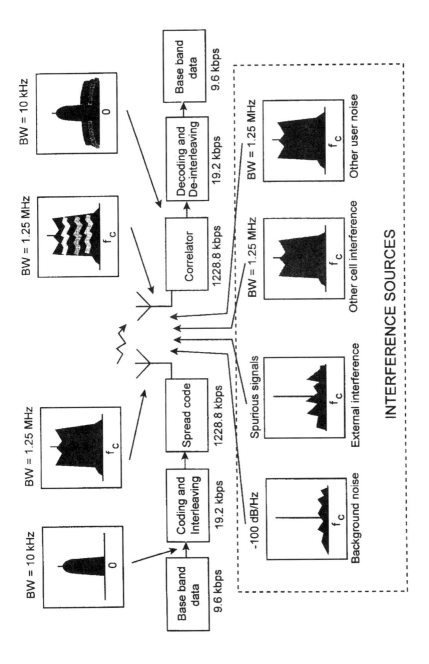

Figure 11.8 Sources of interference in a CDMA system [4].

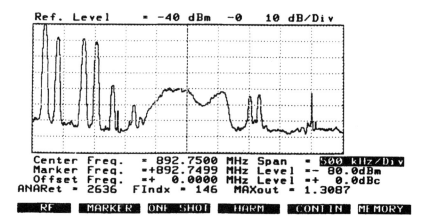

Figure 11.9 Conventional AMPS emission spectra with a CDMA channel. (Source: [3]. Reprinted with the permission of Artech House.)

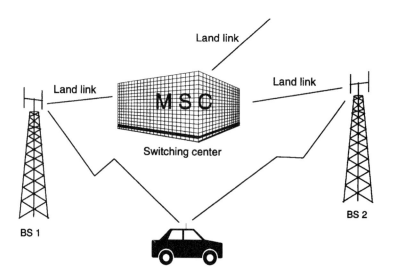

Figure 11.10 Soft hand-over [4].

The various fingers include a delay cell, a variable attenuator, and a phase shifter, their outputs are combined together in an adder. Varying the delay, attenuation and phase in each finger, an optimum combination of signals can be achieved.

Another form of time diversity is the use of error correction codes (FEC) followed by interleaving. The loss of transmitted bits tends to be grouped together in time (error bursts), whereas the majority of the error correction

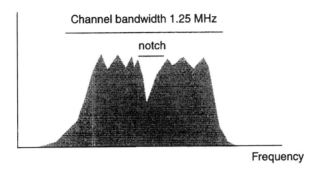

Figure 11.11 Effects of frequency selective fading [4].

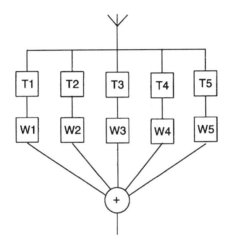

Figure 11.12 RAKE receiver [4].

algorithms function better when the erroneous bits are evenly distributed. Interleaving helps to randomize the times when channel errors appear.

11.3.3 Power Control

In CDMA systems, an unnecessary increase in transmitted power by a mobile affects all other communications, both in its own cell and in adjacent ones. The objective is for all the signals from the mobiles to reach the base station with the same power. The power should also be sufficient to obtain the best performance possible. In these systems, two forms of transmitted power control are used: *open loop* and *closed loop*.

Open loop power control is based on the similarity of path losses between the base-to-mobile path and the reverse direction. The received power at the mobile is used as a reference. If the received signal is low, the MS will transmit with a higher power and, if it is high, the MS will reduce it.

In order to complement open loop control, closed loop power control is used. To implement it, an active feedback loop is set up from the base to the mobile station. The base sends power control messages with a fixed period, which is usually on the order of 1 ms (every 1.25 ms, i.e. 800 times per second in the case of IS-95), advising the mobile to increase or decrease the transmitted power in fixed step increments (1 dB in IS-95).

Since in CDMA systems the mobile station only transmits the required power to maintain the link, the average transmitted power will be far less than that required when operating on an analog system. An analog cellular telephone always transmits sufficient power to overcome fading, despite the fact that no fading occurs for most of the time. Due to this advantage of being able to transmit with very low powers, CDMA systems offer better possibilities for longer terminal battery life and the use of smaller and more economical power amplifiers.

11.4 The CDMA IS-95 Standard

11.4.1 Introduction

In this section a description of the CDMA IS-95 standard, defined to increase analog AMPS capacity, is given. A proposal was brought forward for a new digital cellular system based on CDMA technology. This new system would provide a larger capacity increase than that achieved with the other standards. The IS-95 standard only specifies the air-interface and was released in 1993. This system uses a basic binary rate of 9,600 bps per information channel. This binary rate is increased to 1.228 Mbps (spread spectrum signal) by the channel coding and the spreading processes. This frequency span is equivalent to approximately 42 analog AMPS channels. Thus, this technique does not lend itself to a direct channel replacement of analog channels, unlike in the NADC-TDMA system; instead, large blocks of channels must be replaced at one time.

The digital 1.228-Mbps CDMA signal is transmitted using a filtered QPSK modulation with a channel bandwidth of 1.25 MHz.

At the receiving end, the decoding process recovers the original signal corresponding to a binary rate of 9,600 bps. To do so, the code with which the signal was expanded must be known. When decoding is applied to the coded signal from other users, these signals maintain their bandwidths corresponding to the spread spectrum rate of 1.228 Mbps.

The relationship between the expanded bandwidth, and the basic rate is known as the *processing gain*. In the case of the American CDMA system, this gain is 128 times or, approximately, 21 dB. There are various basic differences between this system and other cellular systems, analog and digital, these are:

- All users share the same frequency. For a completely loaded system, up to 35 communications may be supported on the same frequency. Channels are two-frequency full-duplex with a bandwidth of approximately 1.25 MHz and a 45 MHz separation between transmission directions. The 800-MHz band is used. At base stations, it is only necessary to have one transmitter/receiver per RF frequency used in the cell or sector.
- The various communication channels are differentiated by the code used. It is thus not necessary to carry out any frequency planning in CDMA cellular networks.
- The capacity limit of the system is not strict (i.e., the appearance of new users adds more interference to the system, causing a higher error rate in all communications). Also, CDMA systems benefit from the speech activity cycle, which is approximately 35%. This permits a considerable increase in the system capacity, evaluated for a 100% speech activity cycle.

In order to maximize capacity, the system uses the following:

- Several power setting algorithms to optimize system operation;
- Powerful error correction codes with interleaving;
- Speech activity detection;
- Variable rate speech encoding;
- RAKE receiver techniques.

11.4.2 Speech Coding

Speech is coded prior to being transmitted in order to reduce the required bit rate while maintaining an adequate quality. The speech coder used with the CDMA IS-95 standard produces a maximum binary rate of 8600 bps. After adding a CRC (12 bits per frame) for error detection and tail bits (eight per frame) for the convolutional coder, the binary rate is increased to 9,600 bps. This is the maximum capacity of the channel, and it is not used when the user is not speaking. The speech coder detects speech activity and lowers the transmission rate during periods of silence. The lowest rate is 1,200 bps; intermediate rates of 2,400 and 4,800 bps are used for special purposes. The 2,400-bps rate is used to transmit transients in the background noise and the 4,800-bps rate to mix coded speech and signaling data. In this latter case, the binary speed is 9,600 bps, but half of the bits are assigned to speech and the other half to signaling messages. This procedure is known as "dim and burst signaling."

The MS transmits its power in the form of bursts during low binary rate periods. In this manner, it transmits 1/2, 1/4, or 1/8 of the time. The binary rate is always 9,600 bps during transmission periods, so that the average rate is actually 4,800, 2,400, or 1,200 bps, respectively. This mechanism makes it possible to reduce the average power used and the interference experienced by the other users.

The BS uses a different method to reduce the transmitted power during periods of inactivity. The BS always transmits with a 100% activity cycle at 9,600 bps but uses only 1/2, 1/4, or 1/8 of the power and repeats the information 2, 4, or 8 times. MSs recover the necessary signal-to-noise ratio by combining multiple transmissions. The power transmitted by the base station should also be thoroughly controlled as it causes interference to the mobiles on adjacent cells. The size of the cells primarily depends on the transmitted power and on local terrain irregularity.

11.4.3 The Walsh Matrix

An important feature of this CDMA system is the use of *Walsh codes*. These are based on the Walsh matrix, a square matrix with binary elements, which always has a dimension that is a power of two. It is generated by using the first-order Walsh matrix as a basis, Walsh(1) = W_1 = 0, expanding it as shown in Figure 11.13. One of the properties of the Walsh matrix is that any row is orthogonal to any other row and to the logic not of any other row. Orthogonality means that the scalar product of any two rows is zero. In other words, it means that between two rows, half of the bits coincide, and the other half do not. The CDMA system uses a 64-by-64 bit Walsh matrix for the downlink; that is the maximum number of channels possible in each carrier, although these are not all used simultaneously but rather only around 35 of them.

11.4.4 Coding on the Base-to-Mobile Link

The coding scheme on the base-to-mobile link is illustrated with the aid of Figure 11.14. The binary reference rate is 9,600 bps. The information at the speech coder output is produced in blocks of 20 ms. Channel coding is

$$W_{2n} = \begin{matrix} W_n & W_n \\ W_n & \overline{W_n} \end{matrix}$$

$$W_2 = \begin{matrix} 0 & 0 \\ 0 & 1 \end{matrix}$$

$$W_4 = \begin{matrix} 0 & 0 & 0 & 0 \\ 0 & 1 & 0 & 1 \\ 0 & 0 & 1 & 1 \\ 0 & 1 & 1 & 0 \end{matrix}$$

Figure 11.13 Walsh matrices [4].

performed on each block using a rate 1/2 convolutional coder, which provides two output bits per each input bit. The bits are then interleaved to improve the efficiency of the correction algorithms.

After this operation, the data stream is modified using a *long code*, the only purpose of this is to act as a privacy mask. The long code is a pseudo-random binary sequence (PRBS) that is obtained with a 42-bit long shift register. A *mask* is applied to the pseudo-random generator, which selects a combination of the available bits as a function of the user's electronic serial number (ESN). These bits are modulo-2 added using exclusive-OR gates to generate a single burst of bits at a rate of 1.2288 Mbps. Since for the base-to-mobile link a rate of only 19.2 kbps is needed, only one in each 64 bits generated is actually used.

While setting up a call, each MS is assigned one of the rows in the 64×64 Walsh matrix. This is applied to the coding of each signal bit, so that the binary rate is multiplied by a factor of 64, going from 19.2 kbps to 1.2288 Mbps.

The final stage consists on the conversion of the binary signal into two binary channels necessary to drive the QPSK modulator. The bits are separated into the I and Q branches (in phase and quadrature), and the data on each channel are mixed with a unique *short pseudo-random code* (PRBS) through a XOR operation. These short codes are spreading sequences generated in a similar fashion to that used for the long code (i.e., using shift registers). In this case, the registers are made up of 15 stages, and the sequences at their output have a length equal to 32,767 bits. An additional bit is included to render a 32,768 bit period. Each BTS uses a different phase for the short code in order to differentiate its transmissions from those of the other BTSs.

452 Introduction to Mobile Communications Engineering

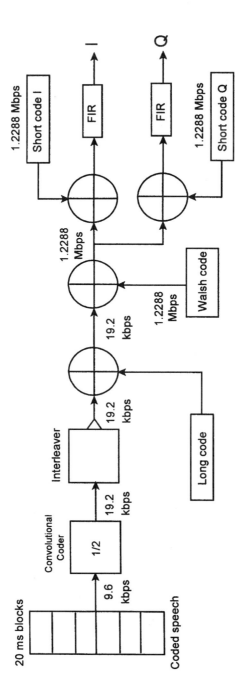

Figure 11.14 Base-to-mobile link [4].

Following the short code stage, two binary channels at a rate of 1.2288 Mbps are obtained. Low-pass filtering is applied to each channel using an FIR filter with a cutoff frequency of approximately 615 kHz. The resultant I and Q signals are converted to analog signals and sent to a linear I/Q modulator.

The multiple base station channels are transmitted by combining the I and Q branches of each communication. Since all users share the composite signal, a *pilot signal* is transmitted that uses row zero in the Walsh matrix (all zeroes). Approximately 20% of the energy in the cell corresponds to this pilot signal. The pilot signal is a coherent phase reference used by mobile stations to demodulate the binary traffic on the channel. This is also the time reference for the correlators. There are 511 different *time offsets* possible, ensuring a unique identification of the different base stations and sectors.

Other channels transmitted from the base station are the PCHs up to a maximum of seven, corresponding to rows 1 to 7 in the 64×64 Walsh matrix. Base stations also transmit a synchronization channel (SCH) which uses row 32. The MS finds BS related information on the synchronization channel, and this makes it possible to align the long-code. Finally, there are the downlink traffic channels (FTC) that correspond to the remaining rows in the Walsh matrix (Figure 11.15).

11.4.5 Coding on the Mobile-to-Base Link

The coding scheme on the mobile-to-base link is illustrated with the aid of Figure 11.16. The mobile does not transmit a pilot signal as this would take up an important fraction of the available power, which is limited in the mobile mainly due to battery capacity. The coding scheme is different in this case, and the demodulation task is rather more difficult at the base station. The same speech coder is used in both directions and the basic binary rate is also 9,600 bps. The signal first passes through a rate 1/3 convolutional coder, which produces an output rate of 28.8 kbps. This output is interleaved, and the bits are grouped together in blocks of six. Each group of bits is a symbol with a value from 0 to 63. Each symbol is assigned one row of the Walsh matrix. All mobile stations may transmit any row in the matrix as a function of their transmitted data.

At this point, the binary rate is 307.2 kbps, but no unique coding has been achieved yet for each mobile. To do so, the long code is used; this increases the binary rate to 1.2288 Mbps. This long code has a time offset that depends on the user's identity. It separates the uplink users in the same cell/sector and in other cells. These bits are sent to the I and Q branches, and later, the same short sequences are applied to them as in the base station. There is yet another

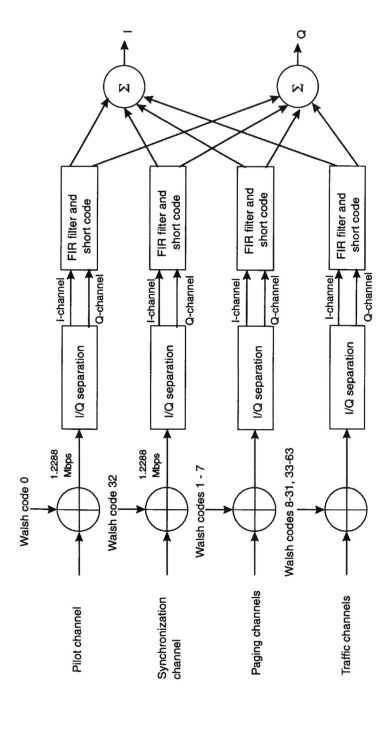

Figure 11.15 Base station channels [4].

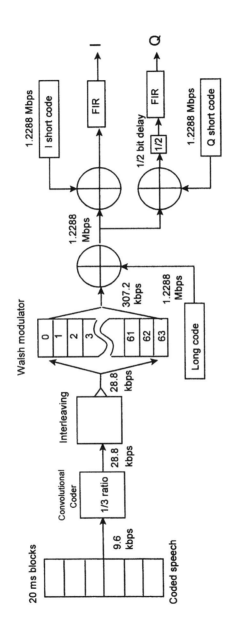

Figure 11.16 Mobile-to-base link [4].

difference. A *half-bit delay* is applied to the Q channel before the FIR filter. In this way an *offset-QPSK* (OQPSK) modulation is generated. This is used to avoid in the signal amplitude zeroes inherent to QPSK modulation, thereby simplifying the design of output amplifiers for mobile stations (Figure 11.17).

Down and up links have different characteristics. In the downlink, a phase reference is present provided by the pilot signal and the orthogonal codes. In the uplink, the signals are simply noncorrelated (non-orthogonal). The base station, however, has the advantage of having several receive antennas available (space diversity). Taking all these factors into consideration, it is the downlink that determines the capacity of the system. Table 11.3 summarizes the applications of the different codes used in CDMA IS-95 systems.

11.4.6 CDMA System Operation

When the mobile station is turned on, it already knows the frequency assigned to the CDMA service. The mobile tunes in to this frequency and searches for the pilot channel. It is normal to find more than one pilot signal, and these are distinguished from each other by the time shift in the short code corresponding to each base station. The mobile station locks to the strongest signal and establishes a frequency and time reference with it. It then demodulates the channel corresponding to row 32 in the Walsh matrix, the SCH. This channel contains the information required by the 42-bit shift register used to generate the long code.

The mobile station may be required to register with the network. This is known as a *start-up registration*. Also, with the change of location area, a

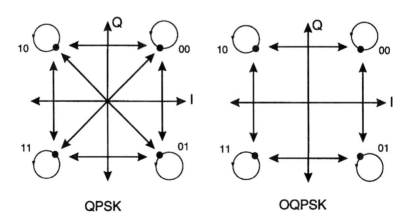

Figure 11.17 Uplink and downlink modulations in the CDMA IS-95 system.

Table 11.3
Codes Used in the CDMA IS-95 System [4]

Parameter	Function	Notes
Frequency	Divides the spectrum into several 1.23-MHz frequency allocations	Forward and reverse links are separated by 45 MHz.
Walsh codes	Separate forward link users of the same cell and they are used as a part of the modulation process on the reverse link	Assigned by BS for the downlink. Walsh code 0 is always the pilot channel. Walsh code 32 is always the sync. channel.
Long code	Separate reverse link users of the same cell	Depends on time and user ID. The long code is composed of a 42-bit long PRBS generator and a user specific mask.
Short codes (also called the I and Q spreading sequences)	Separate cell sites or sectors of cells	The I and Q codes are different but based on 15-bit long PRBS generators. Both codes repeat at 26.667 ms intervals. BSs are differentiated by time offsets of the short sequence.

new registration process occurs to update the position of the mobile in the fixed infrastructure of the network.

When the user wishes to make a call, he/she keys in the wanted number and presses the send button. The MS will attempt to contact the BTS with a signal known as an *access probe* via an access channel. To do this, it uses a long code based on cell parameters. It is possible that several mobiles attempt to gain access at the same time, in which case a collision will occur. Each MS, in case it does not receive an acknowledgment from the base (via the paging channel), will wait for a random period of time before trying again. After establishing contact, the base station will assign a downlink traffic channel with a row number in the Walsh matrix. At this point, the mobile station changes its long code mask for one based on its ESN identifier and starts the conversation mode.

It is usual for an MS communicating with a given base station to detect the pilot signal from another cell with sufficient strength to permit its use. The mobile then requests a soft hand-over. When this function starts, the mobile station receives a Walsh number and a time reference for the different pilot signal corresponding to the other base and will use them in different receiver correlators. The mobile is able to combine signals from both cells. If the signal from the first cell decreases too much, the mobile requests the second cell to end the hand-over process and only uses the signal from the new cell.

At the end of the call, the channels are released. When the mobile station is turned off, it will generate a *turn off registration,* which advises the system that it is no longer available to receive calls.

11.4.7 Comparison of Capacities in Different FDMA, TDMA, and CDMA Systems

When comparing two cellular telephone systems, capacity is the essential parameter. In TDMA and FDMA systems, different frequencies are assigned to adjacent cells. This gives rise to the use of different frequency reuse factors (i.e., only one part of the channels available may be used in each cell). The final capacity of the system will depend on the reuse factor, implementation of the cells (number of sectors per cell) and on their sizes. Nevertheless, a series of simplifications will be assumed to estimate the increase in capacity from one system to another.

A 1.25-MHz bandwidth is considered, which is the width for one IS-95 channel, and various systems, both European and American, are analyzed. These are the American FDMA (AMPS) system with a 30-kHz channel spacing and the European FDMA system (TACS) with a 25-kHz channel spacing, the American TDMA system (30 kHz, three users per channel), GSM, and the American CDMA system. The capacity will be expressed in terms of the maximum number of channels per cell.

In the AMPS system a cluster of seven trisectorized cells is used. The number of AMPS usable channels in the bandwidth of a CDMA channel is 1250 kHz/30 kHz = 41.6 channels. In this comparison 42 channels are assumed. As there are $7 \times 3 = 21$ sectors, the capacity is 42/21 = 2 channels per sector. For the TACS systems, which employ the same reuse pattern, a similar result is obtained. Using a 25-kHz channel spacing instead of 30, a capacity of approximately two channels per sector is also obtained.

The North American Digital Cellular (NADC) has the same number of channels as AMPS, but it offers a number of speech channels three times greater because of it being a three-time slot TDMA system. Assuming a cluster of four trisectorized cells and given that the RF carrier spacing is the same as in AMPS, a capacity of $(42/12) \times 3 \approx 10$ channels per sector is obtained.

For GSM a reuse pattern of four trisectorized cells is normally used. The number of RF channels in the 1.25 MHz band is 1250/200 = 6. An RF carrier in GSM offers eight traffic channels in a TDMA frame. As a consequence, a total of 6 × 8 = 48 channels are available; this corresponds to a capacity of 48/12 = 4 traffic channels per sector.

In a CDMA system, under ideal conditions (i.e., uniform traffic loading, complete synchronization between transmission and reception, perfect power control, and negligible background noise), Viterbi [1] has given the following expression for the reverse link capacity (as we stated before, this link is less critical than the uplink):

$$N \approx \frac{W/R}{E_b/I_o} \frac{g_v \cdot g_A}{1 + f} \qquad (11.6)$$

where

N is the number of offered channels per trisectorized cell;
W/R is the processing gain;
E_b/I_o is the bit energy to interference power density ratio;
g_v is the voice activity gain, due to discontinuous transmission which reduces the overall average interference;
g_A is the antenna gain in a three-sector antenna cell;
f is the external-to-internal cell interference ratio.

Using typical values [1]: $W/R = 128$ ($PG = 21$ dB), $E_b/I_o = 5$–7 dB, $g_v = 2.67$ (4.2 dB), $g_A = 2.4$ (1 dB loss from the ideal gain) and $1 + f = 1.6$, a value of $N = 102$ is obtained which corresponds to 34 users per sector. This is a theoretical result. A more practical value is around 60% of that (i.e., close to 20 channels per sector).

It must be emphasized that the above results are rough estimations. They only intend to reflect the progress from analog to digital TDMA and CDMA systems. In the case of CDMA a much more detailed analysis is needed to refine and validate the coarse approximation that has been presented here. The interested reader may find more information in CDMA specific texts such as [1, 2, 5].

11.5 Cordless Systems

11.5.1 Introduction

Cordless telecommunications systems are considered as a complement to mobile cellular radio systems in many areas of application and, therefore, as a potential market for tens of millions of units worldwide. Apart from the well-known setup of a base plus a portable terminal orientated toward *domestic applications* (cordless telephones), there are two additional directions in which cordless systems can be applied:

- Regional and national networks of telephone booths, allowing the holder to have a handy terminal suitable for making calls without having to use conventional payphones. This type of service is known as *telepoint*.
- *Microcellular systems* (permanent or temporary) in such facilities as large industrial complexes, factories, offices, leisure centers and building sites. These networks are generally known as wireless private branch exchanges (WPABX) or business cordless telecommunication systems (BCTSs).

Analog cordless telephones under the CEPT standard T/R 24-03, better known as CT1, operate on the 900-MHz band and are orientated toward domestic applications. They have been allocated 40 duplex channels, and each piece of equipment dynamically selects a free channel. Originally, the CEPT was responsible for establishing procedures and standards for telecommunications in Europe. Since 1988, these tasks are carried out by ETSI.

Despite the use of these simple "cordless" systems, which may be considered as first-generation, it was decided that the future cordless telecommunications standard should benefit from the latest developments in digital technology and from the synergy with other pan-European developments such as the GSM system. This gave rise to the DECT standard, which will be reviewed in Section 11.5.4. First, however, it is important to define exactly what cordless systems are and what their differences with cellular systems are.

11.5.2 Advanced Cordless Telecommunications Systems

Several services and applications [6] are already available that fall under the category of "cordless," and these go one step beyond the single terminal system. One feature common to all these applications is their relatively short radio range, typically of less than 300m. These cordless services and applications include the following:

- *Basic cordless telephone system.* This is a telephone whose cord is replaced by a radio link. This is the case of the current CT systems. Some implementations also offer the possibility to intercommunicate between the handheld and the fixed unit. It is expected that this application will continue to be important in the second generation of cordless systems, particularly if the new equipment is more economical in comparison with the current CT1s.

- *Telepoint.* Another important application of second-generation CTs is the telepoint for public access (Figure 11.18(a)). Telepoints are base stations located in public places such as train stations, shopping arcades, motorway service stations, and airports. Anybody with a suitable terminal may access the PSTN via a telepoint. It is assumed that the user does not move significantly during a conversation, so that hand-over service to other telepoint stations need not be provided. A telepoint system needs to access a central database to authenticate the user and generate charging information. In many cases, database access is performed via the telephone network itself.

The advantage for users is that they do not need coins or telephone cards, as calls are invoiced on the home or office number. As telepoints are much easier to operate and maintain than conventional telephone

(a)

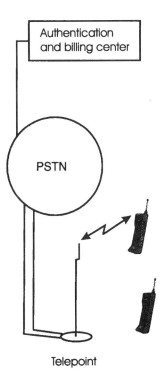

Figure 11.18 (a) Telepoint system. (b) WPABX configuration [6].

(b)

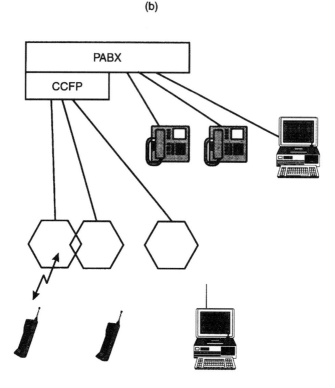

Figure 11.18 (continued).

booths, tariffs may be lower, except for the existence of an initial subscription fee. This service, however, entails three main disadvantages:

The system only allows user-originated calls. If the telepoint user wants to receive communications, his/her terminal should be *associated with a paging terminal.* Current paging systems, however, covering large areas (national coverage, for example), do not provide a swift response and the combination telepoint handheld plus pager does not provide a valid solution for user terminated calls.

The telepoint service should cover a reasonably wide area. It is assumed that it will be available in the main train stations and airports. This will not attract sufficient customers. In order to obtain profits to recover investment, the telepoint operator should provide a service with wide coverage, which may even mean national coverage.

Many telepoints may be installed in noisy places (railway or tube stations, for example). To solve this drawback, it may be necessary to also install telephone booths.

Despite the disadvantages, this system could compete with cellular networks, although the commercial experience with it has not been successful.

It should also be taken into account that the telepoint service is public and, therefore, a common standardized air-interface needs to be defined to permit functional operability between the equipment from different manufacturers.

- *Micro/picocellular systems.* In micro/picocellular systems, coverage is achieved by splitting the service area in cells similar to those in cellular networks. Nevertheless, since cordless networks normally operate in indoor environments, the radio link experiences difficult propagation conditions. This implies that cell sizes should be very small, with diameters even down to 10m. Cells may even be as small as a room. Micro/picocellular systems are suitable for applications in two main areas: Industrial plants and office blocks. As shown in Figure 11.18(b), for some telephones, the cable may be replaced by a radio link using cordless terminals. These communicate with base stations which are connected to the so-called common control fixed point (CCFP). The CCFP performs all the necessary functions in a radio network, such as hand-over. The PABX/CCFP combination is normally termed WPABX. In these systems, the user may originate or receive calls with a handheld terminal throughout the coverage area. Since most internal traffic in the "wired" (PABX) network is nowadays data, these systems will have to offer a wide range of data together with voice services. In order to support advanced paging functions or office electronic systems, there will be a demand for SMSs. It is also possible that other services typical of the ISDN will be offered.

  Micro/picocellular systems may be started up as a temporary solution if the wired network is not available or is too costly. Examples of this are building sites or events such as important sports competitions, exhibitions, or conventions.

  The difference between these cordless micro/picocellular type systems and the cordless domestic systems is that the hand-held terminal may access more than one base. This means that these systems should offer functions normally available in cellular systems, particularly call hand-over.

11.5.3 Cordless Telecommunications and Cellular Radio

Cordless and cellular systems [6] differ in several aspects, particularly in terms of service and market segments. It should also be noted that a cordless system is purely and simply a cordless access technique to the conventional telephone network, whereas cellular systems are complete telephone networks. These differences lead to technical peculiarities in the areas of application.

Cordless telecommunication services. Cordless systems are mainly orientated toward indoor communications. They differ from cellular systems in several aspects, namely the following:

- The total telephone traffic in an office block may present densities of up to 10,000 E/km^2 on one single floor. This may give rise to a total of several 100,000 E/km^2 at very localized points. Obviously, traffic distribution must be evaluated in three dimensions. Modern cellular systems, on the other hand, are designed for maximum two-dimensional traffic densities on the order of 1,000 E/km^2.
- The range of radio links (i.e., the distance between mobile handheld terminals and the fixed station) is very short, normally less than 300m. In cellular systems, the base stations may cover up to several tens of kilometers. This also has an impact on terminal battery lives.

The cordless telecommunications market. Two basic differences are noted, in this aspect, between cordless and cellular radio telecommunications [6]:

- Cordless systems are sold to customers as a package (i.e., handheld terminals and the fixed infrastructure are provided together). The customer may not have experts available with a knowledge in radio planning or adaptation of an existing network to a product suitable for installations both with a small number of terminals and with large cordless extensions. So, extreme flexibility is needed when assigning radio resources (radio channels). This can only be achieved by using dynamic channel allocation (DCA). Also, in the neighboring environment, there will be systems from different manufacturers that should be able to operate with overlapping coverage areas without any mutual interference. On the other hand, there are two markets for cellular networks: the professional market for providers that own their network and have the assistance of their own technicians and the consumer terminal market. Many manufacturers of cellular products target one of these two markets exclusively.
- Contrary to the telepoint case, microcellular systems, as noted above, will not provide public services but rather will serve a closed user group. There is no need for interoperability. The only compatibility required is that which addresses interference control (common air-interface). This gives manufacturers a good deal of freedom to imple-

ment their own designs and ideas. The effect this has on the market is to increase competition in offering systems with better features, services and applications.

Technical differences. The differences described in the two previous points lead to some important technical consequences:

- Short transmission ranges inside buildings give rise to completely different propagation characteristics in cordless systems to those found in other systems. Typical *delay spreads* are considerably smaller in indoor environments than in outdoor, long-range communications. This feature makes the use of equalizers at the receiver unnecessary.
- The spectral efficiency of cordless systems in terms of Erlang/MHz/km^2, will be very high due to the short range of the radio links. Even in the case of selecting modulation and coding schemes requiring high interference protection rations (R_p), they will yield systems with spectral efficiencies an order of magnitude higher than those of the most up-to-date digital cellular systems.
- The liberalization of the CT systems market will produce environments with a strong likelihood of interference. This rules out the possibility of using fixed frequency planning as in cellular systems. This calls for the use of dynamic channel allocation techniques.
- Cellular systems permit the user to make calls while moving at a relatively high speed (e.g., traveling in a car or in a fast inter-city train). Cordless systems are designed to establish communications while the user is walking. The DECT system falls into this second category.

11.5.4 The Digital Enhanced Cordless Telecommunication (DECT) System

This section describes the so-called DECT system, highlighting its most relevant features. This was originally a pan-European cordless digital system developed by ETSI. The DECT system only permits radio access between a handheld terminal and one or several bases, without defining routing aspects as is the case of complete networks. Thus, this is a *radio access system*. Unlike cordless systems, a cellular system (e.g., GSM) defines a complete network structure with switching offices and databases (HLR, VLR) for mobility management.

Since the demand for cordless systems appeared, a series of technical solutions have arisen that have attempted to meet market demands. Nevertheless, none of the solutions are pan-European, but rather they offer partial solutions of a domestic nature. Table 11.4 shows the characteristics of earlier

Table 11.4
Different Cordless Standards

Parameter	CT0	CT1	CT2	CT3	DECT
Technology	Analog	Analog	Digital	Digital	Digital
Transmitted power (mW)	< 10	< 10	< 10	< 80	<100
Multiple access	Fixed freq.	FDMA	FDMA	TDMA	TDMA
Duplexing	FDD	FDD	TDD	TDD	TDD
Radio-channel bandwidth (kHz)	12,5 / 25	25	100	1,000	1,730
No. of traffic channels per radio channel	1	1	1	8	12
Voice coder	—	—	ADPCM	ADPCM	ADPCM
Binary rate in voice channel (kbps)	—	—	32	32	32
Voice channel modulation	FM/PM	FM	GFSK	GMSK	GMSK
Transmission rate	—	—	72	640	1,152
Working band (MHz)	< 50	900	864–868	900	1,800
Main application	Residential	Residential	Telepoint	Business	All

standards together with those of DECT. A brief review of several systems reveals the fragmented nature of the market:

- CT0. These are systems using low frequency bands (below 50 MHz). Although they have only been authorized in a few European countries, a large number can be found on sale in almost all the European countries.

- CT1. This is an analog CEPT standard, operating on the 900-MHz band. This system has been adopted in several countries. The CT1 system has been allocated 40 channels on the 914–915 MHz band for the mobile-to-base link and 959–960 MHz for the base-to-mobile link. This system uses frequency modulation and channels are assigned to each call in a dynamic manner so that, for each new user in the system, its terminal automatically selects one of the free channels at the time of establishing the connection. The most relevant problems with this first generation of cordless telephones lie in their limited coverage range (a few tens of meters) and, basically, their great vulnerability to the effects of interference caused by other users as a result of the limited number of radio channels available.

- CT2. This is a digital system using an FDMA multiple access technique. The main objective of this Britain-originated system was its application in telepoint networks. The CT2 system uses 40 radio channels with a 100-kHz bandwidth each. The band used is 864.05–868.05 MHz and, unlike the CT1 system, which uses different transmission sub-bands for the mobile-to-base and base-to-mobile link (FDD), this system uses a time division multiplex method (TDD) to separate both transmission directions. The fact that the voice signal is transmitted in digitized form facilitates adopting time multiplexing. The TDD scheme used defines a 2 ms frame in which two bursts of 72 kbps are transmitted, one of 66 bits (0.917 ms) in each direction, although, at the request of one of the users, the burst length may be increased to 68 bits. The rest of the frame time is used as guard time between bursts to avoid possible overlapping. Voice digitization is carried out by means of a simple coding system (adaptive differential pulse code modulation (ADPCM)) with a rate of 32 kbps. The modulation method used is binary FSK with a modulation index which may range from 0.4 to 0.7. The data to be transmitted are previously filtered by a Gaussian filter, thus minimizing interference in adjacent channels. For this reason, this type of modulation is known as Gaussian frequency shift keying (GFSK).

- CT3 is also a TDMA-based digital system, operating on the 900-MHz band. This system has been especially designed for business applications and is Swedish in origin.

Both the British and the Swedish administrations attempted to make CEPT accept their proposed CT systems as a European standard for digital cordless telecommunications systems. But the CEPT considered that neither of these two standards fully complied with the requirements and diversity of applications expected for such systems. For this reason, it was decided, similarly to what occurred with GSM, to standardize a new cordless digital telephone communications system of a pan-European nature. This standard is called DECT. It is evident that such a diversity of incompatible systems aimed at meeting the very same communication needs did not at all help to create and develop competitive equipment and services. On the other hand, a common standard such as DECT is the solution for bringing down the barriers of market fragmentation.

Description of the DECT System

The DECT system is a technical concept that attempts to cover the whole range of domestic "cordless" applications, telepoint and wireless PABXs. In other words, a system that allows the use of any telephony service in the home, at the office and in public places, using a DECT terminal (i.e., via radio links).

In its domestic profile, DECT equipment consists of a base unit connected to a telephone line and a hand-held unit. Both components may communicate with each other via a radio link with a range of approximately 100m. As shown in Figure 11.19, this application is simply a replacement of the conventional telephone line/cord by a radio link, although some more sophisticated functions may also be made available, such as a multiple handheld telephone system serving all the members of a family, with the capacity to transfer calls between them.

The second main application of DECT systems is in public access via radio, the so-called telepoint service. In this case, several base stations would normally be connected to several telephone lines, located in public places such as train stations, airports, or shopping malls and business areas. Users can make telephone calls via radio using their hand-held terminals, if they are within the system's service area, as defined by their base stations.

The third use of DECT systems is in business applications or, as this is normally termed, WPABX. To a large extent, this is similar to a cellular system (i.e., a number of base stations strategically installed throughout an office block, for example, providing communications via radio to hand-helds). This would be, in effect, an indoor cellular system with cells distributed in a three-dimensional way and in a relatively small volume, which, together with the necessarily close proximity of the base stations, would lead to a picocellular system. With this configuration, two phenomena occur. First, it is highly likely that a user in movement will cross over several cells while holding a conversation. Second, the number of simultaneous communications expected in an office environment would be large (due to the high traffic density). These conditions necessitate a technology capable of carrying out numerous call hand-overs swiftly, silently and efficiently.

Apart from call hand-overs, a roaming service must also be available (i.e., to carry out incoming and outgoing calls anywhere within the service area). Wireless PABXs also provide functions inherent to the modern digital PABXs, such as call on hold, call transfer/forward, and abbreviated dialing.

Figure 11.19 schematically shows several possible DECT configurations for different applications where it can be seen how the cells for different applications overlap.

The capacity of DECT is not limited, however, to speech communication applications. The introduction of DECT coincides with the widespread introduction of ISDN and the popularization of speech and data services in the office. DECT systems support both voice and non-voice services. More specifically, all the services associated with ISDN's basic access (2B+D) are possible within the DECT system, including circuit and packet switching.

Finally, provision has been made for the development of applications in such a manner that DECT may be used as an extension to other public and

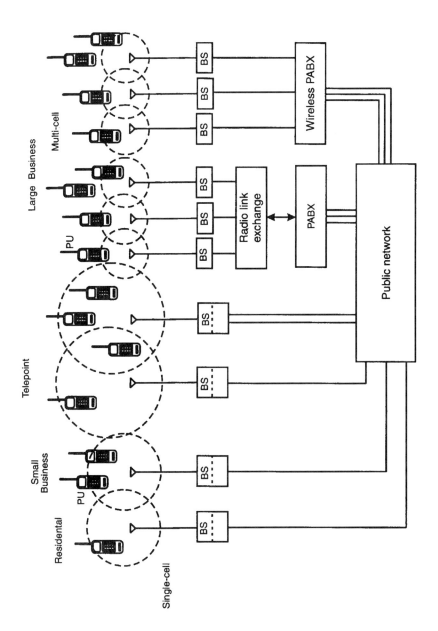

Figure 11.19 Application areas for DECT.

private networks (e.g., GSM, PSTN, ISDN, or X.25). DECT is also being used for wireless local loop applications for rural telephony.

The DECT Network

Figure 11.20 shows a complete network structure where DECT is introduced. This figure clarifies the relationships, interfaces, and compatibilities within the DECT network and with associated networks. Four functional domains are defined in the model: global network, extended service networks, DECT networks, and end system. The tree structure shown illustrates how each network may communicate with several domains and how direct communication between two or more domains in the same category is not possible. To do so, they should be connected via a higher level network.

The global network is a general purpose network that will mainly provide address translation between connected networks. From the viewpoint of DECT, overall networks may be, for example, the telephone network PSTN, ISDN, a packet switching network, X.25, or a GSM network. The extended services network offers services higher to those on the global network, such as PABX, ISPBX, LAN, and telepoint.

The DECT network interconnects the extended services network to the end system. DECT is, therefore, an intermediary system with no application process content. It comprises the fixed DECT system and the cordless DECT system, which are described as follows:

- The fixed DECT system is responsible for providing cordless access to the extended services network and/or to the global network via a radio link and a fixed infrastructure. It should be noted that the fixed DECT system does not include switching entities, except those required to carry out call hand-overs or multiple calls. This means that the DECT system will not be able to perform connections between two hand-helds on the DECT network. All switching functions are performed outside the DECT network.

- The cordless DECT system is normally the hand-held part of the DECT network and is responsible for routing functions to the end system.

The end system is able to connect, via the DECT network, to the extended services network. Normally, the cordless DECT system and the end system are implemented on a single piece of equipment (the hand-held), although they may be implemented separately, as is the case, for example, with a speech or data terminal or an LAN.

Other Mobile Radio Systems 471

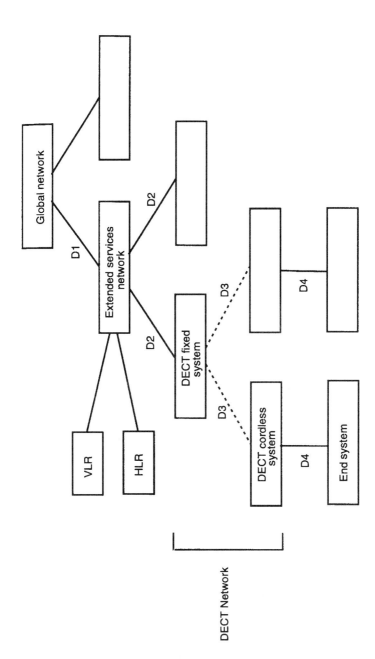

Figure 11.20 Location of the DECT system in an overall system.

There are two functional entities related to mobility management functions that are external to the DECT network: the HLR and the VLR. Both will preferably reside on the extended services network.

Figure 11.20 also shows reference points D1, D2, D3, and D4, which separate the different networks. Point D2 should support the signaling required for call routing, mobility management, and user information transport. The air-interface coincides with point D3. Finally, D4 may or may not be a physical interface (for example, when the cordless DECT system and the final system are integrated in one single hand-held unit).

11.5.5 Technical Principles of the DECT System

The DECT standard includes two levels of specification, known as coexistence specifications (CX) and the common interface specification (CI). Common to both levels is a time division multiple access, time division duplex multicarrier system (MC/TDMA/TDD). This is explained in some detail below. The frequency band allocated to DECT is from 1.88 to 1.9 GHz. The specification of the common interface is not compulsory, except for telepoint applications. It includes the definition of a minimum set of air-interface parameters, which ensure interoperation between terminals from different manufacturers and is considered compulsory. When defining these two DECT specification levels, numerous options are left open so that the different manufacturers may implement tailor-made interfaces, services and functions, while maintaining a common air-interface.

The radio link specification follows a layer based structure. Following this approach, the signaling protocols are divided into 4 layers: physical layer (PHL), medium access control layer (MAC), data link control layer (DLC), and network layer (NWL). Figure 11.21 shows the correspondence with the OSI layered model, which is as follows: OSI layer 1 corresponds to DECT physical layer and part of the MAC layer. The other part of the MAC and the DLC layer correspond to OSI layer 2. The DECT network layer corresponds to OSI layer 3.

The main objective of the *physical layer* is to establish a set of radio bit streams. The tasks performed in the physical layer are described as follows:

1. To modulate and demodulate radio carriers with a bit stream to create an RF channel;
2. To create physical channels with fixed throughput;
3. To monitor the radio environment to achieve activation of physical channels on request of MAC, recognition of calling physical channel, synchronization between transmitters and receivers and to inform

Other Mobile Radio Systems 473

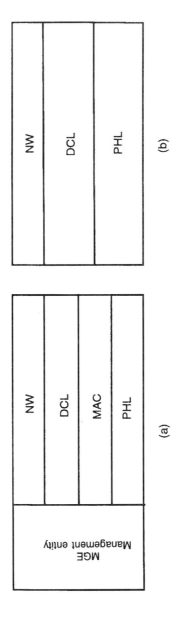

Figure 11.21 Structuring into layers or functional levels of the DECT system: (a) layered structure of DECT and (b) OSI model.

the management entity of the status (e.g., field strength) of physical channels.

The main objective of the MAC layer is to select a suitable physical channel and to transfer the information in a reliable manner through it. This is achieved by multiplexing four *logical channels*: signaling, user information, paging, and broadcast, on one or more physical channels. The broadcast channel is used to provide hand-helds with network information to allow their efficient access to the network. By means of the paging channel, the handheld terminals may be paged from the network. Also, signaling data protection is achieved within the MAC.

The data link control layer performs, among others, the functions of maintenance of the data link while changing cells (intercell hand-over), HDLC protocol and encryption. The main function of the network layer is to route calls within the network and to the outside world. Finally, the management entity (MGE), which interfaces to the four layers, carries out management functions that require information from more than one layer.

As stated earlier, a TDMA/TDD access technique is used. In TDMA, different users share an RF carrier. Each user is assigned different time slots (TS) within a frame. In this way, each user has access to the entire channel bandwidth, but only for a small percentage of the total time. Figure 11.22 shows the DECT TDMA/TDD structure. The TDMA frame is divided into 24 time slots (TS0 to TS23). The length of the TDMA frame is 10 ms, so that each slot lasts 10/24 = 0.417 ms. The first 12 time slots are grouped together in the first half of the TDMA frame and allocated to communications from the base stations to the portable units (PU) and the remaining 12 time slot group occupies the second half of the TDMA frame and is devoted to the opposite transmission direction (portable units-to-base stations).

Thus, a physical duplex channel is formed by the sequence every 10 ms of a pair of time slots located in the same position within the two halves for the TDMA frame. Therefore, the TDMA/TDD structure will contain 12 physical duplex channels.

Each slot is used for the transmission of bursts, each of them containing 420 bits. The length of the burst is 52.1-μs shorter than the relevant time slot. This guard time allows for small timing errors, propagation dispersion, and transmitters ramping.

The user's logical channels are assembled onto one or more physical channels. In Figure 11.22 also the multiplex format for normal telephony operation is shown. In this case, the 420 bits of the bursts are split into 32 synchronization bits (SYNC) and 388 signaling and data bits (DATA).

Other Mobile Radio Systems 475

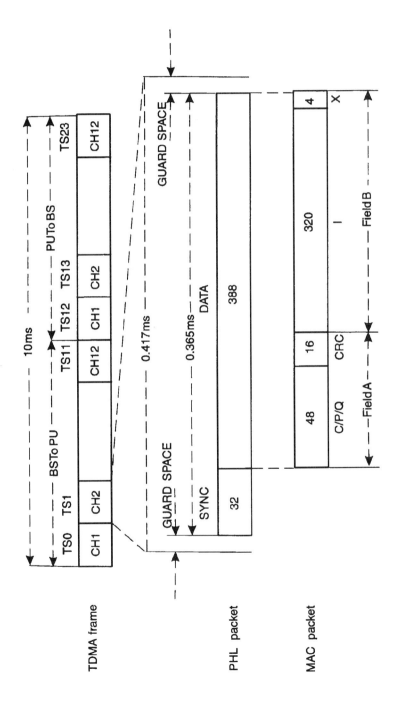

Figure 11.22 DECT system frame structure.

The SYNC channel is always present and transmits the same 32 bits, thus allowing resynchronization in each burst.

The DATA channel is divided into two fields: *field A* with 64 bits, and *field B* with 324 bits. Field A assembles the *signaling* (C), *paging* (P) and *broadcast* (Q) channels which share 48 bits upon demand, while a minimum capacity for each channel is simultaneously guaranteed. The remaining 16 bits of field A are used to protect the data with cyclic redundancy checking (CRC). The signaling gross binary rate (i.e., the throughput) of field A, is 6.4 kbps.

Field B contains 320 bits per burst for the user information channel (I). It offers, therefore, a throughput of 32 kbps. It must be stressed that it is possible to join up to 12 time slots together to create high capacity channels of up to 384 kbps both ways (12 × 32 kbps). The remaining 4 bits (X bits) in field B are used to detect interference within the burst, thus allowing for unsynchronized systems to operate in the same area. Figure 11.22 illustrates the I *information channel* for telephony applications where all speech coded information (ADPCM, 32 kbps) is sent unprotected.

The frame structure explained above results in an overall data rate of 1,152 kbps. In fact, taking into account that the 52.1 μs of guard time is approximately equivalent to 60 bits, the peak binary rate of the TDMA frame is

$$(60 + 32 + 388) \text{ bits/TS} \times 24 \text{ TS}/10 \text{ ms} = 1{,}152 \text{ kbps}$$

This bit stream of 1,152 kbps will be the signal used to modulate the corresponding RF carrier using GMSK modulation with a relative bandwidth BT = 0.5. Table 11.5 summarizes the technical characteristics of the DECT system.

DECT also has a multicarrier configuration. The carrier spacing is derived from the nominal bit rate (1,152 kbps) to be transmitted resulting in a separation of carriers of 1,728 kHz. The total number of carriers and, hence, the bandwidth required for the system, is calculated taking into account traffic density parameters. To cope with the maximum expected telephone traffic in an office building (10,000 Erlang/km^2/floor), DECT was provided with 10 RF carriers, which means that the total number of channels available is

$$10 \text{ carriers} \times (12 \text{ channels/carrier}) = 120 \text{ channels}$$

The total bandwidth required to accommodate the 10 carriers is approximately 20 MHz (10 carriers × 1,728 kHz/carrier plus guard bands).

Table 11.5
DECT System Characteristics

GMSK modulation with BT = 0,5	
Multicarrier TDMA/TDD	
Dynamic channel assignment	
Frequency band	1.88–1.9 GHz
Bandwidth	20 MHz
RF carrier spacing	1,728 MHz
No. of RF carriers	10
No. of duplex channels per carrier	12
Frame interval	10 ms
Nominal bit interval	1,152 kbps
Traffic binary rate	32 kbps
Signaling binary rate	6.4 kbps

11.5.6 Basic Operation of the DECT System

The operation of DECT is designed to allow high capacity, while maintaining high speech quality. It is based on the combination of the TDMA technique and the *dynamic channel allocation* procedure (DCA). In DECT, physical channels are assigned to communications in a decentralized manner. For each call, the hand-held chooses the channel that best suits the wanted connection. DECT base stations consist of one single radio transceiver (Tx-Rx) that can change frequency with each time slot. This simple but flexible base station can thus operate on each time frame on all the 12 duplex channels, each channel working independently on any of the 10 carriers.

Each base station is always active on at least one channel. Every active channel broadcasts system information and base station identification. This allows any hand-held unit within the coverage range of the base station to identify the system (without performing any transmission). When the terminal recognizes a parent system, the receiver locks to any active channel of the nearest base station (i.e., with the strongest field). In this idle locked state, the hand-held unit periodically listens for possible paging calls from the system.

When the hand-held wishes to access the system either to make a call or to answer a paging message, it looks for a free channel and sends the appropriate burst, thus accessing the network.

Finally, it is important to pay attention to the *call hand-over* technique used where seamless or soft hand-over is employed. Hand-over is performed in a decentralized manner. This feature in DECT is controlled by the terminal. While the terminal communicates via the original link, it scans the other channels and keeps record of the free channels and identities of the base stations with their respective status (field strength), thus being ready to perform a hand-over, if necessary.

The hand-over process is initiated as soon as another channel has a stronger field than the current one. The old link is maintained on one slot in the portable unit, while the new link is set up in parallel on another time slot. When the new link is established, the new base station requests the switching from the old to the new radio link. As it can be observed, DECT does not depend at all on the old channel to quickly set up the new one. The procedure to set up calls or hand them over ensures stable DCA, high capacity, and high link quality.

11.5.7 Other Information of Interest

By way of a conclusion to this section, Table 11.6 summarizes and compares the technical characteristics of the GSM cellular system with those of the cordless DECT system.

Finally, typical expected traffic densities are given in Table 11.7 for the cordless systems for various application environments.

Table 11.6
Relative Characteristics of the GSM and DECT Systems

System	DECT	GSM
Duplex separation	In time	In frequency
Multiple access	12-channel TDMA	8-channel TDMA
Frame interval	12 ms	4.615 ms
No. of time slots	2 × 12/frame	8/frame
Voice codec	32 kbps ADPCM	13 kbps
Frequency bands	1800 MHz	900 MHz
RF binary rate	1024 kbps	271 kbps
Carrier spacing	≈ 1,3 MHz	200 kHz
Error protection*	None	Convolutional + block
Interleaving*	None	Eight frames
Slow frequency hopping	None	1 hop/frame
Equalization	None	16 μs delay
Max. peak power	0.3W	5W hand-helds
Range	300m	32 km
System delay*	12 ms	57.5 ms
Control dedicated channels	none	Dedicated TSs on dedicated carrier
Signaling and traffic multiplexing	within each burst	26-frame multiframe

*For voice traffic.

Table 11.7
Typical Expected Traffic Densities for CT Systems

Residential area			
Suburban	150 Erlang/km^2		
Urban	200 Erlang/km^2		
Offices and business areas			
Building density	Low	Medium	High
No. of stories	3	6	20
Area per terminal	40 m^2	20 m^2	20 m^2
Traffic density	3,000 Erlang/km^2	30,000 Erlang/km^2	120,000 Erlang/km^2
Telepoint	40,000 Erlang/km^2		

11.6 The Digital Trunking System TETRA

11.6.1 Introduction

TETRA is a new European Standard for PAMR trunked systems. In fact, from the beginning it has been designed as a trunked radio platform that effectively and economically supports its shared usage by several organizations, while maintaining privacy and mutual security and providing integrated voice and data services. Virtual networking inside a TETRA network enables each organization to operate independently but still enjoy the benefits of a large, high-functionality system with efficient resource employment.

TETRA supports voice, circuit switched data, and packet switched data services with a wide selection of data transmission rates and error protection levels. The following features of TETRA can be highlighted:

- Fast call set-up time (typically 300 ms);
- Full duplex voice and data (V+D) communications for telephone type individual calls;
- High data transfer rates up to 28.8 kbps;
- Secure communications;
- Direct mode operation (DMO) between radio units (RU);
- Packet data optimized (PDO) operation;
- A wide range of call services such as individual, group, broadcast, priority and emergency calls;
- Call barring;
- Late entry: latecomers may join a call in progress;

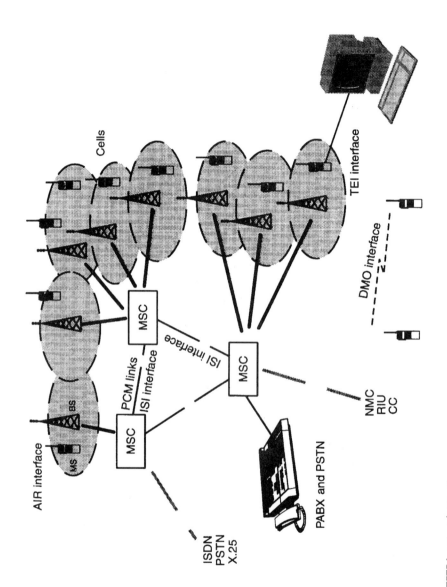

Figure 11.23 TETRA network structure.

- Ambiance listening. The dispatcher may turn on the transmitter of a radio unit (RU) without any indication being provided on the RU (a feature that can be used in hijack situations);
- Interworking and interoperability;
- A wide range of gateways for interworking with external telephone networks such as the PSTN or the ISDN.

In order to assure a multivendor environment with the benefit of economies of scale TETRA has been specified with several open interfaces:

1. Air interface for the interoperability between terminals from different manufacturers;
2. Terminal equipment interface, which enables the independent development of mobile applications;
3. Intersystem interface for interconnection of TETRA networks from different suppliers.

11.6.2 TETRA Telecommunications Services

The TETRA standard defines the basic services for voice and data shown in Figure 11.24.

Teleservices provide complete communication capability between users, including all terminal functions. In the TETRA standard, teleservices cover voice communication services. A bearer service provides communication capability between standard terminal network interfaces. TETRA bearer services are defined for data transfer. Tables 11.8 and 11.9 list these services.

In addition, the standard defines supplementary services for very flexible system application. These services modify or supplement the above mentioned services. Table 11.10 lists some of TETRA's supplementary services.

11.6.3 Harmonized TETRA Frequencies

TETRA technology is independent of the frequency band used. Nevertheless, the harmonized use of frequencies will allow substantial economies of scale. Nowadays, there is a common band for emergency and public safety services in Europe. In fact, the North Atlantic Treaty Organization (NATO) has given up 20 MHz between 380 and 400 MHz for these services. National authorities have allocated two sub-bands of 5 MHz each in these 20 MHz for public safety TETRA networks. Also, a process of allocating frequencies for commercial TETRA has just begun its deployment in several countries. Various plans exist

482 Introduction to Mobile Communications Engineering

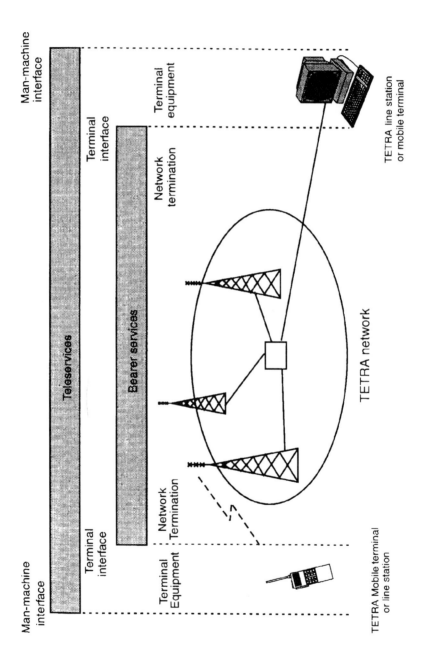

Figure 11.24 TETRA services.

Table 11.8
TETRA Teleservices

Individual call
Group call
Acknowledged group call
Broadcast call

Table 11.9
TETRA Bearer Services

Circuit mode data at 7.2/14.4/21.6/28.8 kbps
Circuit mode protected data at 4.8/9.6/14.4/19.2 kbps
Circuit mode heavily protected data at 2.4/4.8/7.2/9.6 kbps
Connection-oriented packet data
Connectionless packet data

Table 11.10
TETRA Supplementary Services

Call authorized by dispatcher
Area selection
Access priority
Late entry
Pre-emptive priority call
Discrete listening
Ambiance listening
Dynamic group number assignment
Calling line identification present
Connected line identification present
Call barring
Call hold
Advice of charge
Call report
...

to implement TETRA networks in the 410–430 MHz frequency band. Other frequencies foreseen for commercial TETRA applications in Europe are in the following bands: 450–460/460–470 MHz and 870–876/915–921 MHz.

11.6.4 TETRA Functional Configuration

As shown in Figure 11.23, a TETRA network is built around switching centers, MSCs, connected together in a meshed structure. MSCs are linked both to other networks (public or private) and to base stations. There also exists the possibility to directly connect PABXs and local line terminals to BSs.

Other elements of importance are the recording information unit (RIU), the network management center (NMC), and the customer care (CC).

Also in Figure 11.23, the direct mode operation (DMO) is illustrated. This feature enables the direct link between mobile radios without the need for network infrastructure. Also repeater and gateway functions are defined to extend the coverage of hand portable radios in both direct mode and network operation.

11.6.5 TETRA Signal Processing

In this section various relevant signal processing features in the TETRA system are briefly presented.

Coding

As is shown later, TETRA offers four voice/data channels in a 25-kHz RF channel. This means that a capacity equal to double that of analog PAMR systems channelized at 12.5 kHz is offered. This higher efficiency is achieved through the use of advanced voice codecs that combine voice quality acceptable for PAMR applications with a binary rate of only 4.8 kbps.

Besides, due to the fact that PAMR users often have to work in noisy environments, with backgrounds of sirens, loudspeaker announcements, vehicular traffic noise and so on, the TETRA voice codec has been designed to provide good speech quality in harsh situations. The codec used is of the adaptive code excited linear predictive (ACELP) type.

A reduced BER is achieved by making use of interleaved channel coding. A double coding scheme is used: an outer error detecting code and an inner error correcting convolutional code. The channel code bits are interleaved and scrambled before their transmission.

A voice traffic channel operates in 60-ms frames. In each frame, $60 \times 4.8 = 288$ bits are delivered. The block code adds four redundancy bits. The resulting 292 bits are applied to a rate-compatible punctured convolutional coder (RCPC) of rate 299/432, which produces 432 output bits. Afterward, these bits go through interleaving and scrambling processes and are then transmitted as TDMA bursts.

Encryption

TETRA is a high-security technology that includes encryption of voice, data, signaling, and user identities. Two encryption mechanisms have been defined:

1. Air interface encryption, which ciphers the radio path between the base station and the mobiles.
2. End-to-end encryption, for the most critical applications, where ciphering is required for the transmission from one terminal through the network to another terminal (for example in security applications).

Modulation

An important feature in TETRA is that it uses a 25-kHz channel spacing similar to that used by analog PAMR systems. In this way, TETRA system frequency plans can be made compatible with existing PAMR plans. However, as a TETRA radio channel provides four traffic channel capacity (voice and/or data), a multilevel modulation scheme must be used. On the other hand, the requirement was set for a simple modulation type to allow the use of inexpensive receivers. This leads to the use of differential modulation schemes. Finally, strict adjacent channel protection must be provided.

All these considerations have led to the selection of a linear, multilevel, differential modulation method. In principle, a four phase modulation scheme like QPSK could be considered. However, the spectrum of this modulation presents high sidelobes, thus causing significant adjacent channel interference effects. A modification of this modulation scheme called O-QPSK has been selected. With this modulation, a half-bit interval offset is applied to smooth out phase transitions and thus reduce sidelobe levels. This technique has the disadvantage that requires a coherent receiver hence increasing the complexity of the receiver. The type of O-QPSK modulation selected for TETRA is quadratic, differential $\pi/4$-DPSK.

The advantages of the selection of $\pi/4$-DQPSK modulation are the following:

- High spectral efficiency (2 bits/s/Hz), since two bits per symbol are transmitted;
- Transitions do not give rise to amplitude nulls, thus facilitating the design of linearized RF amplifiers;
- Given that information is transmitted by means of phase transitions, it is not necessary to know the absolute value of the signal phase. This means that simple demodulators can be used, except when an equalizer is required at the receiver.

The main drawback of this modulation type comes from the need to use linear power amplifiers; this means that the full power capability of the amplifier can not be explored.

Air Interface

The TETRA air interface for V+D operation uses a TDMA multiple-access scheme with four time slots per TDMA frame. This is further structured into an 18-frame multiframe (Figure 11.25). In circuit mode, voice and data operation, the transmitted information corresponding to an 18-frame multiframe is compressed and conveyed within 17 TDMA frames, thus allowing the 18th frame to be used for control signaling without interrupting the flow of data. This 18th frame is called the *control frame* and provides the basis for associated signaling.

Each time slot has a duration of 85/6(14.17) ms and a capacity of 510 bit periods, so that the transmission rate is 510/(85/6) = 36 kbps. The TDMA frame has a period of 4 × 85/6 = 56.67 ms. In the downlink, half time slots (of 255 bits) can also be used.

Each RF carrier offers four physical channels, but this number can be varied as subslots or groups of slots (up to four) can be used to convey information. For example, if three slots of a carrier are used to transmit a high data rate signal, this carrier will have only two physical channels: the three-slot channel and the other consisting of the remaining fourth slot in the frame. The frames of the uplink and downlink are offset by two time slots to enable full duplex transmission without the need of a duplexer.

In the packet data optimized (PDO) mode of operation, the multiple access technique is different: statistical multiple access (STMA) and statistical multiplexing (STM) are used in the uplink and downlink, respectively.

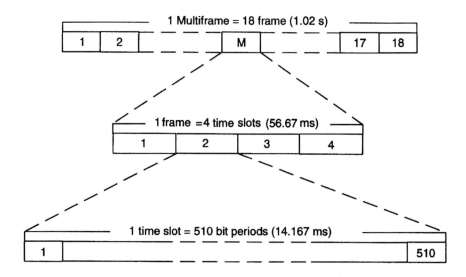

Figure 11.25 TETRA TDMA frame and multiframe structure.

TETRA V+D Channels

As in GSM, for TETRA, logical and physical channels have been defined. Logical channels are classified as control and traffic channels. Control channels (CCH) exclusively transport signaling and packet mode data. There are five types classified as follows:

1. Broadcast control channel (BCCH), used to broadcast general network information;
2. Linearization channel (LCH), a time period reserved for MSs and BSs to linearize their transmitters;
3. Signaling channel (SCH), used for the interchange of common signaling messages. Thus, they are shared by all MSs although, in each case, only convey specific messages from one MS or a group of MSs;
4. Access assignment channel (AACH), that indicates, for each traffic channel, the assignment of the uplink/downlink time slots;
5. Stealing channel (STCH), used to interchange urgent information related to a given call. Hence it is part of a TCH. The signaling capacity is achieved by stealing information bits from traffic bursts and replacing them with signaling bits.

Traffic channels transport circuit-switched voice or data information bits. They can be subdivided into speech channels, TCH/S, used exclusively for voice and mixed channels for voice and data transmission. These latter ones can be further classified according to their binary rates: TCH/7.2, TCH/4.8, and TCH/2.4. Greater rates can be achieved by grouping together time slots up to four, so that the maximum rates are variable in the range from $4 \times 2.4 = 9.6$ kbps up to $4 \times 7.2 = 28.8$ kbps.

Physical Channels

A physical channel is made up of a radio channel (two frequencies, for the uplink/downlink) and a time slot.

Each cell is fitted with one or several two-frequency radio channels. The higher frequency of each pair is used on the downlink while the lower frequency is used on the uplink. Among the RF carriers assigned to a given cell there is one whose first time slot carries the control channels. This carrier is called the *main carrier* in the cell.

Three physical channel types have been defined:

1. Control physical channel (CP);
2. Traffic physical channel (TP);
3. Unallocated physical channel (UP).

The physical channel type used in a given time slot is indicated in the AACH logical channel transmitted within this slot.

Bursts

TETRA V+D control and traffic messages are inserted into TDMA time slots and frames in the form of bursts or discontinuous bit packets. In one time slot, up to 510 bits can be transmitted. The burst *active period* is the fraction of the time slot in which bit transmission takes place. The rest is called *guard time*.

In TETRA, seven burst types with different structures and contents have been defined. These are applicable to the uplink, downlink, and are used for different purposes. Uplink bursts are classified as follows:

- Control burst (CB);
- MS linearization burst (LB);
- Normal burst (NUB).

Downlink bursts are listed as follows:

- Normal continuous burst (NCB);
- Normal discontinuous burst (NDB);
- Continuous synchronization burst (SB);
- Discontinuous synchronization burst (SB).

All bursts (except for NDB and SB) have variable length guard periods. Guard periods, located at the beginning of the burst, have been envisaged to facilitate the transmitter linearization and power ramping. Guard periods at the end of the burst are used to minimize the collisions between bursts from adjacent time slots. All bursts include a *training sequence* to help the channel decoding in the receiver.

References

[1] Viterbi, A. J., *CDMA Principles of Spread Spectrum Communications*, Reading, MA: Addision-Wesley Publishing Company, 1995.

[2] Ojanpera, T., and R. Prasad, *Wideband CDMA Third-Generation Mobile Communications,* Norwood, MA: Artech House, 1998.

[3] Redl, S. M., M. K. Weber, and M. W. Oliphant, *An Introduction to GSM,* Norwood, MA: Artech House, 1995.

[4] Whipple, D. P., "The CDMA Standard," *Applied Microwave & Wireless,* Winter 1994, pp. 24–39.

[5] Prasad, R., *CDMA for Wireless Personal Communications,* Norwood, MA: Artech House, 1996.

[6] Ochsner, H., DECT-Digital European Cordless Telecommunications, IEEE Vehicular Tech. Conf., VTC '89.

Selected Bibliography

www.dect.ch (DECT Forum)
scouria@ccnga.uwaterloo.ca, *DECT. The Standard Explained,* DECT Forum, February 1997.
www.tetramou.com
ETSI Technical specifications, *Terrestrial Trunking Radio,* www.etsi.fr.

12

Future Mobile Communications Systems

12.1 Introduction

It is expected that, during the first decade of the 21st century, digital mobile communications systems of the current generation will have stimulated the market in such a way that it will not be possible to satisfy users' requirements in terms of capacity, services, coverage, and quality.

This will be the appropriate moment to introduce the so-called third-generation (3G) systems whose main characteristics will be their vocation for universality and their multifunction capability. These types of systems will make extensive use of digital techniques and the technology developed during this decade. Given that the main feature of mobile communications systems is their capability to set up calls while the user is on the move or traveling, it is obvious that the final objective of these types of systems will be to provide the user with a light, inexpensive personal telephone terminal that can be used anywhere by means of terrestrial or satellite radio links. These are called personal communications systems, and they are the basis of the ITU's concept of universal personal telecommunications (UPT).

These systems are based on the fact that most communications are initiated or addressed to a specific person. The basic objective of these systems is to allow each person to be able to communicate anytime and anywhere.

In fact, this would fulfill the utopia of total communication. Although such a utopia is probably not achievable in its entirety, it is expected that, for the first time in history, technology puts at the disposal of mankind the means to get close to the intended objective.

It must be stressed that an important driving force in the evolution of mobile communications systems are the requirements set by the users; this trend will continue in the future. Along this line and for the next 10–15 years it can be stated that mobile communications systems will make possible the following:

- The enhancement of the services offered include multimedia and other wideband services, together with the extension of services to address the majority of the population;
- The development of both adequate fixed and mobile networks to provide these evolved/enhanced services;
- The availability of light terminals adaptable to the wanted service; this adaptability will also be extended to the radio interface.

Another guiding line toward third-generation mobile systems is that mobile networks will more and more trend to emulate and offer similar services to those provided by fixed telephone networks with the added value of mobility. It is already common to hear about the mobile access to the information highway. This entails the need to provide high binary rates to download huge Internet files. Networks will no longer be symmetric. The uplink can carry a low binary rate while the downlink provides high data rates to fit the above-mentioned application.

Personal mobility in future networks will only be reached by following some key trends, namely:

- An evolutionary approach from second to third generation, allowing a smooth and progressive transition;
- A path from standalone to integrated solutions;
- A convergence of networks, applications and services;
- An introduction of the UPT functionality.

The transition from the first to the second generation involved breaking up with existing systems, given that the second generation brought new digital technologies, services, and facilities. However, with consolidated second-generation systems providing supranational coverage, the transition to the third generation must be approached in an evolutionary way as, in fact, has been the case with the progressive upgrading of services and facilities seen for second-generation systems. One trend is, for example, keeping the core network and its associated signaling system while the radio interface is replaced. Backward compatibility must be guaranteed with currently existing systems with the use of multi-mode user terminals during the transition phase.

To establish a service with universal coverage capable of supporting high user densities it is necessary to set up a cellular network with a hierarchical structure including megacells (satellite coverage) and picocells (indoor coverage).

The growing demand for mobile services, in addition to being quantitatively greater (higher traffic density), is qualitatively more demanding (involving new communications scenarios, such as rural areas or sea environments, and higher personalization, among other demands). On the other hand, the easy and swift deployment of radio communication systems makes the possibility of setting up mobile resources to provide services to fixed and mobile users in new and developing countries that lack the minimum telecommunication infrastructure more and more likely.

For all this, and due to the fast evolution of mobile communication systems, the International Telecommunication Union asked its Radiocommunications Sector (ITU-R) to start the study and standardization process for a global system with the same characteristics, performance, and facilities as those of second-generation systems, both PAMR and PMT, and new added features. Such system was called in the beginning Future Public Land Mobile Telecommunications System (FPLMTS). Recently, the name has been changed to International Mobile Telecommunications (IMT-2000).

IMT-2000 can be defined as a third-generation mobile system that tries to unify the currently existing systems in a single radio infrastructure able to offer a wide range of services at the beginning of the 21st century in multiple operational environments.

Since the evolution of mobile communications aims at the concepts of globalization and personalization, IMT-2000 aims, as its basic principle, toward the specification of a global network that allows the establishment of communications between personal terminals.

At the same time in Europe, a similar project is being carried out, the Universal Mobile Telecommunications System (UMTS). It is the fruit of the R&D work carried out under the European Union funded RACE and ACTS programs.

This chapter provides a general overview of the most important projects under way to set up 3G mobile communications systems, namely, IMT-2000 and UMTS. The description presented reflects the situation at the time of this writing. It is acknowledged that the situation may change and certainly will during the 1999–2000 period. The material presented here is thus provisional.

12.2 IMT-2000

12.2.1 Introduction

IMT-2000 is a system concept that embraces the so-called worldwide third-generation mobile systems, which are scheduled to start service around the year 2000, subject to market needs. IMT-2000 systems will make available to users who are on the move or whose position may change, irrespective of their location (national or international roaming), a wide range of telecommunication services supported by fixed networks (i.e., PSTN/ISDN) and other services specific to mobile users.

A range of mobile terminal types that are able to access terrestrial or satellite networks, these terminals being designed for mobile or fixed use, is being considered. Key features of IMT-2000 are the following:

- Aims to provide these services over a wide range of user densities and geographic coverage areas;
- Features a high degree of commonality of design worldwide;
- Strives for compatibility of services within IMT-2000 and with fixed networks;
- Uses a small pocket-sized terminal with worldwide roaming capability;
- Makes an efficient and economical use of the radio spectrum;
- Admits the provision of service by more than one network in any area of coverage;
- Provides an open architecture that will permit the easy introduction of technology advances and equipment interoperability.

Special attention has been paid to the possibility to use IMT-2000 to provide communication services in rural, remote areas, as well as in small and developing countries for which these systems could provide a quick solution to fixed and mobile telecommunication needs.

The close relation between the terrestrial and the satellite components of IMT-2000 will allow the development of some services initially via satellite when no terrestrial infrastructure exists. When this is insufficient, it will later pass to the terrestrial mode in a slow, smooth way as the introduction of terrestrial infrastructure allows.

The IMT-2000 studies carried out in the framework of the ITU-R are performed by Task Group 8/1. These studies progress in a coordinated way with those carried out on other ITU bodies that take care of the fixed telecommunication network. The outcome of these studies has been the preparation of a set of recommendations (see Table 12.1).

A design objective of IMT-2000 is that the number of radio interfaces should be kept to a minimum, and, if more than one interface is required, those should have a high degree of commonality between them. These interfaces will serve different radio operation environments. It is assumed that a number of radio transmission technologies (RTT) may meet the requirements for the radio interfaces.

The material presented next has been extracted from the recommendations listed in Table 12.1, especially, ITU-R M.687.

12.2.2 Frequency Bands

IMT-2000 will operate worldwide in bands allocated by ITU's Radio Regulations as follows: 1,885–2,025 and 2,119–2,200 MHz with the satellite component limited to 1,980–2,010 and 2,170–2,200 MHz.

12.2.3 Basic Objectives of IMT-2000

The basic objectives of IMT-2000 can be classified into three broad groups:

1. General objectives:
   - To provide, as far as practical, services with a quality of service comparable to that of fixed networks;

Table 12.1
ITU-R Recommendations for IMT-2000

No.	Title
M.687	Concepts and objectives of IMT-2000
M.816	Framework for services supported on IMT-2000
M.817	IMT-2000 network architecture
M.818	Satellite operation within IMT-2000
M.819	IMT-2000 for developing countries
M.1034	Requirements for the radio interface(s) for IMT-2000
M.1036	Spectrum considerations for implementation of IMT-2000 in the bands 1885–2025 MHz and 2110–2200 MHz
M.1078	Security principles for IMT-2000
M.1079	Speech and voice band data performance requirements for IMT-2000
M.1167	Framework for the satellite component of IMT-2000
M.1168	Framework for IMT-2000 management
M.1223	Evaluation of security mechanisms for IMT-2000
M.1224	Vocabulary of terms for IMT-2000
M.1225	Guidelines for evaluation of radio transmission technologies for IMT-2000
M.1311	Framework for modularity within IMT-2000

- To provide for the continuing flexible extension of service provision, subject to the constraints of radio transmission, spectrum efficiency, and system economics;
- To adopt a phased approach for the definition of IMT-2000, with the first phase (Phase 1) including those services supported by user bit rates up to approximately 2 Mbps and phase 2 augmenting phase 1 with new services, some of which may require higher bit rates;
- To accommodate a variety of mobile terminals ranging from those that are small enough to be easily carried by a person (personal pocket radio) to those mounted in a vehicle;
- To admit the provision of services by more than one network in any area of coverage;
- To provide a modular structure that will allow the system to start from a configuration as small and simple as possible and grow as needed, both in size and complexity within practical limits.

2. Technical objectives:
   - To support integrated communication and signaling;
   - To provide high levels of security (voice and data services) allowing for the provision of end-to-end encryption for voice and data services;
   - To provide service flexibility that permits the optional integration of services such as mobile telephone, dispatch, paging and data communication, or any combination thereof;
   - To support terminal equipment interfaces (and procedures) defined for the fixed public networks that allow the alternative use of terminal equipment in the fixed public networks;
   - To allow the connection of PABXs or small rural exchanges to mobile stations.

3. Operational objectives:
   - To provide for the required user authentication and billing functions;
   - To provide for unique user identification and numbering in accordance with appropriate ITU-T Recommendations;
   - To provide for a unique equipment identification scheme;
   - To enable each mobile user to request particular services and initiate and receive calls, as desired, with calls for a given mobile user, incoming or outgoing at the same mobile termination, having the ability to be simultaneously multiplexed and associated to different services (i.e., advanced voice and data services including multimedia);

- To aid the emergency services by providing additional emergency call information as far as possible (e.g., user identity and location information) and other information that may be required by national or local authorities.

12.2.4 Radio Interfaces

The family of radio interfaces is the following:

- R1: Radio interface between a mobile station (MS) and the base station (BS);
- R2: Radio interface between a personal station (PS) and the personal base station (CS);
- R3: Radio interface between the satellite and the mobile Earth station (MES) (may also allow for the automatic routing of traffic between terrestrial and mobile satellite systems);
- R4: An additional radio interface used for alerting (e.g. paging) in case of a call terminated at an IMT-2000 terminal.

A possible scenario for the evolution and implementation of personal communications within IMT-2000 is given in Figures 12.1 and 12.2.

Taking into account the characteristics of areas where IMT-2000 is introduced, the system structure should be optimized according to the geographical coverage areas and traffic conditions. Therefore, plural air interfaces should be allowed in the design of systems.

The radio interface should be designed to allow different applications to use the same interface where this can be shown to be technically and economically feasible. If the same radio interface cannot be used for all applications, then the individual interfaces should have a maximum commonality to allow interworking with the minimum extra complexity.

In order to improve performance, systems may adapt such parameters as channel bit rate, bandwidth, frequency/time/coding arrangements, diversity techniques, and multipath equalization to actual propagation, interference, and traffic conditions, subject to cost and power consumption considerations. Voice and data services will be a basic part of IMT-2000 and a flexible multiple access scheme is desirable to handle the wide range of traffic densities and services offered.

When considered with frequency reuse factors, the modulation techniques should achieve efficient use of the spectrum. Technologies having transmission efficiencies greater than 1 bit/s/Hz should be used for IMT-2000.

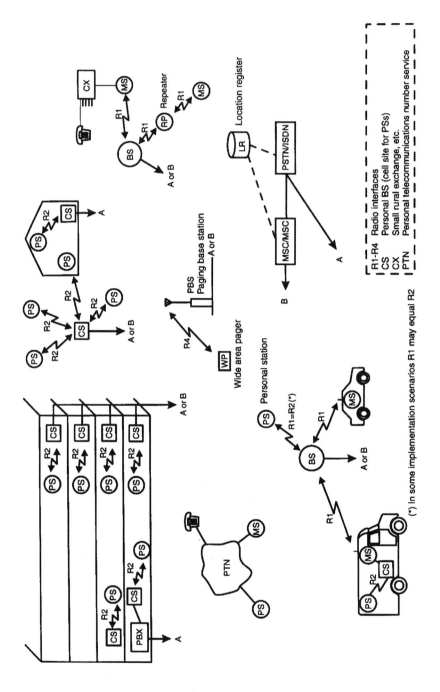

Figure 12.1 Scenario for personal communications within IMT-2000 (terrestrial component) (ITU).

Future Mobile Communications Systems 499

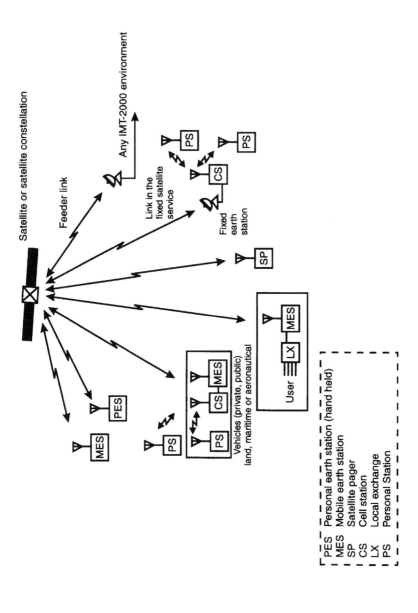

Figure 12.2 Scenario for personal communications within IMT-2000 (satellite component). (ITU).

12.2.5 Traffic Densities and Spectrum Requirements

The maximum demand for "personal" telecommunication services is in large cities where different categories of traffic can be found (i.e., that generated by mobile stations (MS), vehicle-mounted or portable and personal stations (PS), outdoor or indoor).

In Recommendation M.687, a detailed traffic forecast is given. This leads to the figures of Tables 12.2 and 12.3 (also contained in the recommendation). From these it follows that the minimum spectrum bandwidth required is 227 MHz (111+56 for MS and 27+3 and 24+6 for PS).

A speech coding rate of 8 kbps was assumed for mobile stations in Tables 12.2 and 12.3, since speech coders with lower bit rates, quality, and transmission delays comparable to those of the PSTN may not be available in the foreseeable future. Higher rate speech coders have been assumed for use in inexpensive personal stations. It should be noted that the current fixed networks use codecs with bit rates of 64 kbps and mobile applications range from 32 down to

Table 12.2
General Characteristics of Personal Communications (High-Density Area) Voice Service Traffic Demands and Spectrum Requirement (ITU)

Specifications	MS Outdoor Interface R1	PS Outdoor Interface R2	PS Indoor Interface R2
Radio coverage (%)	90	>90	99
Base station antenna height (m)	50	<10	<3[1]
Base station installed: indoors/outdoors	No/yes	Yes/yes[2]	Yes/yes[2]
Traffic density (Erlang/km^2)	500	1,500	20,000[1]
Cell area (km^2)	0.94	0.016	0.0006
Blocking probability (%)	2	1	0.5
Cluster size (cell sites × sectors/site)	9	16	21 (3 floors)
Duplex bandwidth per channel (kHz)	25	50	50
Traffic per cell (Erlang)	470	24	12
Number of channels per cell	493	34	23
Bandwidth (MHz)	111	27	24
Station[3]	Vehicle mounted or portable		
Volume (cm^3)		<200	<220
Weight (g)		<200	<200
Highest power	5W	50 mW	10 mW

Notes: (1) Per floor; (2) usual case; (3) a range of terminal types available to suit operational and user requirements.

Table 12.3
Spectrum Estimation for Non-Voice Services (ITU)

	MS Outdoor Interface R1		PS Outdoor Interface R2		PS Indoor Interface R2	
	Circuit switched	Packet switched	Circuit switched	Packet switched	Circuit switched	Packet switched
Traffic density (Erlang/km^2)	45	37	(1)	150	2,000$^{(2)}$	2,500$^{(2)}$
Duplex bandwidth per channel (kHz)	100	50	50	50	50	50
Bandwidth (MHz)	56		3		6	

Notes: (1) Insignificant; (2) per floor.

approximately 10 kbps. It is believed that the choice of access scheme (FDMA, TDMA, or CDMA) will not substantially affect the overall estimation.

12.2.6 Radio Transmission Technology Evaluation

A team of ITU members, representatives of regional and national standardization bodies, members of the worldwide telecommunication industry, regulatory bodies, and operators are cooperating toward the development of IMT-2000 as indicated in Figure 12.3.

According to this plan, before March 1999, the key decisions concerning the radio transmission technologies (RTT) associated to IMT-2000 must be adopted, and toward 2000 the drafting of the relevant ITU-R recommendation

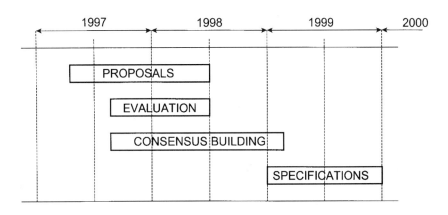

Figure 12.3 Steps toward IMT-2000.

must be accomplished. Table 12.4 enumerates the minimum performance capabilities of the RTT for IMT-2000.

Recommendation M.1225 provides guidelines for both the procedures and the criteria to be used in evaluating RTTs for a number of test environments, chosen to simulate closely the more stringent radio operating conditions. The main evaluation criteria are the following:

- Spectrum efficiency;
- Technology complexity-incidence on cost of installation and operation;
- Quality;
- Flexibility of radio technologies;
- Implications on network interface;
- Hand-portable performance optimization capability;
- Coverage/power efficiency.

The test environments foreseen are:

- Indoor office (o);
- Outdoor-to-indoor and pedestrian (p);
- Vehicular;
- Satellite.

The recommendation contains also the guidelines for evaluating the RTTs by independent evaluation groups, the evaluation methodology and the template for the description of the RTTs.

The IMT-2000 evaluation groups are the following:

- Europe: ETSI Project DECT, ETSI SMG2, European Space Agency;
- Brazil: ANATEL;

Table 12.4
IMT-2000 Minimum Performance Objectives (ITU)

Scenario	Indoor Office	Outdoor to Indoor/ Pedestrian	Vehicular	Satellite
Binary rate	2.048 Mbps	384 kbps	144 kbps	9.6 kbps
BER	10^{-6}	10^{-6}	10^{-6}	10^{-6}

- Canada: Canadian Evaluation group;
- USA: TIA/TR-45 International Standards Develop; Group & T1P1/TR46 international Standards Ad-Hoc;
- Australia: Cooperative Research Centre for Broadband;
- China: China Evaluation Group;
- India: Design Domain International;
- Japan: Association of Radio Industries and Business (ARIB);
- Korea: TTA Evaluation Ad-Hoc;
- Malaysia: ITU;
- New Zealand: Radio Spectrum Management;
- Others: Inmarsat, Globalstar.

As of June 30, 1998, the ITU had received several RTT proposals (10 for terrestrial systems and five for satellite). These proposals are listed in Table 12.5. Seven of the terrestrial proposals are CDMA-based. Two of them stand out at the time of this writing:

- W-CDMA (ETSI, ARIB, T1P1 and TIA/TR46.1);
- CDMA 2000 (TR 45.5 and TTA).

12.3 The UMTS Concept

12.3.1 Introduction

The work toward the European third-generation mobile system UMTS started in Europe by extensive research activities within the framework of the European Union ACTS program. Currently, it is being driven by standardization activities carried out within ETSI's Special Mobile Group (SMG). Their developing parties state that the aim of UMTS is to lead the personal communications user into the information society of the 21st century. UMTS will deliver information directly to people and provide them with access to new and innovative services. It will offer broadband multimedia-oriented personalized communications to the mass market, regardless of location, network, or terminal.

UMTS will combine new technologies with the evolution of existing networks (for example, ISDN and GSM). Also, it will have high interworking capabilities with these networks, especially with enhanced GSM, as well as interoperability for global roaming.

Table 12.5
Proposals for RTT in IMT-2000

Proposal	Description	Environment				Source
		Indoors	Pedestrian	Vehicular	Satellite	
DECT	Digital enhanced cordless telecommunications	x	x	—	—	ETSI Project (EP) DECT
UWC-136	Universal wireless communications	x	x	x	—	USA, TIA TR45.3
WIMS W-CDMA		x	x	x	—	USA, TIA TR46.1
TD-SCDMA	Time division synchronous CDMA	x	x	x	—	China Academy of Telecommunication Technology (CATT)
W-CDMA	Wideband CDMA	x	x	x	—	Japan, ARIB
CDMA II	Asynchronous DS-CDMA	x	x	x	—	Republic of Korea, TTA
UTRA	UMTS terrestrial radio access	x	x	x	—	ETSI SMG2
NA: W-CDMA	North American: Wideband CDMA	x	x	x	—	USA, T1P1-ATIS
cdma2000	Wideband CDMA (IS-95)	x	x	x	—	USA, TIA TR45.5
CDMA I	Multiband synchronous DS-CDMA	x	x	x	—	Republic of Korea, TTA
SAT-CDMA	49 LEO satellites on seven orbital planes at 2,000 km	—	—	—	x	Republic of Korea, TTA
SW-CDMA	Satellite wideband CDMA	—	—	—	x	ESA
SW-CTDMA	Satellite wideband hybrid CDMA/TDMA	—	—	—	x	ESA
ICO RTT	10 MEO satellites in two orbital planes	—	—	—	x	ICO Global Communications
Horizons	Horizons satellite system	—	—	—	x	Inmarsat

UMTS will have both terrestrial and satellite components that will enable service access in a very wide range of radio environments from megacells (satellite) through to macro, mini, micro, to picocells.

As a consequence, UMTS must offer universal coverage; that is, it must have connectivity capacity over large geographic areas (as a minimum over Europe, although potentially over the entire world). Universality also implies the availability of UMTS services in multiple environments (such as rural, residential, indoor, and business areas). Hence, the terminal in this future communication system must adapt automatically its technical characteristics (including transmission rate, modulation type, and power) to the propagation conditions found in the different operational scenarios (indoors, outdoors) and as a function of the services (voice, data, messaging) demanded by the user requires.

The introduction of UMTS faces an important challenge due to the existence of the GSM network with a large coverage and capacity with low-cost terminals which offers new services such as generalized packet radio services (GPRS), high-speed circuit-switched data (HSCSD), and enhanced data rates over GSM evolution (EDGE). This is why the deployment of UMTS must be carried out in a progressive way using as much as possible, in a first stage, the existing GSM core network with modifications that do not require large investments from the operators.

For some time, the GSM network will cope with the telephony service and the basic data services. Possibly, the starting points for the deployment of UMTS will be those areas where GSM networks are saturated and advanced high binary rate services are required that GSM cannot provide. This deployment will be facilitated if GSM/UMTS roaming is allowed to achieve seamless service between UMTS and GSM, and, if affordable, dual mode GSM/UMTS terminals are available.

As a consequence, on the network side, UMTS will be built on the GSM structure using the MAP network protocol and therefore will support GSM features, thus assuring a seamless upgrading of GSM and backward compatibility. Cost-effective evolution requires that as much as possible of the core network from GSM is retained. Multimode terminals will ensure user-friendly, cost-efficient access in all environments, while supporting GSM basic services.

Since the mobile communications market is one of the most promising markets today, the European Parliament and Council recommended the establishment of an organization called "UMTS Forum." The forum will contribute to the development of a common vision for the UMTS and the drafting of a European policy on mobile and personal communications.

It will advise and issue recommendations to the European Commission, European Radiocommunication Committee (ERC), European Telecommunications Office, European Telecommunications Standards Institute, and National Regulatory Authorities.

The main milestones for UMTS development and putting into operation, are the following:

- 30 June 1997: ERC Decision on UMTS Core band;
- 30 November 1997: UMTS additional frequency spectrum demand included on the agenda of WRC '99; suggestions for UMTS regulatory framework from the EC;
- 31 December 1997: Identification of candidates for additional UMTS frequency spectrum (Extended band, approximately 185 MHz);
- First quarter 1998: ETSI freezing of basic UMTS parameters; discussions on licensing conditions;
- Second quarter 1998: Operator's commitment;
- Spring 2000: WRC '00 Recommendation on UMTS Extended band; ETSI UMTS Phase 1 standard;
- 1 October 2000: ERC Decision on additional UMTS frequency spectrum;
- 2001: Pre-operational UMTS trials;
- 2002: Partial availability of UMTS core band; Commercial UMTS operation;
- 2005: Availability of UMTS core band;
- 2008–2010: Availability of UMTS extended band.

12.3.2 UMTS Objectives

The main objectives foreseen for UMTS Phase 1 are the following:

1. General objectives:
   - High capacity to support large user densities;
   - Support a wide variety of services with different binary rates and modes, both packed-switched and circuit-switched;
   - Support of multimedia applications, with a wide range of data rates from 144–512 kbps for wide area mobility to 2 Mbps for restricted mobility;
   - Good performance for fair competition with other networks;
   - High speech quality using low bit rates;
   - High spectral efficiency.

2. Operational objectives:
   - Must be capable of supporting handy phones that can be used throughout Europe and, possibly, worldwide, at home, in cities, in rural areas, within offices, etc.;
   - Must be capable of supporting a wide range of terminals from the basic handy phone to sophisticated terminals capable of supporting a wide range of services or a specific selection of these;
   - Dual mode/band terminals for roaming between UMTS/GSM networks;
   - Virtual home environment for service creation and service portability;
   - Seamless indoors to outdoors and for outdoor coverage;
   - Operation in unlicensed cordless environments.
3. Core network aspects:
   - Evolution of GSM and ISDN;
   - Mobile/fixed convergence elements.
4. Economic issues:
   - Manufacturer's competition;
   - Economies of scale.

12.3.3 Spectrum Requirements

The following items have been identified concerning the spectrum needs for UMTS:

- The bandwidth must be in the order of 2×20 MHz per license. With a minimum of two licenses in any given geographical area, this means a total of 2×40 MHz by 2002.
- The spectrum must be available early on to allow both technical and market trials. At least the core band needs to be fully available through Europe, even if extension bands may be specific to individual countries.
- There is a need to allocate an additional 20 MHz as start-up band for non-public, nonlicensed, in-building, low mobility systems. This spectrum will be required from 2002 to help the starting of the market for multimedia terminals and to stimulate a demand for public UMTS access.

According to the two types of operations foreseen: TDD, FDD (see Section 12.3.4), the European Radiocommunications Committee (ERC) has established two sets of frequency allocations for terrestrial UMTS, as follows:

1. Unpaired bands (U) for TDD applications, in which each radio channel consists of only one carrier. Two sub-bands with a total bandwidth of 35 MHz have been reserved, 1,900–1,920 MHz and 2,010–2,025 MHz.
2. Paired bands (P) for FDD applications. In this case each radio channel consists of two carriers with a duplex spacing of 90 MHz. Also, two sub-bands with a total bandwidth of 60 MHz have been reserved, 1,920–1,980 MHz and 2,110–2,170 MHz.

These bands have been further divided in four segments for different uses. Also two additional segments are provided for UMTS FDD satellite component.

In Table 12.6 the frequencies, segments and intended uses are indicated. The DECT application has been included because it uses segment no. 1.

Refarming of used frequency bands like 900 MHz and 1,800 MHz will be possible. This will give the regulatory bodies more flexibility to issue licenses, besides those on the new 2-GHz band. Refarming capabilities allow introduction of additional third-generation multimedia services to the existing GSM networks without additional spectrum, thus enabling low entry costs for UMTS services using current GSM topology.

12.3.4 Radio Interface

Currently, in ETSI, a standards development process is in progress for the UMTS terrestrial radio interface which will be called UMTS terrestrial radio access (UTRA). It is being designed to enable both time division duplex (TDD) and frequency division duplex (FDD) modes of operation.

Table 12.6
UMTS Frequency Allocations

Segment	Band (MHz)	Type	Access	Use
1	1,885–1,900	U	TDD	DECT
2	1,900–1,920	U	TDD	
3	1,920–1,980	P	FDD	WAM with segment 6
4	1980–2,010	P	FDD	MS with segment 7
5	2,010–2,025	U	TDD	
6	2,110–2,170	P	FDD	WAM with segment 3
7	2,170–2,20	P	FDD	MS with segment 4

Notes: P: Paired band; U: unpaired band; WAM: wide area mobility; MS: mobile satellite.

The UTRA should support a range of maximum user bit rates according to the user's current environment, as follows:

1. Rural outdoor, available throughout the operator's service area, with the possibility of large cells: at least 144 kbps (ISDN 2B+D channel), with the possibility of achieving 384 kbps (ISDN H0 channel) at the maximum speed of 500 km/h.
2. Suburban outdoor, available with complete coverage of a suburban or urban area, using smaller macrocells, or microcells, at least 384 kbps (goal to achieve 512 kbps) at the maximum speed of 120 km/h.
3. Indoor/low-range outdoor, available indoors and localized coverage outdoors, at least 2 Mbps (1.92 Mbps "provided" by the ISDN H12 channel), at the maximum speed of 10 km/h.

The TDD mode becomes more and more important because the spectrum demand in the future will be highly asymmetric to support enhanced data services. The demand on the downlink will be much higher than on the uplink. Flexible allocation of bandwidth for the uplink and the downlink is possible using the TDD mode, thus the available spectrum can be used in a more efficient way. TDD is considered a good candidate for unlicensed cordless and public wireless local loop applications.

For the radio interface various alternatives were proposed that later were reduced to two:

- W-CDMA;
- TD-CDMA.

In the following sections, both access techniques are described briefly.

W-CDMA

The W-CDMA scheme has been developed as a joint effort between ETSI and the Japanese ARIB. This system includes:

- Wideband CDMA carrier to offer a high degree of frequency diversity and high bit-rates;
- Flexible physical layer for implementation of UMTS services, with support for large range of varying bit-rates with high granularity;

- Built-in support for co-existence and efficient hand-overs with GSM;
- Feasible implementation from day one of UMTS, with possibility for performance enhancement using more demanding features like adaptive antennas and multi-user detection in the future.

Table 12.7 summarizes the key technical characteristics of the W-CDMA radio-interface.

There are some performance-enhancement features that can be applied to the W-CDMA system:

1. Transmitter diversity: Orthogonal transmit diversity, where the data stream is split into several streams which are sent through different antennas. This scheme can be used in the downlink to get quality and capacity gains without increasing the mobile complexity.
2. Receiver structures: W-CDMA is designed to work without requiring receivers for joint detection of multiple user signals. However, the potential capacity gains of such receivers have been recognized and taken into account in the design.
3. Adaptive antennas: Adaptive antennas are recognized as a way to enhance capacity and coverage of a CDMA system. Solutions employing these kinds of antennas are supported in the W-CDMA concept and adaptive antenna issues have been included in the design of the downlink common physical channels.

There exist two basic physical channels in WCDMA: the dedicated physical data channel and the dedicated physical control channel. The data

Table 12.7
Key Technical Characteristics of the W-CDMA Radio Interface

Multiple-access scheme	DS-CDMA
Duplex scheme	FDD / TDD
Chip rate	4.096 Mcps (expandable to 8.192 Mcps and 16.384 Mcps)
Carrier spacing (4.096 Mcps)	Flexible in the range 4.4–5.2 MHz (200 kHz carrier raster)
Frame length	10 ms
Interbase station synchronization	FDD mode: No accurate synchronization needed; TDD mode: synchronization needed
Multirate/variable-rate scheme	Variable-spreading factor and multicode
Channel coding scheme	Convolutional coding (rate 1/2-1/3) Optional outer Reed-Solomon coding (rate 4/5)
Packet access	Dual mode (common and dedicated channel)

channel is used to carry dedicated data generated at layer 2 and above (i.e. the dedicated logical channels). The control channel carries control information generated at layer 1. The control information consists of known pilot bits to support channel estimation for coherent detection, transmit power-control commands and optional (variable-length) rate information. The rate information informs the receiver about the instantaneous rate of the different services and how services are multiplexed on the dedicated physical data channels.

The frame length on the physical channels is 10 ms, and each frame is divided into 16 slots of 0.625 ms each, corresponding to one power-control period. In the downlink, the dedicated physical control and data channels are time-multiplexed within the slots, with one power-control command per slot. In the uplink, control and data are code-multiplexed and transmitted in parallel.

In both uplink and downlink, the dedicated physical control and data channels are spread to the chip-rate using orthogonal variable-rate spreading factor codes. These channelization codes have varying spreading factors to carry varying bit-rate services (i.e. the channelization codes are of different lengths to match different user bit rates, with spreading factors from 4 up to 256). Using different channelization codes, several data and control channels of different rates can be spread to the chip-rate and still be orthogonal after spreading. Hence, multicode transmission can be employed for the highest bit-rates, typically above 384 kbps, and several services of different rates can be transmitted in parallel with maintained orthogonality.

In the uplink and downlink QPSK modulation is used, with root-raised cosine pulse-shaping filters (roll off 0.22 in the frequency domain).

WCDMA offers three basic service classes with respect to forward error correction coding: standard services with convolutional coding only (BER $\approx 10^{-3}$), high-quality services with additional outer Reed-Solomon coding (BER $\approx 10^{-6}$), and services with service-specific coding where WCDMA layer 1 does not apply any prespecified channel coding. The latter class can be used to enable other coding schemes such as turbo-coding. Rate 1/2 or 1/3 convolutional codes are used, with block interleaving over one or several frames depending on delay requirements. The additional Reed-Solomon code employed is of rate 4/5 and is followed by symbol-wise interframe block interleaving.

After channel coding and service multiplexing, the total bit rate is rather arbitrary. Rate matching is used to match the coded bit-rate to the limited set of possible bit-rates of a dedicated physical data channel. In the uplink, puncturing and repetition are employed to match the rate, while in the downlink, puncturing and repetition for the highest rate are used together with discontinuous transmission for the lower rates.

Using the above-mentioned coding, interleaving and rate matching techniques the WCDMA concept has shown that rates of at least 2 Mbps can be achieved using a 4.096 Mcps carrier. Also, low bit-rates as well as high bit-rates can be supported efficiently, with high bit-rate granularity.

The WCDMA system operates with a frequency reuse of one. Soft hand-over enables this and provides capacity and coverage gains compared to hard hand-over. Seamless interfrequency hand-over is needed for operation in hierarchical cell structures and hand-over to other systems (e.g., GSM).

The FDD mode assumes asynchronous base stations. To enable asynchronous operation a fast cell search scheme has been defined. In the cell search procedure the mobile station acquires two synchronization codes broadcast by the base station, from which the mobile can determine the scrambling code and frame synchronization of the base station.

Due to the varying characteristics of packet data traffic in terms of packet size and packet intensity, a dual-mode packet-transmission scheme is used for WCDMA. With this scheme, packet transmission can either take place on a common fixed-rate channel or on a dedicated channel, with an adaptive choice of method based on the packet traffic characteristics. Small infrequent packets are typically transmitted on the common channel, while larger, more frequent packets are transmitted on a dedicated channel.

TD-CDMA

The TD-CDMA mode is based on an eight-slot time frame structure very similar to that in GSM with a carrier spacing of 1.6 MHz. The duration of the TDMA frame as well as time slots are exactly the same for both GSM and TD-CDMA. This allows the building of TD-CDMA easily on top of proven GSM technology and ensures an easy hand-over between GSM and UMTS. For each carrier within each time slot there is the possibility of multiplexing a maximum of 8 to 12 physical channels using different code sequences. This reduced number of sequences permits the implementation/realization of multiuser receivers at a reasonable cost with the technology currently available.

In Table 12.8 the key technical characteristics of TD-CDMA are summarized.

A physical channel corresponds thus to a code sequence-time slot pair within a carrier (or within a frequency hopping pattern, in case FH is used). The transmitted bursts in a physical channel include a training sequence to allow the estimation of the channel response at the receiver. There exist two types of bursts that may be distinguished by their training sequence lengths and the maximum allowable time dispersion. The TDMA structure brings a great flexibility at the time of establishing a correspondence between logical and physical channels. The wanted multiplicity of binary rates for the traffic channels is achieved by assigning a variable number of time slots and/or code

Future Mobile Communications Systems

Table 12.8
TD-CDMA Key Technical Characteristics

Multiple access method	TDMA and CDMA (FDMA inherent)
Duplex method	FDD and TDD
Channel spacing	1.6 MHz
Carrier chip / bit rate	2.167 Mcps
Time slot structure	8 slots/TDMA frame
Spreading	Orthogonal, 16 chips/symbol
Frame length	4.615 ms
Multirate concept	multi-slot and multi-code
FEC codes	rates 1/8 . . . 1 (convolutional, punctured)
Interleaving	inter-slot interleaving
Modulation	QPSK / 16-QAM
Burst types	Two different burst types: burst 1, for long delay-spread environments and burst 2, for short delay-spread environments
Pulse shaping	GMSK basic impulse
Detection	Coherent, based on midamble
Power control	Slow
Hand-over / IF hand-over	Mobile-assisted hard hand-over
Channel allocation	Slow and fast DCA supported
Intracell interference	Suppressed by joint detection
Intercell interference	Like in other clustered systems

sequences to each user, as well as different coding rates (between 1/8 and 1) and modulation types (QPSK and 16-QAM with Gaussian pulses).

The possibility of using time division duplex, TDD, makes TD-CDMA a suitable technique for use in the unpaired bands allocated to UMTS. Due to the multiuser/joint detection scheme foreseen, the requirements on uplink power control are quite relaxed and the benefits of CDMA without the intracell interference can be utilized. A slow control power mechanism at carrier level has been defined in much the same way as in GSM. Furthermore, soft hand-over, which implies additional infrastructure costs, is not needed.

However, surely, the most important advantage of TD-CDMA is its high degree of compatibility with GSM, thus allowing the realization of dual mode terminals at a reasonable price with the possibility of performing hand-overs between both systems.

TD-CDMA permits asymmetric operation, which is very convenient for some applications such as Internet access and some degree of on demand assignment.

In January 1998, the Special Mobile Group of ETSI reached an agreement over the radio interface for UMTS. This interface, UTRA, is dual. Its specifications are listed as follows:

- In the paired frequency bands, W-CDMA will be used;
- In the unpaired frequency bands, TD-CDMA will be used.

It is expected that this combination of CDMA and TDMA will achieve the following:

- High spectrum efficiency;
- Service flexibility concerning data rates and applications, with transmissions up to 2 Mbps right from the beginning;
- Data rates of 144 kbps for full mobility, 384 kbps for limited mobility and 2,048 kbps for low-mobility users, respectively;
- Investment savings by smooth migration from GSM to UMTS, while maintaining GSM functionalities and using the existing equipment as much as possible;
- Seamless applications and flexible traffic handling.

An adequate standards development must be carried out to allow the coexistence of both access technologies in such a way that harmonization with GSM is eased and it will be possible, in the future, to manufacture dual FDD/TDD terminals at a reasonable cost.

Selected Bibliography

ETSI Technical Report TR 101.398 - UMTS, High level requirements relevant for the definition of the UMTS Terrestrial Radio Access concept (UMTS 21.02 Version 3.0.1), 1998.

ETSI Technical Report TR 101.146 UMTS Terrestrial Radio Access (UTRA), Concept evaluation, (UMTS 30.06 Version 3.0.0), 1997.

About the Authors

José M. Hernando received his M.S. degree in Telecommunications Engineering in 1967 and his Ph.D. in 1970, both from Madrid Polytechnic University. From 1967 to 1969, he worked at ITT Laboratories of Spain. Afterwards, he joined IBERIA Airlines of Spain, where he was responsible for the design and project implementation of the PMR airport and air-to-ground mobile radiocommunication networks. In 1975, he was appointed assistant professor in the Radiocommunications Laboratory of the Telecommunication School of Madrid Polytechnic University. From 1977 until nowadays, he has been serving as a full professor at the university. Dr. Hernando also works as a Senior Consultant for private companies and public administration. He has published several papers on radiocommunications and is the author of four textbooks on the subject. He currently participates in the Study Groups of the Radiocommunications Sector of the International Telecommunications Union devoted to radio propagation and mobile systems.

Fernando Pérez-Fontán received his M.S. degree in Telecommunications Engineering in 1982 and his Ph.D. in 1992 from Madrid Polytechnic University. In 1988, following several years in industry, he joined the Communications Technologies Department of the University of Vigo, where he teaches a Mobile Communications course. He has published a number of papers in international journals and conferences and has led various mobile satellite projects for the European Space Agency. Currently, he is chairman of Working Group 1.b (mobile propagation) within Euro-COST 255 (Radiowave Propagation Modeling for New SatCom Services at Ku-Band and Above).

Index

π/4-DQPSK modulation, 436, 485

Absorption losses, 100
Access Assignment Channel (AACH), 487
Access burst, 408
Access Grant Channel (AGCH), 402
Adaptive differential pulse code
　　　modulation (ADPCM), 467
Adaptive time alignment, 406, 413–15
Advanced mobile phone service (AMPS),
　　　323
　adjacent channel interference
　　　reduction, 351–54
　allocated band, 349
　base station location tolerance, 334
　cell splitting, 354–55
　channels, 429
　cochannel reuse ratio, 335–36
　deployment plan, 361
　extended (EAMPS), 428, 429
　features, 347–51
　maximum cell radius, 334–35
　minimum cell radius, 335
　network roll-out, 351–60
　network rollout plan example, 358–60
　operation, 336–38
　overlaid cell concept, 355–58
　saturation, 428
　schematic representation, 337
　ST frequency, 349
　system description, 336–38
　system parameter selection criteria,
　　　334–36
　system spectrum allocation, 350
　See also Cellular networks
Amplitude
　response, 93–94
　variations, 23
Antenna feeder, 197
Antenna gain, 421
　base station, 206
　isotropic, 421
　usage correction, 196
Antenna multicouplers, 267–69
　block diagram, 268
　characteristics, 269
　defined, 267
　elements, 267
　See also Antenna systems
Antennas
　directive, 358
　isotropic, 147
　omni, 358, 371
Antenna switch (AS), 28
Antenna systems, 257–73, 288
　complete, 272–73
　connection options, 258
　duplexers, 269
　feeders, 269–72
　illustrated, 273

introduction, 257
multicouplers, 267–69
Tx-Rx, 273
See also Base stations
Area coverage
multichannel operation, 53
overlap/capture areas, 57
radiocommunication system, 207
synchronous/quasi-synchronous
   operation, 56–58
techniques, 52–58
voting systems, 53–56
Area coverage probability, 200, 201, 202
defined, 202
fringe coverage probability relation, 203
illustrated, 201
Asynchronous duplex data transmission services, 424
Attenuation
linear, 193
propagation, 223
wall, 255
Audio tone signaling system, 43
Authentication, 388–90
challenge-response method, 389
MS, 390
process, 389, 390
Automatic number indication (ANI), 51
Average fade duration (afd), 76, 79–81
computing, 79
for different fade depths, 81
expression, 80
illustrated, 77
for Rayleigh case, 79
See also Second-order statistics

Bandpass cavities, 260–62
arrangement, 262
features, 260
role, 260
See also Transmitter combiners
Bandpass duplexers, 270
Band-reject cavities, 262–65
combiner illustration, 264
defined, 262
See also Transmitter combiners
Bandwidth, coherence, 89, 108
Base station controllers (BSCs), 381, 384

Base station identity codes (BSIC), 417
Base stations
antenna gain, 206
antenna multicouplers, 267–69
antenna systems, 257
cellular components, 333
components, 272
coverage range, 209
creates urban microcells, 252
defines macrocells, 252
in dispatch system, 58
diversity at, 86
duplexers, 38, 269
effective antenna height, 125, 158
elements, 282–85
engineering, 257–73
feeders, 269–72
four-channel, 286–87
full-duplex systems and, 39
functions, 287–88
height gain factor, 131, 133
location tolerance, 334
nominal transmit power, 379
radio link controlled, 42
remote control functions, 287
schematic diagram, 286
spatial correlation at, 87
transmitter combiners, 257–67
trunked system, 282–88
Base station subsystem (BSS), 381
BSCs, 381, 384
BTSs, 381, 384
update intervals, 388
See also GSM
Base transceiver stations (BTSs), 381, 384
Basic losses
computation, 209
defined, 14, 123
expression, 208
Bistatic radar equation, 102–3
Bit error rate (BER), 90
Blank-and-burst, 345
Breakpoint, 240, 245
Broadcast Control Channel (BCCH), 402, 432, 487
Built-up area losses
JRC model, 165–69
quantification of, 136–41

Bursts, 405–8
    abbreviated, 435
    access, 408
    defined, 406
    downlink, 434, 435, 488
    dummy, 406–8
    frequency correction, 406
    NADC, 433
    normal, 406
    synchronization, 406
    TETRA, 488
    types of, 407
    uplink, 434–35, 488
Calls
    data, 295
    emergency, 295
    establishing, to mobile, 405
    fixed network originated, 340–41
    fixed network terminal to mobile terminal, 393
    group, 308
    include, 296
    mobile-originated, 341, 404
    MS to fixed network terminal, 393
    MS to MS, 394
    stages, 393
    types of, 44
    voice, 295
    See also Hand-over
Call setup, 338–40
    GSM, 393–97, 403–8
    incoming call, 394–97
    mobile-originated call, 393–94
    process, 403–8
Carrier/bearer services
    defined, 423
    list of, 424
    See also GSM
Carrier sense multiple access (CSMA) systems, 60
Carrier-to-interference ratio, 331, 367
    calculating, 34–36
    exceeding threshold level, 212
    expression, 35–36
    at fringe of interfered cell, 367
    link budge, 237–38
    in logarithmic units, 212
    See also Interference

Carrier-to-multipath ratio, 84
    in microcell/macrocell, 244
    Rice distribution characterized by, 242, 243
Carson rule, 27
Cavity-isolator combiners, 265–67
    defined, 265
    for four Tx, 265
    insertion loss, 265, 266
    See also Transmitter combiners
CDMA, 25, 439–88
    adjacent cells, 442
    advantage, 442
    American system, 449
    AMPS emission spectra, 446
    capacity, 442
    direct-sequence (DS-CDMA), 441
    diversity mechanisms, 444–47
    FDMA/TDMA comparison, 458–59
    frequency reuse, 443
    frequency selective fading, 447
    full-duplex operation, 442
    interference sources, 445
    multipath and, 444
    orthogonal codes, 441
    power control, 447–48
    soft hand-over, 446
    speech coding, 449–50
    system operation, 456–58
    TD-CDMA, 512–14
    Walsh matrix, 450, 451
    W-CDMA, 509–12
CDMA IS-95, 25, 448–59
    codes, 457
    coding on base-to-mobile link, 450–53
    coding on mobile-to-base link, 453–56
    defined, 448
    requirements, 449
    speech coding, 449–50
    system operation, 456–58
    uplink/downlink modulations in, 456
    Walsh matrix, 450, 451
Cell radius, 334–35
    maximum, 334–35
    minimum, 335
Cells
    cochannel, 330, 367
    coexistence, with different sizes, 357

minimum radius of, 359
number of, in cluster, 330
number of channels in, 365
omni, 353
See also Clusters
Cell splitting, 354–55
defined, 327, 354
illustrated, 356
minimum cell radius and, 359
Cellular
concept, 323–71
geometry, 327–32
grid, 331
GSM structure, 379–80
layout, 323
Cellular networks
base stations, 332, 333
blocking probability, 363
data links, 333
deployment plan, 361
deployment stages, 328–29
dimensioning of, 362–66
functions, 360–62
interference, 368
noise limits, 368
operation fundamentals, 333
reuse index, 364, 366
structure, 360–62
traffic density, 366
voice links, 333
CEPT standard, 466, 467
Channel coding
CDMA, 450–51
GSM, 409–10
NADC, 439
Channel distortion, 95
Channel reuse, 292
Channels
Access Assignment Channel (AACH), 487
Access Grant Channel (AGCH), 402
AMPS, 429
Broadcast Control Channel (BCCH), 402, 487
control, 380, 400, 401, 402–3
Fast Associated Control Channel (FACCH), 403, 432
Frequency Correction Channel (FCCH), 402

frequency selective, 93
full-duplex, 39
half-duplex, 37–39
illustrated types, 29–30
intermodulation-free, 225, 227
Linearization Channel (LCH), 487
logical, 431–32
multipath, 95–96
number per cell, 365
Paging Channel (PCH), 402, 418
physical, 487–88
propagation, 98
Random Access Channel (RACH), 402
single-frequency simplex, 28–34
Slow Associated Control Channel (SACCH), 403, 432
Stand Alone Dedicated Control Channel (SDCCH), 402–3
Stealing Channel (STCH), 487
Synchronization Channel (SCH), 402, 453
traffic, 348, 364–65, 380, 401
two-frequency simplex, 36–37
types of, 27–39
wideband, 112–15
Channel sets, assigning, 329
Channel simulator, 97
Channel sounding, 114
Channel spacing, 27
GSM, 379
NADC, 427–28
two-frequency systems as function of, 218
Closed loop power control, 447, 448
Clusters
with 120-degree sectors, 354
defined, 325
illustrated, 331
number of, 325
number of cells in, 330
of omni cells, 353
repeated, 329
size for different propagation exponents, 370
total surface, 364
See also Cells
Cochannel cells, 367
location of, 330
See also Cells

Cochannel interference, 367–71
  compatibility, 229–33
  transmitter, 327
  See also Interference
Cochannel reuse ratio, 335–36
  effects, 335
  values, 336
Cochannel tiers, 367, 369
Coherence bandwidth, 108
Combiners. See Transmitter combiners
Common control fixed point (CCFP), 463
Complex envelope, 90
Configuration assistance, 288
Continuous current system, 43
Continuous tone controlled signaling
    system (CTCSS), 45–47, 294
  circuitry, 47
  defined, 45
  shared repeater, 47–48
  standardized tones, 48
  See also Signaling
Control channels, 401
  common (CCCH), 402, 418
  dedicated, 402–3
  digital (DCC), 432
  general information, 402, 403
  GSM, 380
  multiframes, 400
  TETRA, 487
Control channel system codeword
    (CCSC), 298
Control system, 338–46
  call setup, 338–40
  cell and-over, 341–45
  fixed network originated calls, 340–41
  mobile-originated calls, 341
  signaling tone (ST), 345–46
  supervision audio tone (SAT), 346
Conversion losses
  for class C transmitter, 221
  defined, 220
Cordless systems, 459–79
  basic, 460–61
  cellular radio and, 463–65
  DECT, 460, 465–79
  introduction to, 459–60
  micro/picocellular, 463
  standards, 466
  telecommunications market, 464–65

telecommunications services, 464
telepoint, 461–63
Correction factors, 132–36
  isolated mountain, 135–36, 146, 147
  mixed see-land path, 135, 143–44
  open-area, 134, 141
  sloping terrain, 134, 142
  street orientation, 133, 139
  suburban area, 133–34, 140
  undulating terrain, 135, 145
  urban area, 133
Correlator, 441
COST 207 project, 107
  channel model, 115–19
  Doppler spectra, 115–19
  PDPs, 117
COST 231 model, 122, 176–79
  applicability, 178
  calculations, 179
  defined, 176
  general expression, 177
  geometry, 176
  Hata, 159
  street/path geometry parameters,
    176–77
Coverage
  area, 229
  evaluation, 237
  indoor, 252–55
  mobile station evaluation, 207–11
  study elements, 211
Coverage calculations spreadsheet, 155
  for four quadrants, 157
  for urban area, 156
Coverage computations, 192–207
  antenna effective length, 193
  area coverage probability, 200
  correction parameter, 297
  effective area, 193
  fringe coverage probability, 200
  linear attenuation, 193
  median necessary field strength, 192,
    193
  minimum necessary received field
    strength, 192
  usage correction parameter, 196
Coverage contours, 154, 202
  computation, 211
  for four quadrants, 155
  for locations, 204

Coverage range, 208
　base station, 209
　maximum, 209
Cumulative distribution function (CDF)
　defined, 73
　obtaining, 79
Data calls, 295
DECT system, 460, 465–79
　application area, 468, 469
　call hand-over, 477–78
　capacity, 468
　characteristics, 477
　coexistence specifications (CX), 472
　common interface specification (CI), 472
　cordless, 470
　CT0, 466
　CT1, 466
　CT2, 466–67
　CT3, 467
　DCA, 477
　description, 467–70
　equipment, 468
　fixed, 470
　frame structure, 475
　GSM vs., 478
　layers, 473
　location, in overall system, 471
　MAC layer, 474
　multicarrier configuration, 476
　network, 470–72
　operation of, 477–78
　physical layer, 472–74
　technical principles, 472–76
　See also Cordless systems
Degradation
　due to fast signal variations, 197
　experienced by base station receiver, 199
　experienced by mobile receiver, 198
　noise, 420
Delay
　interval, 108
　spreads, 107, 108, 465
　window, 107
Demand assignment, 281
Demand assignment multiple access (DAMA), 278

Depolarization, 251–52
Deterministic scattering model, 100–106
Diffracted rays, 182
　Keller law for, 181
　shadow region, 184
Diffraction
　coefficient, 182
　geometry, 185
　losses, 100
Diffuse scattered components, 102
Digital color code (DCC), 346
Digital control channel (DCC), 432
Digital enhanced cordless telecommunication system. See DECT system
Digitally controlled squelch (DCS), 45, 48–49
　advantage, 49
　data structure, 49
　decoder, 48–49
　defined, 48
　See also Signaling
Dimensioning
　of cellular networks, 362–66
　defined, 317
　direct case, 319
　inverse case, 319–21
　parameters, 317
　of trunked systems, 317–21
Directive antennas, 358
Direct ray, 172
Discontinuous reception (DRX), 418
Discontinuous transmission (DTX), 417
Dispatch system, 58–59
　channels, 58–59
　elements, 58
　illustrated, 58
　operation characteristics, 59
Distributions
　global, 74
　parameters, 73–74
　random FM, 82
　triangular, 108, 110
Doppler shift, 15, 65, 67, 103
　GSM and, 381
　illustrated, 18
　magnitude, 15
　maximum possible, 103

measuring, 89
of n-th plane wave, 67
spectra, 21, 22
of transmitted carrier frequency, 17
Doppler spectra
  classic spectrum (CLASS), 115
  COST 207, 115–19
  for different distributions, 70
  for directive receiving antenna, 72
  Gaussian spectrum 1 (GAUS1), 116
  Gaussian spectrum 2 (GAUS2), 116
  if scatterers are in movement, 71
  for Rayleigh case, 68
  for Rice case, 71, 85, 116
Dual-tone multifrequency (DTMF) signaling, 52
  defined, 52
  frequencies, 53
  See also Signaling
Dummy burst, 406–8
Duplexers, 269
  bandpass, 270
  block diagram, 270
  defined, 269
  VHF, 271
Duplex separation, 378
Dynamic channel allocation (DCA), 464, 477

Effective antenna height
  base station, 125, 158
  calculation, 127
Effective antenna length, 193
Effective area, 193
Effective radiated power (ERP)
  for bases/mobiles, 37
  evaluation of, 207–11
  expression, 207
  maximum, 33
Effective radius, 160, 161
Electronic serial number (ESN), 341, 451, 457
Emergency calls, 295, 423
Empirical propagation models, 244–47
  defined, 170
  parameters for LOS paths, 247
  parameters for NLOS paths, 247
Encryption
  process, 411
  TETRA, 484–85

Enhanced data rates over GSM evolution (EDGE), 505
Environments
  open area, 128–30
  suburban area, 130
  urban area, 130
  wooded area, 130
Equivalent isotropically radiated power (EIRP), 207
  evaluation of, 207–11
  expressions, 207
  field strength as function of, 209
Erlang-B tables, 372–73
Erlang-C formula, 318
European Telecommunications Standardization Institute (ETSI), 26
Excess losses, 208
  clutter, 208
  defined, 123
  Okumura model, 150
  urban area, 131, 132
Extended AMPS (EAMPS), 428, 429

Fade duration, average, 76, 79–81
Fade rate, 76
Fading
  fast, 19, 85, 239
  flat, 95, 98
  frequency, 21
  frequency selective, 87–95, 98
  margins, 76
  Rayleigh, 243
  Rice, 83–85, 243
Fast Associated Control Channel (FACCH), 403, 432
Fast fading, 19
  effects of, 85, 346–47
  multipath, 239
Fast variations, 9, 14–24
  defined, 8
  superimposition of, 12, 14
  See also Signal variations
FDMA, 325
  CDMA/TDMA comparison, 458–59
  frequency reuse, 443
Feeders, 269–72
  characteristics, 271
  defined, 269

foam dielectric, 271
mesh coaxial, 270
FFSK modulation, 60, 296
Flat fading, 95, 98
Forward control channel data format, 348
Forward error correction (FEC), 60
Four-ray model, 239
  geometry, 240
  result comparison, 240
  results for urban microcell, 242
Free-space losses, 12, 167–69, 208
  computing, 155
  expression, 147
  JRC model, 159
  See also Total losses
Frequency
  bands, GSM, 378
  correction burst, 406
  inversion operation, 41
Frequency Correction Channel (FCCH), 402
Frequency planning, 225–29, 292
  channel reuse, 292
  frequency sets, 293
  for trunked systems in 400-MHz band, 229
  for trunked systems in 800-MHz band, 228
Frequency reuse
  CDMA, 443
  distance, 212, 213
  FDMA, 443
  GSM, 379–80
  index, 326
  TDMA, 443
  two-cell pattern, 237
  two-frequency simplex channels and, 36
Frequency selective fading, 87–95, 98
Frequency-separation correlation function, 98
Fringe coverage probability, 200, 203
FSK modulation, 60
  continuous tones, 41
  direct, 347
Full-duplex channels
  base stations and, 39
  defined, 39

illustrated, 40
See also Channels
Future mobile systems, 491–514
  IMT-2000, 494–503
  UMTS, 503–14
Future public land mobile
  telecommunications system
  (FPMLTS), 493
Gaussian distribution, 201, 205
Generalized packet radio services (GPRS), 505
Global distribution, 74–75
Global signal variations, 74–75
Global system for mobile
  communications. *See* GSM
Gross duplex data transmission services, 424
Group calls, 308–9
  numbering systems, 308
  signaling sequence, 309
  See also Calls
Groups
  assignment schematic, 231
  channel arrangement in, 231
  defined, 227
GSM, 375–425
  adaptive time alignment, 413–15
  authentication, 388–90
  base station subsystem, 381
  BSC, 384
  BTS, 384
  burst types, 405–8
  call setup/routing, 393–97, 403–8
  call stages, 393
  carrier/bearer services, 423, 424
  cellular structure, 379–80
  channel spacing, 379
  channel types, 401–3
  complexity, 376
  control channels, 380
  data traffic channel, 380
  DECT vs., 478
  development, 375–76
  discontinuous transmission/reception, 417–18
  duplex separation, 378
  EIR, 385
  encryption, 411

equipment identification, 391
evolution, 376–78
first generation, 375–76
frame structure, 399
frequency bands, 378
frequency reuse pattern, 379–80
hand-over, 381, 391–92, 416–17
HLR, 385
integration objective, 378
interface reduction technique, 380–81
interfaces, 380, 381, 382
introduction to, 375–78
link budget data, 419–22
maximum BS EIRP, 379
maximum Doppler shift, 381
maximum equalizable time dispersion, 381
maximum time-slot time advance, 381
mobile stations, 381
modulation, 379, 411–12
MSC, 384–85
multiple access technique, 380
network elements, 383
network identifiers, 385–86
network signaling, 380
network subsystem (NSS), 383
nominal BS transmit power, 379
nominal MS transmit power, 379
operation subsystem (OSS), 383
power control, 415–16
protection ratio, 379
radioelectric functions, 384
registration/location update, 387–88
roaming, 381, 391
second generation, 376
security, 381
services, 422–25
signaling bits coding, 412–13
slow frequency hopping, 418–19
specifications, 378–81
specification structure, 376
subsystems, 381
supplementary services, 423, 424–25
switching subsystem (SSS), 381
teleservices, 422, 423–24
traffic channels, 380
uplink/downlink budgets, 422
VLR, 385

voice link, 408–12
voice traffic channel, 380

Half-duplex channels, 37–39
  advantages, 38
  defined, 37
  disadvantages, 38–39
  operation, 38
  See also Channels
Hand-held terminals, 206–7
Hand-over, 341–45
  acknowledgment, 345
  DECT, 477–78
  defined, 334
  GSM, 381, 391–92, 416–17
  illustrated, 392
  mobile assisted (MAHO), 416
  signaling message interchanges, 344
Hata model, 157–59
  application, 159
  basic loss expression, 157
  corrections, 158
  COST 231, 159
  defined, 157
  propagation formula, 422
  validity limits, 158
Height gain factor
  BS, 131, 133
  MS, 131
High-speed circuit switched data (HSCSD), 505
Highway microcells, 236
  two-ray model results, 241
  See also Microcells
Home location register (HLR), 362
  content, 386
  defined, 385
  routing protocol and, 395–96

Ikegami model, 122, 170–76
  components, 173
  development, 172–76
  direct ray, 172
  geometry, 171
  reflected ray, 172
  See also Physical-geometrical models
IMT-2000, 494–503
  defined, 493
  design objective, 495

evaluation groups, 502–3
features, 494
frequency bands, 495
general objectives, 495–96
introduction to, 494–95
ITU-R Recommendations for, 495
objectives, 495–97
operational objectives, 496–97
performance objectives, 502
personal communications scenario, 498–99
radio interfaces, 497–99
RTT, 501–3
spectrum requirement, 500–501
steps toward, 501
technical objectives, 496
traffic densities, 500–501
See also UMTS
Include calls, 296
Individual calls
 numbering system, 308
 to terminal with different prefix, 314–15
 to terminal with same prefix, 314
Indoor coverage, 252–55
 macrocell models, 252–53
 microcell models, 253–55
 from outdoor bases, 253
 wall attenuation and, 255
Indoor propagation, 248–52
 depolarization, 251–52
 model parameters, 251
 picocell models, 248–52
 variability, 251
Insertion losses, 265, 266
Integrated services digital network (ISDN), 377
Interconnection network, 290–91
 defined, 290
 illustrated, 291
 See also Trunked systems
Interfaces
 GSM, 380, 381, 382, 397–413
 IMT-2000, 497–99
 NADC, 432–36
 TETRA, 481, 486
 UMTS, 508–14
Interference
 CDMA, sources, 445
 cellular network, 368

cochannel, 229–33, 327, 367–71
degradation margin, 421
diversity, 419
GSM reduction technique, 380–81
intermodulation, 213–25
margin, 421
multiple study, 233
near-far, 32
pattern for receiver antenna heights, 151
between single-frequency simplex stations, 34
study by radial lines, 232
study for terrain database surface element, 232
Interference-limited systems, 212–13
Interleaving
 GSM, 410–11
 NADC, 439
Intermodulation, 213–25
 causes, 224
 external, 224
 generated at nonlinear external elements, 223–24
 generated at transmitter, 214–15
 generation mechanism, 224
 interference generation process, 220
 as power ratio, 224
 receiver-generated, 215, 224–25
 sources, 214
 transmitter external elements, 215
 transmitter-generated, 219–23
Intermodulation products, 213, 215–19
 defined, 214
 even, higher-order, 218
 fifth-order, 218
 fourth-order, 218
 frequency ranges for, 219
 generation of, 214–15
 link budget, 221, 222
 second-order, 217
 third-order, 216, 218
International mobile equipment identity (IMEI), 386
International mobile subscriber identity (IMSI), 385, 389
Intersymbol interference (ISI), 90
IS-95 standard. *See* CDMA IS-95
IS-136 standard, 25

Isolated mountain correction factor
   defined, 135–36
   illustrated, 146, 147
   See also Correction factors
Isotropic antennas, 147
ITU-R, 493
   IMT-2000 Recommendations, 495
   Recommendation 370 propagation
      model, 223
   Recommendation M.687, 500
   Recommendation M.1225, 502
   Report 319, 230
   Report 793, 225

JRC model, 159–69
   built-up area losses, 165–69
   defined, 159
   diffraction losses, 164
   insufficient path clearance, 164
   knife-edge, 163
   for more than three knife-edges, 166
   plane Earth loss, 169
   radio paths, 160
   reference loss, 159–60
   sufficient path clearance, 163
   total loss, 169

Keller law, 181
Knife-edge model, 163

Land mobile satellite (LMS), 3
Land usage factor, 165
Larger areas
   defined, 10, 121
   propagation calculations for, 123
   received field in, 122
Level crossing rate (lcr), 76–79
   for different fade depths, 81
   illustrated, 77
   normalized, 78
   for Rayleigh case, 78
   in terms of median signal level, 78
   See also Second-order statistics
Linear attenuation, 193
Linearization Channel (LCH), 487
Line-based remote control, 39–41
   illustrated, 43
   options, 39
Line-of-sight (LOS), 248
   non (NLOS), 248
   obscured (OLOS), 248

Link budgets, 76
   carrier-to-interference ratio, 237–38
   GSM, 419–22
   hand-held vehicle TACS, 352
   for intermodulation product, 221, 222
   in microcell system, 238
   TACS, 349
Link bypassing, 38
Local mean, 10, 170
Locations, 334
Locations variability, 205
   defined, 122
   values, 205
Log-normal distribution, 238

Macrocells
   achievement, 236
   base station defines, 252
   illustrated, 236
   K-factor, 244
   uses, 237
   See also Microcells
Mean opinion scores (MOSs), 192
Mean-square values, 73
Median, 73
Median field strength
   necessary, 192, 193, 206
   received, 130–32
Message broadcasting, 424
Microcell propagation, 235–47
   environments, 239
   four-ray model, 239
   modeling, 239–44
   two-ray model, 239
   urban, 244–47
Microcells
   achievement, 236
   highway, 236, 241
   illustrated, 236
   k-factor, 244
   setup, 237
   sizes, 236
   system link budget, 238
   urban, 242, 244–47, 252
   See also Macrocells
Micro/picocellular systems, 463
Mixed path correction factor
   defined, 135
   illustrated, 143–44
   See also Correction factors

Mobile attenuation code (MAC), 347
Mobile identity number (MIN), 341
Mobile networks
    cochannel interference compatibility, 229–33
    coverage computations, 192–207
    coverage quality objective, 191
    frequency planning, 225–29
    interference-limited system, 212–13
    intermodulation, 213–25
    system engineering, 191–92
    traffic calculations, 191–92
    traffic quality objective, 191
    transmission quality objective, 191–92
Mobile-originated calls, 341
    process, 341
    signaling message interchanges, 343
Mobile relative spot height, 167
Mobile stations
    authenticated, 390
    coverage evaluation, 207–11
    height gain factor, 131
    median necessary field strength, 206
    nominal transmit power, 379
Mobile subscriber roaming number (MSRN), 386
Mobile switching centers (MSCs), 332
    commands, 345
    GSM, 384–85
    LAs and, 388
    switching matrix reconfiguration, 345
Mobile telephone switching office (MTSO), 332
Mobile-terminated calls, 342
Mode, 73
Modulations
    $\pi/4$-DQPSK, 436, 485
    carrier, 61
    direct, 61
    GSM, 379, 411–12
    TETRA, 485
Monitoring and maintenance, 289
MPT13xx standards, 292–316
    access protocol, 298
    call processing, 297
    call types, 295–96
    characteristics, 296–97
    defined, 292

    facilities, 295–96
    introduction to, 292–95
    MAP-27, 294
    MPT-1317, 294
    MPT-1318, 294
    MPT-1323, 294
    MPT-1327, 294, 295, 303, 304
    MPT-1331, 294
    MPT-1343, 294, 303, 304, 305, 306, 307
    MPT-1347, 294
    MPT-1352, 294
    multisite systems, 297–98
    signaling message types, 298–300
    terminal addressing, 303–4
    See also Trunked systems
MSK, 412
Multichannel systems
    frequency assignment methods for, 227–29
    operation, 53
Multipath
    echos, 21, 65, 66, 67
    effect, 4
    phenomenon generation, 14–24
Multipath channel
    function parameters, 95
    schematic representation, 96
Multipath propagation, 65–119
    effects on signal transmission, 98
    nonuniform frequency response, 88–89
    ray tracing and, 184
Multiwall model, 249–50
Mutual assistance concept, 32

Narrowband measurements, 111
Near-far interference, 32
Negative exponential model, 111
Network control centers (NCCs), 282
    base station connection, 291
    block diagram, 285
    tasks, 282
Network subsystem (NSS), 383, 389
Noise degradation, 420
Normalized cochannel reuse distance, 330
North American Cellular Digital System (NADC), 427–39
    $\pi/4$-DQPSK, 436

bursts, 433
channel coding scheme, 439
channel spacing, 427–28
control channel bursts, 434
downlink burst, 434, 435
elements, 431
frames for digital traffic channels, 433
functional entities, 431
interleaving scheme, 439
logical channels, 431–32
modulation, 436–38
overview, 427–30
radio interface, 432–36
speech processing, 438–39
system description, 430–31
traffic channel bursts, 434
uplink burst, 434–35
Obscured LOS (OLOS), 248
Offset QPSK (OQPSK), 456
Okumura model, 124–50
    BS effective height, 124–25
    built-up area losses, 136–41
    calculations using, 146–50
    correction factors, 132–36, 150
    environment types, 128–30
    excess loss, 150
    isolated mountain and path parameter, 125–26
    locations variability, 141–46
    mean slope of terrain, 126
    mixed see-land path parameter, 126–27
    received median field strength calculation, 130–32, 149
    received power values, 146, 150
    terrain features, 124–28
    terrain undulation, 125
    total losses, 150
Omni antennas, 358, 371
Open areas, 128–30
    correction factor, 134, 141
    defined, 129–30
    Hata model correction, 158
Open loop power control, 447–48
Operation subsystem (OSS), 383
Overlaid cell networks
    concept of, 355–58
    defined, 355
    practical implication of, 355

Packet assembler disassembler (PAD), 424
Paging Channel (PCH), 402, 418
Paging messages, 338
Passband cavities, 260–62
    responses of, 263
    role of, 260
Path loss
    basic, 14
    calculation, 14
    expression, 245–46
    factors, 13
    plane Earth model, 148–49
    reference, 14
    signal variations and, 10–14
    slope with distance, 242
    two-slop model for, 245
Phase
    received signal, 75–76
    response, 93–94, 95
    term, 182
    variations, 18, 23
Physical channels, 487–88
Physical-geometrical models, 169–88
    COST 231 (Walfish-Ikegami), 176–79
    defined, 170
    Ikegami, 170–76
    ray-tracing, 179–88
Plane Earth loss, 160
Plane Earth model, 150–57
    defined, 151
    geometry, 150, 151
    path loss, 148
    received power, 149
    receiver, 153
Point-to-point short messages, 423–24
Power control
    CDMA, 447–48
    closed loop, 447, 448
    GSM, 397, 415–16
    open loop, 447–48
Power delay profile (PDP), 98
    averaged, 106
    calculated, 188
    characterization of, 111
    coherence function relationship, 100
    COST 207, 117
    defined, 116

exponential model, 110
measurement, 98, 99, 112–13
related models, 108–11
Prediction errors, 238
Preselector filter, 285
Principle of reciprocity, 86
Private mobile radio (PMR), 25–61
 conventional, 26
 data transmission, 59–61
 introduction to, 25–27
 radio stations, 225
 signaling in, 44–52
Probability density function, 74, 83–84, 243
Processing gain, 440, 449
Projection ratio, 379
Propagation
 areas used in studying, 11
 attenuation, 223
 effects, 4
 indoor, 248–52
 microcell, 235–47
 multipath, 65–119
 in new scenarios, 235–55
Propagation channels
 frequency dispersive, 98
 time dispersive, 98
Propagation loss, 121–88, 208
 basic, 208, 209
 expressions, 13
 introduction to, 121–24
 transmitter-generated intermodulation, 220
Pseudo-noise sequences, 115
Pseudo-random binary sequence (PRBS), 451
Public access mobile radio (PAMR), 26
Public land mobile network (PLMN), 360
Public mobile telephony (PMT), 324
Pulse compression techniques, 113
Pure ALOHA, 60
Push-to-talk (PTT) button, 28

Quasi-synchronous operation, 57–58
 clocks, 57
 defined, 57

Radio access system, 465
Radio station control
 local, 39
 remote, 39–41
 types of, 39–43
 via single-channel radio links, 41
Radio transmission technologies (RTT), 501–3
RAKE receiver, 447
Random Access Channel (RACH), 402
Random access protocol, 301–3
Random frequency modulation (FM), 81–83
 defined, 81
 distribution, 82
 simulated series, 83
 spectrum, 82
Rayleigh
 distribution, 5, 10, 24, 74, 121, 243
 fading, 243
 probability density function, 72
Rayleigh case
 antenna directivity and, 69
 average fade duration (afd) for, 79
 azimuth angle distribution for, 86
 Doppler spectrum for, 68
 level crossing rate (lcr) for, 78
 phase distribution, 76
 RF spectrum for, 68
Ray-tracing
 example using UDB, 179, 180
 models, 179–88
 multipath propagation and, 184
 rules, 181
 See also Physical-geometrical models
Received field strength
 illustrated, 7
 measurements, 7
 median, calculating, 130–32
 minimum necessary, 192
 variations, 6–8
Received signals
 angle of incidence of, 65–69
 envelope statistics, 69–75
 instantaneous power, 73
 phase, 75–76
 spectrum of, 65–69
 variation measurement system, 112
Receiver antenna heights, 151
Receiver-generated intermodulation, 215, 224–25
 characteristics, 226
 characterization, 224

Receiver noise, 156
Receiver systems
   schematic representation of, 194
   sensitivity, 195
Reference losses
   defined, 14
   JRC model, 159–60
Reflected components, 102
Reflected rays
   on buildings, 172
   Snell law and, 181
Reflection coefficients, 182
   building surfaces, 174
   magnitude, 176
Registration, 360
   protocol, 297
   start-up, 456
Remote control
   base station, 42, 287
   configuration, 39
   with fixed radio link, 41
   line-based, 39, 43
Repeaters
   link bypassing, 38
   locations for, 38
   shared, 47–48
Reuse index, 364, 366
Reverse control channel, 348
Rice
   environment, 69
   fading, 83–85, 243
   probability density function, 83–84
Rice distribution, 4, 24, 83, 121
   carrier-to-multipath ratio, 84
   characterized by carrier-to-multipath ratio, 242, 243
   time-series, 5
Roaming, GSM, 381, 391

Scattering
   deterministic model, 100–106
   function, 98, 104
   matrix, 98, 101
Second-order statistics, 76–81
   average fade duration (afd), 76, 79–81
   level crossing rate (lcr), 76–79
Security, GSM, 381
Seizure precursor, 340

Service area
   defined, 229
   illustrated, 229
   in multisite system, 230
Shadowing effects, 4, 238
Shadow region, 184
Shape factor, 246
Shared repeaters, 47–48
   decoders, 48
   efficiency, 47
   user groups for, 48
   See also Repeaters
Short data messages, 296
Short message service (SMS), 432
Signaling
   CTCSS, 45–47
   DCS, 48–49
   digital, 44
   DTMF, 52
   GSM, 380
   squelching systems, 45
   tone sequence, 49–52
   types of, 44
Signaling messages, 298–300
   formats, 346–47
   forward control channel data format, 348
   hand-over, 344
   list of, 300
   mobile-originated call, 343
   mobile-terminated call, 342
   reverse control channel data format, 348
   traffic channel data format, 348
   types of, 300
   See also Trunked systems
Signaling protocols, 301
Signaling sequences, 304–16
   call setup, 304
   group call, 309
   individual call with different prefix, 311
   individual call with same prefix, 310
   short data message, 312
   status message, 313
Signaling tone (ST), 341, 345–46
   for alert confirmation, 346
   AMPS frequency, 349

for call cleardown, 345
for hand-over acknowledgment, 345
for hook flash, 345
TACS frequency, 349
Signal sampling, 111–12
Signal variations
    calculations, 187
    characterization of, 8–10
    decay with distance, 12
    fast, 8, 9, 14–24
    instantaneous speed and, 9
    path loss and, 10–14
    slow, 8, 9, 10–14
Silence identification frames (SIDs), 418
Single-frequency simplex channels, 28–34
    advantages of, 29
    drawbacks, 29–30, 33
    illustration, 31
    interference phenomenon between, 34
    See also Channels
Sloping terrain correction factor
    defined, 134
    illustrated, 142
    See also Correction factors
Slotted ALOHA, 60
Slow Associated Control Channel
    (SACCH), 403, 432
Slow frequency hopping (SFH), 418–19
    defined, 418–19
    illustrated, 419
    interference diversity and, 419
Slow variations, 9, 10–14
    defined, 8
    free-space losses, 12
    See also Signal variations
Small areas, 10, 121
Snell law, 181
Spatial correlation, 85–87
    at base station, 87
    illustrated, 86
Speech processing, 438–39
Spread factor, 182
Spread spectrum (SS) schemes, 441
Squelching
    carrier-activated, 46
    circuit types, 45
    systems, 45
Stand Alone Dedicated Control Channel
    (SDCCH), 402–3

Stationary mobile terminal, 15
Status messages, 316
    definition, 296
    pre-defined meanings, 296
    signaling sequence, 313
Stealing Channel (STCH), 487
Street orientation correction factor
    defined, 133
    illustrated, 139
    See also Correction factors
Subscriber
    administration, 289–90
    charging, 290
Subscriber identity module (SIM),
        386–87
    card, 387
    contents, 386
    parameters, 386–87
Suburban areas
    correction factor, 133–34, 140
    defined, 130
    Hata model correction, 158
Sufficient path clearance, 163
Supervisory audio tone (SAT), 341, 346
    assignments, 346
    missions, 346
Supplementary services
    defined, 423
    list of, 424–25
    See also GSM
Support systems, 288–90
    configuration assistance, 288
    monitoring and maintenance, 289
    subscriber administration, 289
    subscriber charging, 290
    See also Trunked systems
Switching subsystem (SSS), 381
Synchronization burst, 406
Synchronization Channel (SCH), 402,
        453
Synchronous duplex data transmission
        services, 424
Synchronous operation, 56–57
    defined, 56
    drawbacks, 56–57
TD-CDMA, 512–14
    advantages, 513
    asymmetric operation and, 513

bursts, 512
defined, 512
goals, 514
technical characteristics, 513
See also UMTS
TDMA, 25, 325
   FDMA/CDMA comparison, 458–59
   frames, 398, 401
   frequency reuse, 443
   GSM, 380
   mobile digital telephony networks, 326–27
   multi-access structure, 432
   structures, 398–401
   TETRA frame/multiframe structure, 486
   time slot, 398
   time structure, 397
   voice frame interleaving process in, 410
Telephony, 423
Telepoint, 461–63
   advantage, 461–62
   coverage, 462
   defined, 460
   disadvantages, 462–63
   illustrated, 461–62
   receiving communications and, 462
   See also Cordless systems
Teleservices
   defined, 422
   list of, 423–24
   See also GSM
Teletraffic theory, 277
Temporary mobile subscriber identity (TMSI), 385, 390
Terminal addressing, 303–4
Terrain
   blockage, 14
   database grid, 210
   databases, 200
   mean slope of, 126
   Okumura model and, 124–28
   slope parameter definition, 127
   sloping, 134
   undulation, 125, 135
Terrain profiles, 159, 168
   illustrated, 168
   JRC model, 159
   land-usage superposed on, 210

TETRA, 479–88
   air interface, 486
   bearer services, 483
   bursts, 488
   coding, 484
   defined, 479
   direct mode operation (DMO), 484
   encryption, 484–85
   features, 479–81
   functional configuration, 483–84
   harmonized, frequencies, 481–83
   interfaces, 481
   modulation, 485
   network structure, 480
   physical channels, 487–88
   services, 482
   signal processing, 484–88
   supplementary services, 483
   support, 479
   TDMA frame/multiframe structure, 486
   telecommunication services, 481
   V+D channels, 487
Theoretical rays, 170
Third-generation (3G) systems, 491–92
Time division duplex (TDD), 442, 467, 509
Time hopping-SS (TH-SS), 441
Time variability, 205
Tone sequence signaling, 49–52
   in dispatch system, 58
   frequencies, 50
   pager types, 52
   transmission standards, 50
   See also Signaling
Total access cellular system (TACS), 323
   allocated band, 349
   extended (ETACS), 349
   features, 347–51
   link budgets, 349
   maximum ERPs, 351
   ST frequency, 349
   See also Cellular networks
Total losses
   expression of, 153
   JRC model, 169
   Okumura model, 150
   transmitter-generated intermodulation, 221

Touch tone. See Dual-tone
multifrequency (DTMF)
signaling
Traffic channels, 401
data, 380
data format, 348
full-rate, 401, 433
GSM, 380
half-rate, 401, 433
multiframes, 400
NADC, 433
total number offered, 364–65
voice, 380
Transfer characteristic function, 214
Transmitter combiners, 257–67
band-reject cavities, 262–65
cavity-isolator, 265–67
circulator plus hybrid, 259
configurations, 258
coupling loss and, 257
four, 258–59
passband cavities, 260–62
performances, 267
Transmitter filter losses, 35
Transmitter-generated intermodulation, 219–23
conversion loss, 220
coupling loss, 220
propagation losses, 220
total losses, 221
See also Intermodulation
Transversal filter elements, 115
Triangular distributions, 108, 110
Trunked systems, 277–321
access protocol, 298
antenna system, 288
base stations, 282–88
call processing, 297
dimensioning, 317–21
elements, 280–92
frequency plan, 292
illustrated, 279, 283–84
interconnection network, 290–91
introduction to, 277–80
MPT13xx standards, 292–316
multisite, 297–98
NCC, 282
PMR/PAMR, 278

setup options, 280
signaling message types, 298–300
signaling sequences, 304–16
support systems, 288–90
terminal addressing, 303–4
theoretical fundamentals, 278
Trunking gain, 278
Two-frequency simplex channels, 36–37
frequency reuse and, 36
spectrum organization, 37
use of, 37
See also Channels
Two-ray model, 90–95, 239
illustrated, 92
received signal profile, 240
result comparison, 240
results in highway microcell, 241

UMTS, 503–14
challenges, 505
defined, 493, 503–5
development milestones, 506
FDD satellite component, 508
frequency allocations, 508
introduction to, 503–6
objectives, 506–7
on GSM structure, 505
radio interface, 508–14
spectrum requirements, 507–8
TD-CDMA, 512–14
TDD mode, 509
terrestrial radio access (UTRA), 508
user bit rates, 509
W-CDMA, 509–12
See also IMT-2000
Undulating terrain, 125
correction factor, 135, 145
parameter, 125
Universal mobile telecommunications
system. See UMTS
Universal personal telecommunications
(UPT), 491
Urban areas
coverage calculations spreadsheet, 156
defined, 130
excess loss, 131, 132
field strength variations, 171
Hata model corrections, 158
long-term field strength, 171

losses, 148
multiple interactions in, 186
received filed strength in, 135, 136, 137, 138
short-term field strength, 171–72
street orientation correction factor, 133
Urban databases (UDBs)
ray tracing techniques on, 179, 180
use of, 122
Urbanization factor, 167
Urban microcells
base station creates, 252
empirical propagation models for, 244–47
four-ray model results, 242
k-factor, 244
LOS links, 245
NLOS links, 245
See also Microcell propagation; Microcells
Usage correction parameter, 196

Variance, 73
Velocity-weighted algorithm, 7
VHF duplexers, 271
VHF isolator, 261
Visitor location register (VLR), 362
content, 386
defined, 385
Voice activity detector (VAD), 408
Voice calls, 295
Voice coder, 408
Voice messaging, 423
Voltage standing wave ratio (VSWR), 286

Voting systems, 43–46
with auxiliary receivers, 56
defined, 53–54
diagram, 55
signal selector, 54
Walfish-Ikegami model, 176–79
applicability, 178
calculations, 179
defined, 176
general expression, 177
geometry, 176
street/path geometry parameters, 176–77
See also Physical-geometrical models
Wall attenuation, 255
Walsh matrix, 450, 451
W-CDMA, 509–12
development, 509–10
dual-mode packet transmission scheme, 512
frequency reuse, 512
performance-enhancement features, 510
physical channels, 510–11
rate matching, 511
service classes, 511
technical characteristics, 510
See also UMTS
White noise, autocorrelation, 113
Wideband channels
characterization, 95–106
sounding, 112–15
Wooded areas, 130
World Radio Conference (WRC), 27

Recent Titles in the Artech House Mobile Communications Series

John Walker, Series Editor

Advances in Mobile Information Systems, John Walker, editor

An Introduction to GSM, Siegmund M. Redl, Matthias K. Weber, Malcolm W. Oliphant

CDMA for Wireless Personal Communications, Ramjee Prasad

CDMA RF System Engineering, Samuel C. Yang

CDMA Systems Engineering Handbook, Jhong S. Lee and Leonard E. Miller

Cell Planning for Wireless Communications, Manuel F. Cátedra, Jesús Pérez-Arriaga

Cellular Communications: Worldwide Market Development, Garry A. Garrard

Cellular Mobile Systems Engineering, Saleh Faruque

The Complete Wireless Communications Professional: A Guide for Engineers and Managers, William Webb

GSM and Personal Communications Handbook, Siegmund M. Redl, Matthias K. Weber, Malcolm W. Oliphant

GSM Networks: Protocols, Terminology, and Implementation, Gunnar Heine

GSM System Engineering, Asha Mehrotra

Handbook of Land-Mobile Radio System Coverage, Garry C. Hess

Introduction to Mobile Communications Engineering, José M. Hernando, F. Pérez-Fontán

Introduction to Radio Propagation for Fixed and Mobile Communications, John Doble

Introduction to Wireless Local Loop, William Webb

IS-136 TDMA Technology, Economics, and Services, Lawrence Harte, Adrian Smith, Charles A. Jacobs

Mobile Communications in the U.S. and Europe: Regulation, Technology, and Markets, Michael Paetsch

Mobile Data Communications Systems, Peter Wong, David Britland

Mobile Telecommunications: Standards, Regulation, and Applications, Rudi Bekkers and Jan Smits

Personal Wireless Communication With DECT and PWT, John Phillips, Gerard Mac Namee

Practical Wireless Data Modem Design, Jonathan Y.C. Cheah

RDS: The Radio Data System, Dietmar Kopitz, Bev Marks

RF and Microwave Circuit Design for Wireless Communications, Lawrence E. Larson, editor

Spread Spectrum CDMA Systems for Wireless Communications, Savo G. Glisic, Branka Vucetic

Understanding Cellular Radio, William Webb

Understanding Digital PCS: The TDMA Standard, Cameron Kelly Coursey

Understanding GPS: Principles and Applications, Elliott D. Kaplan, editor

Universal Wireless Personal Communications, Ramjee Prasad

Wideband CDMA for Third Generation Mobile Communications, Tero Ojanperä, Ramjee Prasad

Wireless Communications in Developing Countries: Cellular and Satellite Systems, Rachael E. Schwartz

For further information on these and other Artech House titles, including previously considered out-of-print books now available through our In-Print-Forever® (IPF®) program, contact:

Artech House
685 Canton Street
Norwood, MA 02062
Phone: 781-769-9750
Fax: 781-769-6334
e-mail: artech@artechhouse.com

Artech House
46 Gillingham Street
London SW1V 1AH UK
Phone: +44 (0)20 7596-8750
Fax: +44 (0)20 7630-0166
e-mail: artech-uk@artechhouse.com

Find us on the World Wide Web at:
www.artechhouse.com

$\underline{V}$ .25 μV $S = 20 \log .25 \quad 20(\log 25 + \log 10^{-2})$
              $\approx -12 \, db \, \mu V \quad\quad 1\cdot 4 + 2$

$\underline{P},$ In dbw = $10 \log \frac{V^2}{R} = 10 \log \frac{(.25 \times 10^{-6})^2}{50}$

$= 10 \log \frac{.0625 \times 10^{-12}}{50}$

$= 10 \log \frac{6 \times 10^{-14}}{5 \times 10}$

$= 10 \log 1.2 \times 10^{-15}$

$= 10 \cdot (1.2 - 16)$

$= -148 \, dbw$

$= -148 + 10 \log 1000$

$= -118 \, dbm$